Numerical Recipes Example Book (C)

Second Edition

William T. Vetterling

Polaroid Corporation

Saul A. Teukolsky

Department of Physics, Cornell University

William H. Press

Harvard-Smithsonian Center for Astrophysics

Brian P. Flannery

EXXON Research and Engineering Company

CAMBRIDGE
UNIVERSITY PRESS

Published by the Press Syndicate of the University of Cambridge
The Pitt Building, Trumpington Street, Cambridge CB2 1RP
40 West 20th Street, New York, NY 10011-4211, USA
10 Stamford Road, Oakleigh, Melbourne 3166, Australia

First edition originally published 1988
Second edition originally published 1992
Reprinted 1993 (twice)

Printed in the United States of America
Typeset in TEX

The computer programs in this book are available, in C, in
several machine-readable formats. There are also versions
of this book and its software available in FORTRAN,
Pascal, and BASIC programming languages.
   To purchase diskettes in IBM PC or Apple Macintosh
formats, use the order form at the back of the book or write
to Cambridge University Press, 110 Midland Avenue, Port
Chester, NY 10573.
   Unlicenced transfer of Numerical Recipes programs from
the above-mentioned IBM PC or Apple Macintosh
diskettes to any other format or to any computer except a
single IBM PC or Apple Macintosh or compatible for each
diskette purchased, is strictly prohibited. Licenses for
authorized transfers to other computers are available from
Numerical Recipes Software, P.O. Box 243, Cambridge,
MA 12238 (FAX 617 863-1739). Technical questions,
corrections, and requests for information on other available
formats should be directed to this address.

Library of Congress Cataloging-in-Publication Data available.

A catalogue record for this book is available from the British Library.

ISBN 0-521-43720-2 Example book in C (this book)
ISBN 0-521-43108-5 Numerical Recipes in C
ISBN 0-521-43714-8 C diskette (IBM 5.25", 1.2M)
ISBN 0-521-43724-5 C diskette (IBM 3.5", 1.44M)
ISBN 0-521-43715-6 C diskette (Mac 3.5". 800K)

CONTENTS

iii

Preface

This *Numerical Recipes Example Book (C)* is designed to accompany the text and reference book *Numerical Recipes in C: The Art of Scientific Computing*, Second Edition, by William H. Press, Saul A. Teukolsky, William T. Vetterling, and Brian P. Flannery (Cambridge University Press, 1992). In that volume, the algorithms and methods of scientific computation are developed in considerable detail, starting with basic mathematical analysis and working through to actual implementation in the form of C routines. The routines in *Numerical Recipes in C: The Art of Scientific Computing*, numbering more than 300, are meant to be incorporated into user applications; they are procedures (or functions), not stand-alone (`main`) programs.

It often happens, when you want to incorporate somebody else's procedure into your own application program, that you first want to see the procedure demonstrated on a simple example. Prose descriptions of how to use a procedure (even those in *Numerical Recipes*) can occasionally be inexact. There is no substitute for an actual, C demonstration program that shows exactly how data are fed to a procedure, how the procedure is called, and how its results are unloaded and interpreted.

Another not unusual case occurs when you have, for one seemingly good purpose or another, modified the source code in a "foreign" procedure. In such circumstances, you might well want to test the modified procedure on an example known previously to have worked correctly, *before* letting it loose on your own data. There is the related case where procedure source code may have become corrupted, e.g., lost some lines or characters in editing or customization, and a simple revalidation test is desirable.

These are the needs addressed by this *Numerical Recipes Example Book*. Divided into chapters identically with *Numerical Recipes in C: The Art of Scientific Computing*, this book contains C source programs that exercise and demonstrate all of the *Numerical Recipes* procedures and functions. The programs are commented, and each is also prefaced by a short description of what it does, and of which *Numerical Recipes* routines it exercises. In many cases where the demonstration programs require input data, that data is also printed in this book. In some cases, where the demonstration programs are not "self-validating," sample output is also shown.

Necessarily, in the interests of clarity, the *Numerical Recipes* procedures and functions are demonstrated in simple ways. A consequence is that the demonstration programs in this book do not usually test all possible regimes of input data, or even all lines of procedure source code. The demonstration programs in this book were by no means the only validating tests that the *Numerical Recipes* procedures and functions were required to pass during their development. The programs in this book *were* used during the later stages of the production of *Numerical Recipes in C: The Art of Scientific Computing* to maintain integrity of the source code, and in this role

v

were found to be invaluable.

A Note on Numerical Recipes Utility Functions

The programming conventions used in this book are discussed fully in Chapter 1 of *Numerical Recipes in C: The Art of Scientific Computing*. You should have a copy of that book – without it, this one will not be very meaningful. Nevertheless, we review a few important matters here.

The *Numerical Recipes* software collection in C contains about 350 routines. For the most part the routines are self-contained, making reference only to other routines in the package, or to standard C library functions. There are some consistent exceptions, however. (1) When the *Numerical Recipes* functions (or the `main()` programs in this book) encounter an error, they turn matters over to a function `nrerror()`, not a standard C function. (2) When vectors and multidimensional matrices are allocated or deallocated, use is made of a set of utility functions with names like `matrix()`, `vector()`, `free_matrix()`, and `free_vector()`. These Numerical Recipes-specific shared functions are found in a utility file `nrutil.c`, listed in Appendix B. Example routines that use any of these utilities include the header file `nrutil.h` (Appendix A) along with any of the standard library header files which are required. Similar remarks apply to routines using complex arithmetic. These make calls to functions in the file `complex.c`, which is listed in Appendix C. The appropriate header file is called `complex.h` (Appendix A).

All of the examples in this book call at least one, and sometimes more than one, of the *Numerical Recipes* functions. A prototype declaration must generally be given for each of the functions used. To spare you the trouble of searching program listings for the proper declarations, we have compiled an alphabetical list in the form of a header file `nr.h`, reproduced in Appendix A. You may, if you wish, simply include the file `nr.h` at the beginning of your own application program and use the recipes with abandon. That is what we have done in the example routines in this book. In some cases, however, you might want to cull from the file only those declarations which your program requires, and to declare them individually. The Numerical Recipes routines themselves, in which we sought to make the interdependence of routines explicit, are written in this fashion.

A final important point about the header file `nr.h` is that, as furnished on the diskette, it contains both ANSI C and traditional K&R-style declarations. The ANSI forms are invoked if any of the following macros are defined: `__STDC__`, `ANSI`, or `NRANSI`. (The purpose of the last name is to give you an invocation that does not conflict with other possible uses of the first two names.) If you have an ANSI compiler, it is *essential* that you invoke it with one or more of these macros defined. The typical means for doing so is to include a switch like "`-DANSI`" on the compiler command line.

Of course, with an ANSI compiler, you should also be sure to use the ANSI versions of the programs supplied on the *Numerical Recipes C Diskette* and printed in this Example Book. If you have only a traditional Kernighan and Ritchie (K&R) C compiler, then you should be sure to use the alternative K&R program versions that are also supplied on the diskette. These differ slightly from the listings in this book (and the main book), but the two C dialects are close enough that separate program listings are not necessary.

Chapter 1: Preliminaries

The routines in Chapter 1 of Numerical Recipes are introductory and less general in purpose than those in the remainder of the book. This chapter's routines serve primarily to expose the book's notational conventions, illustrate control structures, and perhaps to amuse. You may even find them useful. We hope that you will use badluk *for no serious purpose.*

⋆　⋆　⋆　⋆

Procedure flmoon calculates the phases of the moon, or more exactly, the Julian day and fraction thereof on which a given phase will occur or has occurred. The program xflmoon asks the present date and compiles a list of upcoming phases. We have compared the predictions to lunar tables, with happy results. Shown are the results of a test run, which you may replicate as a check. In this program, notice that we have set ZON (the time zone) to −5.0 to signify the five hour separation of the Eastern Standard time zone from Greenwich, England. Our convention requires you to use negative values of ZON if you are west of Greenwich, as we are. The Julian day results are converted to calendar dates through the use of caldat, which appears later in the chapter. The fractional Julian day and time zone combine to form a correction that can possibly change the calendar date by one day.

Date	Time(EST)	Phase
12 9 1992	7 PM	full moon
12 16 1992	2 PM	last quarter
12 23 1992	8 PM	new moon
12 31 1992	10 PM	first quarter
1 8 1993	8 AM	full moon
1 14 1993	11 PM	last quarter
1 22 1993	1 PM	new moon
1 30 1993	6 PM	first quarter
2 6 1993	7 PM	full moon
2 13 1993	10 AM	last quarter
2 21 1993	8 AM	new moon
3 1 1993	11 AM	first quarter
3 8 1993	5 AM	full moon
3 14 1993	11 PM	last quarter
3 23 1993	2 AM	new moon
3 30 1993	11 PM	first quarter
4 6 1993	2 PM	full moon
4 13 1993	3 PM	last quarter
4 21 1993	6 PM	new moon
4 29 1993	8 AM	first quarter

```
/* Driver for routine flmoon */

#include <stdio.h>
#include "nr.h"

#define ZON (-5.0)

int main(void)
{
    int i,i1,i2,i3,id,im,iy,n,nph=2;
    float timzon=ZON/24.0,frac,secs;
    long j1,j2;
    static char *phase[]={"new moon","first quarter",
        "full moon","last quarter"};

    printf("Date of the next few phases of the moon\n");
    printf("Enter today\'s date (e.g. 12 15 1992)  :  \n");
    scanf("%d %d %d",&im,&id,&iy);
    /* Approximate number of full moons since january 1900 */
    n=(int)(12.37*(iy-1900+((im-0.5)/12.0)));
    j1=julday(im,id,iy);
    flmoon(n,nph,&j2,&frac);
    n += (int) ((j1-j2)/29.53 + (j1 >= j2 ? 0.5 : -0.5));
    printf("\n%10s %19s %9s\n","date","time(EST)","phase");
    for (i=1;i<=20;i++) {
        flmoon(n,nph,&j2,&frac);
        frac=24.0*(frac+timzon);
        if (frac < 0.0) {
            --j2;
            frac += 24.0;
        }
        if (frac > 12.0) {
            ++j2;
            frac -= 12.0;
        } else
            frac += 12.0;
        i1=(int) frac;
        secs=3600.0*(frac-i1);
        i2=(int) (secs/60.0);
        i3=(int) (secs-60*i2+0.5);
        caldat(j2,&im,&id,&iy);
        printf("%5d %3d %5d %7d:%2d:%2d        %s\n",
            im,id,iy,i1,i2,i3,phase[nph]);
        if (nph == 3) {
            nph=0;
            ++n;
        } else
            ++nph;
    }
    return 0;
}
```

The function `julday`, our exemplar of the `if` control structure, converts calendar dates to Julian dates. Not many people know the Julian date of their birthday or any other convenient reference point, for that matter. To remedy this, we offer a list of checkpoints, which appears at the end of this chapter as the file `dates1.dat`. The program `xjulday` lists the Julian date for each historic event for comparison. Then it allows you to make your own choices for entertainment.

```c
/* Driver for routine julday */

#include <stdio.h>
#include <stdlib.h>
#include "nr.h"
#include "nrutil.h"

#define MAXSTR 80

int main(void)
{
    int i,id,im,iy,n;
    char txt[MAXSTR];
    static char *name[]={"","january","february","march",
        "april","may","june","july","august","september",
        "october","november","december"};
    FILE *fp;

    if ((fp = fopen("dates1.dat","r")) == NULL)
        nrerror("Data file dates1.dat not found\n");
    fgets(txt,MAXSTR,fp);
    fscanf(fp,"%d %*s ",&n);
    printf("\n%5s %8s %6s %12s %9s\n","month","day","year",
        "julian day","event");
    for (i=1;i<=n;i++) {
        fscanf(fp,"%d %d %d ",&im,&id,&iy);
        fgets(txt,MAXSTR,fp);
        printf("%-10s %3d %6d %10ld %5s %s",name[im],id,iy,
            julday(im,id,iy)," ",txt);
    }
    fclose(fp);
    printf("\nYour choices: (negative to end)\n");
    printf("month day year (e.g. 1 13 1905)\n");
    for (i=1;i<=20;i++) {
        printf("\nmm dd yyyy ?\n");
        scanf("%d %d %d",&im,&id,&iy);
        if (im < 0) return 1;
        printf("julian day: %ld \n",julday(im,id,iy));
    }
    return 0;
}
```

The next program in *Numerical Recipes* is `badluk`, an infamous code that combines the best and worst instincts of man. We include no demonstration program for `badluk`, not just because we fear it, but also because it is self-contained, with sample results appearing in the text.

Chapter 1 closes with routine `caldat`, which illustrates no new points, but complements `julday` by doing conversions from Julian day number to the month, day, and year on which the given Julian day began. This offers an opportunity, grasped by the demonstration program `xcaldat`, to push dates through both `julday` and `caldat` in succession, to see if they survive intact. This, of course, tests only your authors' ability to make mistakes backwards as well as forwards, but we hope you will share our optimism that correct results here speak well for both routines. (We have checked them a bit more carefully in other ways.)

```c
/* Driver for routine caldat */

#include <stdio.h>
#include <stdlib.h>
#include "nr.h"
#include "nrutil.h"

#define MAXSTR 80

int main(void)
{
    int i,id,idd,im,imm,iy,iyy,n;
    long j;
    char dummy[MAXSTR];
    static char *name[]={"","january","february","march",
        "april","may","june","july","august",
        "september","october","november","december"};
    FILE *fp;

    /* Check whether caldat properly undoes the operation of julday */
    if ((fp = fopen("dates1.dat","r")) == NULL)
        nrerror("Data file dates1.dat not found\n");
    fgets(dummy,MAXSTR,fp);
    fscanf(fp,"%d %*s ",&n);
    printf("\n %14s %43s\n","original date:","reconstructed date");
    printf("%8s %5s %6s %15s %12s %5s %6s\n","month","day","year",
        "julian day","month","day","year");
    for (i=1;i<=n;i++) {
        fscanf(fp,"%d %d %d ",&im,&id,&iy);
        fgets(dummy,MAXSTR,fp);
        j=julday(im,id,iy);
        caldat(j,&imm,&idd,&iyy);
        printf("%10s %3d %6d %13ld %16s %3d %6d\n",name[im],
            id,iy,j,name[imm],idd,iyy);
    }
    fclose(fp);
    return 0;
}
```

Appendix

File dates1.dat:

```
List of dates for testing routines in Chapter 1
16 entries
12 31   -1 End of millennium
01 01    1 One day later
10 14 1582 Day before Gregorian calendar
10 15 1582 Gregorian calendar adopted
01 17 1706 Benjamin Franklin born
04 14 1865 Abraham Lincoln shot
04 18 1906 San Francisco earthquake
05 07 1915 Sinking of the Lusitania
07 20 1923 Pancho Villa assassinated
```

```
05 23 1934 Bonnie and Clyde eliminated
07 22 1934 John Dillinger shot
04 03 1936 Bruno Hauptman electrocuted
05 06 1937 Hindenburg disaster
07 26 1956 Sinking of the Andrea Doria
06 05 1976 Teton dam collapse
05 23 1968 Julian Day 2440000
```

Chapter 2: Linear Algebraic Equations

Numerical Recipes Chapter 2 begins the "true grit" of numerical analysis by considering the solution of linear algebraic equations. This is done first by Gauss-Jordan elimination (gaussj), and then by LU decomposition with forward and backsubstitution (ludcmp and lubksb). Several linear systems of special form, represented by tridiagonal, band diagonal, cyclic, Vandermonde, and Toeplitz matrices, may be treated with procedures tridag, bandec, cyclic, vander, *and* toeplz *respectively. Cholesky decomposition (*choldc *and* cholsl*) is the preferred method for symmetric positive definite systems. QR decomposition (*qrdcmp*) is less efficient than LU decomposition in general; however, the ease with which it is updated (*qrupdt*) if the system is changed slightly makes it useful in certain applications. For singular or nearly singular matrices the best choice is singular value decomposition with backsubstitution (*svdcmp *and* svbksb*). Linear systems with relatively few non-zero coefficients, so-called "sparse" matrices, are handled by routine* linbcg. *A suite of routines for manipulating general sparse matrices,* sprsax, sp; stx, *etc., is provided.*

$$\star \quad \star \quad \star \quad \star$$

gaussj performs Gauss-Jordan elimination with full pivoting to find the solution of a set of linear equations for a collection of right-hand side vectors. The demonstration routine xgaussj checks its operation with reference to a group of test input matrices printed at the end of this chapter as file matrx1.dat. Each matrix is subjected to inversion by gaussj, and then multiplication by its own inverse to see that a unit matrix is produced. Then the solution vectors are each checked through multiplication by the original matrix and comparison with the right-hand side vectors that produced them.

```
/* Driver for routine gaussj */

#include <stdio.h>
#include <stdlib.h>
#include "nr.h"
#include "nrutil.h"

#define NP 20
#define MP 20
#define MAXSTR 80

int main(void)
{
    int j,k,l,m,n;
```

```
float **a,**ai,**u,**b,**x,**t;
char dummy[MAXSTR];
FILE *fp;

a=matrix(1,NP,1,NP);
ai=matrix(1,NP,1,NP);
u=matrix(1,NP,1,NP);
b=matrix(1,NP,1,MP);
x=matrix(1,NP,1,MP);
t=matrix(1,NP,1,MP);
if ((fp = fopen("matrx1.dat","r")) == NULL)
    nrerror("Data file matrx1.dat not found\n");
while (!feof(fp)) {
    fgets(dummy,MAXSTR,fp);
    fgets(dummy,MAXSTR,fp);
    fscanf(fp,"%d %d ",&n,&m);
    fgets(dummy,MAXSTR,fp);
    for (k=1;k<=n;k++)
        for (l=1;l<=n;l++) fscanf(fp,"%f ",&a[k][l]);
    fgets(dummy,MAXSTR,fp);
    for (l=1;l<=m;l++)
        for (k=1;k<=n;k++) fscanf(fp,"%f ",&b[k][l]);
    /* save matrices for later testing of results */
    for (l=1;l<=n;l++) {
        for (k=1;k<=n;k++) ai[k][l]=a[k][l];
        for (k=1;k<=m;k++) x[l][k]=b[l][k];
    }
    /* invert matrix */
    gaussj(ai,n,x,m);
    printf("\nInverse of matrix a : \n");
    for (k=1;k<=n;k++) {
        for (l=1;l<=n;l++) printf("%12.6f",ai[k][l]);
        printf("\n");
    }
    /* check inverse */
    printf("\na times a-inverse:\n");
    for (k=1;k<=n;k++) {
        for (l=1;l<=n;l++) {
            u[k][l]=0.0;
            for (j=1;j<=n;j++)
                u[k][l] += (a[k][j]*ai[j][l]);
        }
        for (l=1;l<=n;l++) printf("%12.6f",u[k][l]);
        printf("\n");
    }
    /* check vector solutions */
    printf("\nCheck the following for equality:\n");
    printf("%21s %14s\n","original","matrix*sol'n");
    for (l=1;l<=m;l++) {
        printf("vector %2d: \n",l);
        for (k=1;k<=n;k++) {
            t[k][l]=0.0;
            for (j=1;j<=n;j++)
                t[k][l] += (a[k][j]*x[j][l]);
            printf("%8s %12.6f %12.6f\n"," ",
                b[k][l],t[k][l]);
        }
```

```
      }
      printf("**********************************\n");
      printf("press RETURN for next problem:\n");
      (void) getchar();
    }
    fclose(fp);
    free_matrix(t,1,NP,1,MP);
    free_matrix(x,1,NP,1,MP);
    free_matrix(b,1,NP,1,MP);
    free_matrix(u,1,NP,1,NP);
    free_matrix(ai,1,NP,1,NP);
    free_matrix(a,1,NP,1,NP);
    return 0;
}
```

The demonstration program for routine ludcmp relies on the same package of test matrices, but just performs an LU decomposition of each. The performance is checked by multiplying the lower and upper matrices of the decomposition and comparing with the original matrix. The array indx keeps track of the scrambling done by ludcmp to effect partial pivoting. We had to do the unscrambling here, but you will normally not be called upon to do so, since ludcmp is used with the routine lubksb, which knows how to do its own descrambling.

```
/* Driver for routine ludcmp */

#include <stdio.h>
#include <stdlib.h>
#include "nr.h"
#include "nrutil.h"

#define NP 20
#define MAXSTR 80

int main(void)
{
    int j,k,l,m,n,dum,*indx,*jndx;
    float d,**a,**xl,**xu,**x;
    char dummy[MAXSTR];
    FILE *fp;

    indx=ivector(1,NP);
    jndx=ivector(1,NP);
    a=matrix(1,NP,1,NP);
    xl=matrix(1,NP,1,NP);
    xu=matrix(1,NP,1,NP);
    x=matrix(1,NP,1,NP);
    if ((fp = fopen("matrx1.dat","r")) == NULL)
        nrerror("Data file matrx1.dat not found\n");
    while (!feof(fp)) {
        fgets(dummy,MAXSTR,fp);
        fgets(dummy,MAXSTR,fp);
        fscanf(fp,"%d %d ",&n,&m);
        fgets(dummy,MAXSTR,fp);
        for (k=1;k<=n;k++)
            for (l=1;l<=n;l++) fscanf(fp,"%f ",&a[k][l]);
        fgets(dummy,MAXSTR,fp);
```

```
for (l=1;l<=m;l++)
    for (k=1;k<=n;k++) fscanf(fp,"%f ",&x[k][l]);
/* Print out a-matrix for comparison with product of
   lower and upper decomposition matrices */
printf("original matrix:\n");
for (k=1;k<=n;k++) {
    for (l=1;l<=n;l++) printf("%12.6f",a[k][l]);
    printf("\n");
}
/* Perform the decomposition */
ludcmp(a,n,indx,&d);
/* Compose separately the lower and upper matrices */
for (k=1;k<=n;k++) {
    for (l=1;l<=n;l++) {
        if (l > k) {
            xu[k][l]=a[k][l];
            xl[k][l]=0.0;
        } else if (l < k) {
            xu[k][l]=0.0;
            xl[k][l]=a[k][l];
        } else {
            xu[k][l]=a[k][l];
            xl[k][l]=1.0;
        }
    }
}
/* Compute product of lower and upper matrices for
   comparison with original matrix */
for (k=1;k<=n;k++) {
    jndx[k]=k;
    for (l=1;l<=n;l++) {
        x[k][l]=0.0;
        for (j=1;j<=n;j++)
            x[k][l] += (xl[k][j]*xu[j][l]);
    }
}
printf("\n%s%s\n","product of lower and upper ",
     "matrices (rows unscrambled):");
for (k=1;k<=n;k++) {
    dum=jndx[indx[k]];
    jndx[indx[k]]=jndx[k];
    jndx[k]=dum;
}
for (k=1;k<=n;k++)
    for (j=1;j<=n;j++)
        if (jndx[j] == k) {
            for (l=1;l<=n;l++)
                printf("%12.6f",x[j][l]);
            printf("\n");
        }
printf("\nlower matrix of the decomposition:\n");
for (k=1;k<=n;k++) {
    for (l=1;l<=n;l++) printf("%12.6f",xl[k][l]);
    printf("\n");
}
printf("\nupper matrix of the decomposition:\n");
for (k=1;k<=n;k++) {
```

```
            for (l=1;l<=n;l++) printf("%12.6f",xu[k][l]);
            printf("\n");
        }
        printf("\n********************************\n");
        printf("press return for next problem:\n");
        (void) getchar();
    }
    fclose(fp);
    free_matrix(x,1,NP,1,NP);
    free_matrix(xu,1,NP,1,NP);
    free_matrix(xl,1,NP,1,NP);
    free_matrix(a,1,NP,1,NP);
    free_ivector(jndx,1,NP);
    free_ivector(indx,1,NP);
    return 0;
}
```

Our example driver for `lubksb` makes calls to both `ludcmp` and `lubksb` in order to solve the linear equation problems posed in file `matrx1.dat` (see discussion of `gaussj`). The original matrix of coefficients is applied to the solution vectors to check that the result matches the right-hand side vectors posed for each problem. We apologize for using routine `ludcmp` in a test of `lubksb`, but `ludcmp` has been tested independently, and anyway, `lubksb` is nothing without this partner program, so a test of the combination is more to the point.

```
/* Driver for routine lubksb */

#include <stdio.h>
#include <stdlib.h>
#include "nr.h"
#include "nrutil.h"

#define NP 20
#define MAXSTR 80

int main(void)
{
    int j,k,l,m,n,*indx;
    float p,*x,**a,**b,**c;
    char dummy[MAXSTR];
    FILE *fp;

    indx=ivector(1,NP);
    x=vector(1,NP);
    a=matrix(1,NP,1,NP);
    b=matrix(1,NP,1,NP);
    c=matrix(1,NP,1,NP);
    if ((fp = fopen("matrx1.dat","r")) == NULL)
        nrerror("Data file matrx1.dat not found\n");
    while (!feof(fp)) {
        fgets(dummy,MAXSTR,fp);
        fgets(dummy,MAXSTR,fp);
        fscanf(fp,"%d %d ",&n,&m);
        fgets(dummy,MAXSTR,fp);
        for (k=1;k<=n;k++)
            for (l=1;l<=n;l++) fscanf(fp,"%f ",&a[k][l]);
```

```
          fgets(dummy,MAXSTR,fp);
          for (l=1;l<=m;l++)
              for (k=1;k<=n;k++) fscanf(fp,"%f ",&b[k][l]);
          /* Save matrix a for later testing */
          for (l=1;l<=n;l++)
              for (k=1;k<=n;k++) c[k][l]=a[k][l];
          /* Do LU decomposition */
          ludcmp(c,n,indx,&p);
          /* Solve equations for each right-hand vector */
          for (k=1;k<=m;k++) {
              for (l=1;l<=n;l++) x[l]=b[l][k];
              lubksb(c,n,indx,x);
              /* Test results with original matrix */
              printf("right-hand side vector:\n");
              for (l=1;l<=n;l++)
                  printf("%12.6f",b[l][k]);
              printf("\n%s%s\n","result of matrix applied",
                  " to sol'n vector");
              for (l=1;l<=n;l++) {
                  b[l][k]=0.0;
                  for (j=1;j<=n;j++)
                      b[l][k] += (a[l][j]*x[j]);
              }
              for (l=1;l<=n;l++)
                  printf("%12.6f",b[l][k]);
              printf("\n*********************************\n");
          }
          printf("press RETURN for next problem:\n");
          (void) getchar();
      }
      fclose(fp);
      free_matrix(c,1,NP,1,NP);
      free_matrix(b,1,NP,1,NP);
      free_matrix(a,1,NP,1,NP);
      free_vector(x,1,NP);
      free_ivector(indx,1,NP);
      return 0;
}
```

Procedure `tridag` solves linear equations with coefficients that form a tridiagonal matrix. We provide at the end of this chapter a second file of matrices `matrix2.dat` for the demonstration driver. In all other respects, the demonstration program `xtridag` operates in the same fashion as `xlubksb`.

```
/* Driver for routine tridag */

#include <stdio.h>
#include <stdlib.h>
#include "nr.h"
#include "nrutil.h"

#define NP 20
#define MAXSTR 80

int main(void)
{
    unsigned long k,n;
```

```
    float *diag,*superd,*subd,*rhs,*u;
    char dummy[MAXSTR];
    FILE *fp;

    diag=vector(1,NP);
    superd=vector(1,NP);
    subd=vector(1,NP);
    rhs=vector(1,NP);
    u=vector(1,NP);
    if ((fp = fopen("matrx2.dat","r")) == NULL)
        nrerror("Data file matrx2.dat not found\n");
    while (!feof(fp)) {
        fgets(dummy,MAXSTR,fp);
        fgets(dummy,MAXSTR,fp);
        fscanf(fp,"%ld ",&n);
        fgets(dummy,MAXSTR,fp);
        for (k=1;k<=n;k++) fscanf(fp,"%f ",&diag[k]);
        fgets(dummy,MAXSTR,fp);
        for (k=1;k<n;k++) fscanf(fp,"%f ",&superd[k]);
        fgets(dummy,MAXSTR,fp);
        for (k=2;k<=n;k++) fscanf(fp,"%f ",&subd[k]);
        fgets(dummy,MAXSTR,fp);
        for (k=1;k<=n;k++) fscanf(fp,"%f ",&rhs[k]);
        /* carry out solution */
        tridag(subd,diag,superd,rhs,u,n);
        printf("\nThe solution vector is:\n");
        for (k=1;k<=n;k++) printf("%12.6f",u[k]);
        printf("\n");
        /* test solution */
        printf("\n(matrix)*(sol'n vector) should be:\n");
        for (k=1;k<=n;k++) printf("%12.6f",rhs[k]);
        printf("\n");
        printf("actual result is:\n");
        for (k=1;k<=n;k++) {
            if (k == 1)
                rhs[k]=diag[1]*u[1]+superd[1]*u[2];
            else if (k == n)
                rhs[k]=subd[n]*u[n-1]+diag[n]*u[n];
            else
                rhs[k]=subd[k]*u[k-1]+diag[k]*u[k]
                    +superd[k]*u[k+1];
        }
        for (k=1;k<=n;k++) printf("%12.6f",rhs[k]);
        printf("\n");
        printf("**********************************\n");
        printf("press return for next problem:\n");
        (void) getchar();
    }
    fclose(fp);
    free_vector(u,1,NP);
    free_vector(rhs,1,NP);
    free_vector(subd,1,NP);
    free_vector(superd,1,NP);
    free_vector(diag,1,NP);
    return 0;
}
```

The demonstration program **xbanmul** forms a banded matrix, multiplies if by a vector using **banmul**, and compares the answer to the result of carrying out a full matrix multiplication.

```
/* Driver for routine banmul */

#include <stdio.h>
#include "nr.h"
#include "nrutil.h"

#define NP 7
#define M1 2
#define M2 1
#define MP (M1+1+M2)

int main(void)
{
    unsigned long i,j,k;
    float **a,**aa,*ax,*b,*x;

    a=matrix(1,NP,1,MP);
    aa=matrix(1,NP,1,NP);
    ax=vector(1,NP);
    b=vector(1,NP);
    x=vector(1,NP);
    for (i=1;i<=M1;i++) for (j=1;j<=NP;j++) a[j][i]=10.0*j+i;
    /* Lower band */
    for (i=1;i<=NP;i++) a[i][M1+1]=i;
    /* Diagonal */
    for (i=1;i<=M2;i++) for (j=1;j<=NP;j++) a[j][M1+1+i]=0.1*j+i;
    /* Upper band */
    for (i=1;i<=NP;i++) {
        for (j=1;j<=NP;j++) {
            k=i-M1-1;
            if (j>=LMAX(1,1+k) && j<=LMIN(M1+M2+1+k,NP))
            aa[i][j]=a[i][j-k];
            else aa[i][j]=0.0;
        }
    }
    for (i=1;i<=NP;i++) x[i]=i/10.0;
    banmul(a,NP,M1,M2,x,b);
    for (i=1;i<=NP;i++) {
        for (ax[i]=0.0,j=1;j<=NP;j++) ax[i] += aa[i][j]*x[j];
    }
    printf("\tReference vector\tbanmul vector\n");
    for (i=1;i<=NP;i++) printf("\t%12.4f\t%12.4f\n",ax[i],b[i]);
    free_vector(x,1,NP);
    free_vector(b,1,NP);
    free_vector(ax,1,NP);
    free_matrix(aa,1,NP,1,NP);
    free_matrix(a,1,NP,1,MP);
    return 0;
}
```

The sample program **xbandec** forms a random banded matrix $\mathbf{A}$, a random vector $\mathbf{x}$, and calculates $\mathbf{b} = \mathbf{A} \cdot \mathbf{x}$. It then supplies $\mathbf{A}$ and $\mathbf{b}$ to **bandec** and **banbks** and compares the solution to the saved copy of $\mathbf{x}$.

```
/* Driver for routine bandec */

#include <stdio.h>
#include "nr.h"
#include "nrutil.h"

int main(void)
{
    float d,**a,**al,*b,*x;
    unsigned long i,j,*indx;
    long idum=(-1);

    a=matrix(1,7,1,4);
    x=vector(1,7);
    b=vector(1,7);
    al=matrix(1,7,1,2);
    indx=lvector(1,7);
    for (i=1;i<=7;i++) {
        x[i]=ran1(&idum);
        for (j=1;j<=4;j++) {
            a[i][j]=ran1(&idum);
        }
    }
    banmul(a,7,2,1,x,b);
    for (i=1;i<=7;i++) printf("%ld %12.6f %12.6f\n",i,b[i],x[i]);
    bandec(a,7,2,1,al,indx,&d);
    banbks(a,7,2,1,al,indx,b);
    for (i=1;i<=7;i++) printf("%ld %12.6f %12.6f\n",i,b[i],x[i]);
    free_lvector(indx,1,7);
    free_matrix(al,1,7,1,2);
    free_vector(b,1,7);
    free_vector(x,1,7);
    free_matrix(a,1,7,1,4);
    return 0;
}
```

mprove is a short routine for improving the solution vector for a set of linear equations, providing that an LU decomposition has been performed on the matrix of coefficients. Our test of this function is to use ludcmp and lubksb to solve a set of equations specified by the initializations at the beginning of the program. The solution vector is then corrupted by the addition of random values to each component. mprove works on the corrupted vector to recover the original. Note the use of the utility routine convert_matrix() to change a conventionally defined matrix to the format used in *Numerical Recipes*.

```
/* Driver for routine mprove */

#include <stdio.h>
#include "nr.h"
#include "nrutil.h"

#define N 5
#define NP N

int main(void)
{
```

```
int i,j,*indx;
long idum=(-13);
float d,*x,**a,**aa;
static float ainit[NP][NP]=
    {1.0,2.0,3.0,4.0,5.0,
    2.0,3.0,4.0,5.0,1.0,
    1.0,1.0,1.0,1.0,1.0,
    4.0,5.0,1.0,2.0,3.0,
    5.0,1.0,2.0,3.0,4.0};
static float b[N+1]={0.0,1.0,1.0,1.0,1.0,1.0};

indx=ivector(1,N);
x=vector(1,N);
a=convert_matrix(&ainit[0][0],1,N,1,N);
aa=matrix(1,N,1,N);
for (i=1;i<=N;i++) {
    x[i]=b[i];
    for (j=1;j<=N;j++)
        aa[i][j]=a[i][j];
}
ludcmp(aa,N,indx,&d);
lubksb(aa,N,indx,x);
printf("\nSolution vector for the equations:\n");
for (i=1;i<=N;i++) printf("%12.6f",x[i]);
printf("\n");
/* now phoney up x and let mprove fix it */
for (i=1;i<=N;i++) x[i] *= (1.0+0.2*ran3(&idum));
printf("\nSolution vector with noise added:\n");
for (i=1;i<=N;i++) printf("%12.6f",x[i]);
printf("\n");
mprove(a,aa,N,indx,b,x);
printf("\nSolution vector recovered by mprove:\n");
for (i=1;i<=N;i++) printf("%12.6f",x[i]);
printf("\n");
free_matrix(aa,1,N,1,N);
free_convert_matrix(a,1,N,1,N);
free_vector(x,1,N);
free_ivector(indx,1,N);
return 0;
}
```

The pair svdcmp, svbksb is tested in the same manner as ludcmp, lubksb. That is, svdcmp is checked independently to see that it yields proper decomposition of matrices. Then the pair of programs is tested as a unit to see that it provides correct solutions to some linear sets. (Note: Because of the order of programs in *Numerical Recipes*, the test of the pair in this case comes first).

Driver xsvbksb brings in matrices a and right-hand side vectors b from ma-trx1.dat. Matrix a, itself, is saved for later use. It is copied into matrix u for processing by svdcmp. The results of the processing are the three arrays u, w, v which form the singular value decomposition of a. The right-hand side vectors are fed one at a time to vector c, and the resulting solution vectors x are checked for accuracy through application of the saved matrix a.

```
/* Driver for routine svbksb, which calls routine svdcmp */

#include <stdio.h>
#include <stdlib.h>
#include "nr.h"
#include "nrutil.h"

#define NP 20
#define MP 20
#define MAXSTR 80

int main(void)
{
    int j,k,l,m,n;
    float wmax,wmin,*w,*x,*c;
    float **a,**b,**u,**v;
    char dummy[MAXSTR];
    FILE *fp;

    w=vector(1,NP);
    x=vector(1,NP);
    c=vector(1,NP);
    a=matrix(1,NP,1,NP);
    b=matrix(1,NP,1,MP);
    u=matrix(1,NP,1,NP);
    v=matrix(1,NP,1,NP);
    if ((fp = fopen("matrx1.dat","r")) == NULL)
        nrerror("Data file matrx1.dat not found\n");
    while (!feof(fp)) {
        fgets(dummy,MAXSTR,fp);
        fgets(dummy,MAXSTR,fp);
        fscanf(fp,"%d %d ",&n,&m);
        fgets(dummy,MAXSTR,fp);
        for (k=1;k<=n;k++)
            for (l=1;l<=n;l++) fscanf(fp,"%f ",&a[k][l]);
        fgets(dummy,MAXSTR,fp);
        for (l=1;l<=m;l++)
            for (k=1;k<=n;k++) fscanf(fp,"%f ",&b[k][l]);
        /* copy a into u */
        for (k=1;k<=n;k++)
            for (l=1;l<=n;l++) u[k][l]=a[k][l];
        /* decompose matrix a */
        svdcmp(u,n,n,w,v);
        /* find maximum singular value */
        wmax=0.0;
        for (k=1;k<=n;k++)
            if (w[k] > wmax) wmax=w[k];
        /* define "small" */
        wmin=wmax*(1.0e-6);
        /* zero the "small" singular values */
        for (k=1;k<=n;k++)
            if (w[k] < wmin) w[k]=0.0;
        /* backsubstitute for each right-hand side vector */
        for (l=1;l<=m;l++) {
            printf("\nVector number %2d\n",l);
            for (k=1;k<=n;k++) c[k]=b[k][l];
            svbksb(u,w,v,n,n,c,x);
```

```
                printf(" solution vector is:\n");
                for (k=1;k<=n;k++) printf("%12.6f",x[k]);
                printf("\n original right-hand side vector:\n");
                for (k=1;k<=n;k++) printf("%12.6f",c[k]);
                printf("\n (matrix)*(sol'n vector):\n");
                for (k=1;k<=n;k++) {
                    c[k]=0.0;
                    for (j=1;j<=n;j++)
                        c[k] += a[k][j]*x[j];
                }
                for (k=1;k<=n;k++) printf("%12.6f",c[k]);
                printf("\n");
            }
        printf ("***********************************\n");
        printf("press RETURN for next problem\n");
        (void) getchar();
    }
    fclose(fp);
    free_matrix(v,1,NP,1,NP);
    free_matrix(u,1,NP,1,NP);
    free_matrix(b,1,NP,1,MP);
    free_matrix(a,1,NP,1,NP);
    free_vector(c,1,NP);
    free_vector(x,1,NP);
    free_vector(w,1,NP);
    return 0;
}
```

Companion driver `xsvdcmp` takes the matrices a from `matrx3.dat` The rather pathological test matrices for `xsvdcmp` are given in the Appendix as file `matrx3.dat`. The routine reads each matrix a in and passes a copy of each to `svdcmp` for singular value decomposition into u, w, and v. Then u, w, and the transpose of v are multiplied together. The result is compared to a saved copy of a.

```
/* Driver for routine svdcmp */

#include <stdio.h>
#include <stdlib.h>
#include "nr.h"
#include "nrutil.h"

#define NP 20
#define MP 20
#define MAXSTR 80

int main(void)
{
    int j,k,l,m,n;
    float *w,**a,**u,**v;
    char dummy[MAXSTR];
    FILE *fp;

    w=vector(1,NP);
    a=matrix(1,MP,1,NP);
    u=matrix(1,MP,1,NP);
    v=matrix(1,NP,1,NP);
    /* read input matrices */
```

```
if ((fp = fopen("matrx3.dat","r")) == NULL)
    nrerror("Data file matrx3.dat not found\n");
while (!feof(fp)) {
    fgets(dummy,MAXSTR,fp);
    fgets(dummy,MAXSTR,fp);
    fscanf(fp,"%d %d ",&m,&n);
    fgets(dummy,MAXSTR,fp);
    /* copy original matrix into u */
    for (k=1;k<=m;k++)
        for (l=1;l<=n;l++) {
            fscanf(fp,"%f ",&a[k][l]);
            u[k][l]=a[k][l];
        }
    /* perform decomposition */
    svdcmp(u,m,n,w,v);
    /* write results */
    printf("Decomposition matrices:\n");
    printf("Matrix u\n");
    for (k=1;k<=m;k++) {
        for (l=1;l<=n;l++)
            printf("%12.6f",u[k][l]);
        printf("\n");
    }
    printf("Diagonal of matrix w\n");
    for (k=1;k<=n;k++)
        printf("%12.6f",w[k]);
    printf("\nMatrix v-transpose\n");
    for (k=1;k<=n;k++) {
        for (l=1;l<=n;l++)
            printf("%12.6f",v[l][k]);
        printf("\n");
    }
    printf("\nCheck product against original matrix:\n");
    printf("Original matrix:\n");
    for (k=1;k<=m;k++) {
        for (l=1;l<=n;l++)
            printf("%12.6f",a[k][l]);
        printf("\n");
    }
    printf("Product u*w*(v-transpose):\n");
    for (k=1;k<=m;k++) {
        for (l=1;l<=n;l++) {
            a[k][l]=0.0;
            for (j=1;j<=n;j++)
                a[k][l] += u[k][j]*w[j]*v[l][j];
        }
        for (l=1;l<=n;l++) printf("%12.6f",a[k][l]);
        printf("\n");
    }
    printf("*********************************\n");
    printf("press RETURN for next problem\n");
    (void) getchar();
}
fclose(fp);
free_matrix(v,1,NP,1,NP);
free_matrix(u,1,MP,1,NP);
free_matrix(a,1,MP,1,NP);
```

```
        free_vector(w,1,NP);
        return 0;
}
```

A *cyclic tridiagonal* matrix is just like a tridiagonal matrix, but with additional nonzero elements in the top right and bottom left corners. Sample program xcyclic generates a random cyclic tridiagonal matrix and right-hand side vector. It then calls cyclic to solve the resulting linear system. The solution is checked by also solving the system with ludcmp and lubksb and printing out the fractional discrepancy in the answer.

```
/* Driver for routine cyclic */

#include <stdio.h>
#include "nr.h"
#include "nrutil.h"

#define N 20

int main(void)
{
    float alpha,beta,d,*a,*b,*c,*r,*x,**aa;
    int i,j,*indx;
    long idum=(-23);

    indx=ivector(1,N);
    a=vector(1,N);
    b=vector(1,N);
    c=vector(1,N);
    r=vector(1,N);
    x=vector(1,N);
    aa=matrix(1,N,1,N);
    for (i=1;i<=N;i++)
        for (j=1;j<=N;j++) aa[i][j]=0.0;
    for (i=1;i<=N;i++) {
        b[i]=ran2(&idum);
        aa[i][i]=b[i];
        r[i]=ran2(&idum);
    }
    for (i=1;i<N;i++) {
        a[i+1]=ran2(&idum);
        aa[i+1][i]=a[i+1];
        c[i]=ran2(&idum);
        aa[i][i+1]=c[i];
    }
    alpha=ran2(&idum);
    aa[N][1]=alpha;
    beta=ran2(&idum);
    aa[1][N]=beta;
    cyclic(a,b,c,alpha,beta,r,x,N);
    ludcmp(aa,N,indx,&d);
    lubksb(aa,N,indx,r);
    for (i=1;i<=N;i++) printf("%4d  %12.6e\n",i,(x[i]-r[i])/(x[i]+r[i]));
    free_matrix(aa,1,N,1,N);
    free_vector(x,1,N);
    free_vector(r,1,N);
```

```
        free_vector(c,1,N);
        free_vector(b,1,N);
        free_vector(a,1,N);
        free_ivector(indx,1,N);
        return 0;
}
```

Next follow the routines for testing the sparse matrix routines. The first, xsprsin, takes the matrix defined in Equation (2.7.27) of *Numerical Recipes*, prints out the corresponding ija and sa vectors (which you can check against Equation 2.7.28), and then reconstructs the matrix from these vectors.

```
/* Driver for routine sprsin */

#include <stdio.h>
#include "nr.h"
#include "nrutil.h"

#define NP 5
#define NMAX (2*NP*NP+1)

int main(void)
{
    unsigned long i,j,msize,*ija;
    float **a,**aa,*sa;
    static float ainit[NP][NP]={
        3.0,0.0,1.0,0.0,0.0,
        0.0,4.0,0.0,0.0,0.0,
        0.0,7.0,5.0,9.0,0.0,
        0.0,0.0,0.0,0.0,2.0,
        0.0,0.0,0.0,6.0,5.0};

    ija=lvector(1,NMAX);
    sa=vector(1,NMAX);
    aa=matrix(1,NP,1,NP);
    a=convert_matrix(&ainit[0][0],1,NP,1,NP);
    sprsin(a,NP,0.5,NMAX,sa,ija);
    msize=ija[ija[1]-1]-1;
    sa[NP+1]=0.0;
    printf("index\tija\t\tsa\n");
    for (i=1;i<=msize;i++) printf("%lu\t%lu\t%12.6f\n",i,ija[i],sa[i]);
    for (i=1;i<=NP;i++) for (j=1;j<=NP;j++) aa[i][j]=0.0;
    for (i=1;i<=NP;i++) {
        aa[i][i]=sa[i];
        for (j=ija[i];j<=ija[i+1]-1;j++) aa[i][ija[j]]=sa[j];
    }
    printf("Original Matrix\n");
    for (i=1;i<=NP;i++) {
        for (j=1;j<=NP;j++) printf("%5.2f\t",a[i][j]);
        printf("\n");
    }
    printf("Reconstructed Matrix\n");
    for (i=1;i<=NP;i++) {
        for (j=1;j<=NP;j++) printf("%5.2f\t",aa[i][j]);
        printf("\n");
    }
    free_convert_matrix(a,1,NP,1,NP);
```

```
        free_matrix(aa,1,NP,1,NP);
        free_vector(sa,1,NMAX);
        free_lvector(ija,1,NMAX);
        return 0;
}
```

Program `xsprsax` takes the same sparse matrix and multiplies it by the vector (1,2,3,4,5) using `sprsax`. It compares the result to that of a full matrix multiplication.

```
/* Driver for routine sprsax */

#include <stdio.h>
#include "nr.h"
#include "nrutil.h"

#define NP 5
#define NMAX (2*NP*NP+1)

int main(void)
{
    unsigned long i,j,msize,*ija;
    float **a,*sa,*ax,*b;
    static float ainit[NP][NP]={
        3.0,0.0,1.0,0.0,0.0,
        0.0,4.0,0.0,0.0,0.0,
        0.0,7.0,5.0,9.0,0.0,
        0.0,0.0,0.0,0.0,2.0,
        0.0,0.0,0.0,6.0,5.0};
    static float x[NP+1] = {0.0,1.0,2.0,3.0,4.0,5.0};

    ija=lvector(1,NMAX);
    ax=vector(1,NP);
    b=vector(1,NP);
    sa=vector(1,NMAX);
    a=convert_matrix(&ainit[0][0],1,NP,1,NP);
    sprsin(a,NP,0.5,NMAX,sa,ija);
    msize=ija[1]-2;
    sprsax(sa,ija,x,b,msize);
    for (i=1;i<=msize;i++)
        for (ax[i]=0.0,j=1;j<=msize;j++) ax[i] += a[i][j]*x[j];
    printf("\tReference\tsprsax result\n");
    for (i=1;i<=msize;i++) printf("\t%5.2f\t\t%5.2f\n",ax[i],b[i]);
    free_convert_matrix(a,1,NP,1,NP);
    free_vector(sa,1,NMAX);
    free_vector(b,1,NP);
    free_vector(ax,1,NP);
    free_lvector(ija,1,NMAX);
    return 0;
}
```

Program `xsprstx` works in exactly the same way as the previous program to test `sprstx` by computing the product of the transpose of the sparse matrix with the vector:

```
/* Driver for routine sprstx */

#include <stdio.h>
```

```
#include "nr.h"
#include "nrutil.h"

#define NP 5
#define NMAX (2*NP*NP+1)

int main(void)
{
    unsigned long i,j,msize,*ija;
    float **a,*sa,*ax,*b;
    static float ainit[NP][NP]={
        3.0,0.0,1.0,0.0,0.0,
        0.0,4.0,0.0,0.0,0.0,
        0.0,7.0,5.0,9.0,0.0,
        0.0,0.0,0.0,0.0,2.0,
        0.0,0.0,0.0,6.0,5.0};
    static float x[NP+1] = {0.0,1.0,2.0,3.0,4.0,5.0};

    ija=lvector(1,NMAX);
    ax=vector(1,NP);
    b=vector(1,NP);
    sa=vector(1,NMAX);
    a=convert_matrix(&ainit[0][0],1,NP,1,NP);
    sprsin(a,NP,0.5,NMAX,sa,ija);
    msize=ija[1]-2;
    sprstx(sa,ija,x,b,msize);
    for (i=1;i<=msize;i++)
        for (ax[i]=0.0,j=1;j<=msize;j++) ax[i] += a[j][i]*x[j];
    printf("\tReference\tsprstx result\n");
    for (i=1;i<=msize;i++) printf("\t%5.2f\t\t%5.2f\n",ax[i],b[i]);
    free_convert_matrix(a,1,NP,1,NP);
    free_vector(sa,1,NMAX);
    free_vector(b,1,NP);
    free_vector(ax,1,NP);
    free_lvector(ija,1,NMAX);
    return 0;
}
```

Program `xsprstp` simply prints out the transpose of our sample sparse matrix after calling `sprstp`:

```
/* Driver for routine sprstp */

#include <stdio.h>
#include "nr.h"
#include "nrutil.h"

#define NP 5
#define NMAX (2*NP*NP+1)

int main(void)
{
    unsigned long i,j,*ija,*ijat;
    float **a,**at,*sa,*sat;
    static float ainit[NP][NP]={
        3.0,0.0,1.0,0.0,0.0,
        0.0,4.0,0.0,0.0,0.0,
```

```
        0.0,7.0,5.0,9.0,0.0,
        0.0,0.0,0.0,0.0,2.0,
        0.0,0.0,0.0,6.0,5.0};

    ija=lvector(1,NMAX);
    ijat=lvector(1,NMAX);
    sa=vector(1,NMAX);
    sat=vector(1,NMAX);
    at=matrix(1,NP,1,NP);
    a=convert_matrix(&ainit[0][0],1,NP,1,NP);
    sprsin(a,NP,0.5,NMAX,sa,ija);
    sprstp(sa,ija,sat,ijat);
    for (i=1;i<=NP;i++) for (j=1;j<=NP;j++) at[i][j]=0.0;
    for (i=1;i<=NP;i++) {
        at[i][i]=sat[i];
        for (j=ijat[i];j<=ijat[i+1]-1;j++) at[i][ijat[j]]=sat[j];
    }
    printf("Original Matrix\n");
    for (i=1;i<=NP;i++) {
        for (j=1;j<=NP;j++) printf("%5.2f\t",a[i][j]);
        printf("\n");
    }
    printf("Transpose\n");
    for (i=1;i<=NP;i++) {
        for (j=1;j<=NP;j++) printf("%5.2f\t",at[i][j]);
        printf("\n");
    }
    free_convert_matrix(a,1,NP,1,NP);
    free_matrix(at,1,NP,1,NP);
    free_vector(sat,1,NMAX);
    free_vector(sa,1,NMAX);
    free_lvector(ijat,1,NMAX);
    free_lvector(ija,1,NMAX);
    return 0;
}
```

The next program, xsprspm, demonstrates the use of sprspm. Two sparse matrices $\mathbf{A}$ and $\mathbf{B}$ are defined, and then the matrix $\mathbf{A} \cdot \mathbf{B}$ is computed. Since sprspm actually calculates $\mathbf{A} \cdot \mathbf{B}^T$, we first form $\mathbf{B}^T$ by a call to sprstp. Only a given sparsity pattern is output by sprspm. In the example, we specify this to be the tridiagonal part of $\mathbf{A} \cdot \mathbf{B}$. As usual, we check the answer by carrying out a full matrix multiplication.

```
/* Driver for routine sprspm */

#include <stdio.h>
#include "nr.h"
#include "nrutil.h"

#define NP 5
#define NMAX (2*NP*NP+1)

int main(void)
{
    unsigned long i,j,k,*ija,*ijb,*ijbt,*ijc;
    float *sa,*sb,*sbt,*sc,**a,**b,**c,**ab;
    static float ainit[NP][NP]={
        1.0,0.5,0.0,0.0,0.0,
```

```
            0.5,2.0,0.5,0.0,0.0,
            0.0,0.5,3.0,0.5,0.0,
            0.0,0.0,0.5,4.0,0.5,
            0.0,0.0,0.0,0.5,5.0};
    static float binit[NP][NP]={
            1.0,1.0,0.0,0.0,0.0,
            1.0,2.0,1.0,0.0,0.0,
            0.0,1.0,3.0,1.0,0.0,
            0.0,0.0,1.0,4.0,1.0,
            0.0,0.0,0.0,1.0,5.0};

    ija=lvector(1,NMAX);
    ijb=lvector(1,NMAX);
    ijbt=lvector(1,NMAX);
    ijc=lvector(1,NMAX);
    sa=vector(1,NMAX);
    sb=vector(1,NMAX);
    sbt=vector(1,NMAX);
    sc=vector(1,NMAX);
    c=matrix(1,NP,1,NP);
    ab=matrix(1,NP,1,NP);
    a=convert_matrix(&ainit[0][0],1,NP,1,NP);
    b=convert_matrix(&binit[0][0],1,NP,1,NP);
    sprsin(a,NP,0.5,NMAX,sa,ija);
    sprsin(b,NP,0.5,NMAX,sb,ijb);
    sprstp(sb,ijb,sbt,ijbt);
    /* specify tridiagonal output, using fact that a is tridiagonal */
    for (i=1;i<=ija[ija[1]-1]-1;i++) ijc[i]=ija[i];
    sprspm(sa,ija,sbt,ijbt,sc,ijc);
    for (i=1;i<=NP;i++) {
        for (j=1;j<=NP;j++) {
            ab[i][j]=0.0;
            for (k=1;k<=NP;k++) {
                ab[i][j]=ab[i][j]+a[i][k]*b[k][j];
            }
        }
    }
    printf("Reference matrix:\n");
    for (i=1;i<=NP;i++) {
        for (j=1;j<=NP;j++) printf("%5.2f\t",ab[i][j]);
        printf("\n");
    }
    printf("sprspm matrix (should show only tridiagonals):\n");
    for (i=1;i<=NP;i++) for (j=1;j<=NP;j++) c[i][j]=0.0;
    for (i=1;i<=NP;i++) {
        c[i][i]=sc[i];
        for (j=ijc[i];j<=ijc[i+1]-1;j++) c[i][ijc[j]]=sc[j];
    }
    for (i=1;i<=NP;i++) {
        for (j=1;j<=NP;j++) printf("%5.2f\t",c[i][j]);
        printf("\n");
    }
    free_convert_matrix(b,1,NP,1,NP);
    free_convert_matrix(a,1,NP,1,NP);
    free_matrix(ab,1,NP,1,NP);
    free_matrix(c,1,NP,1,NP);
    free_vector(sc,1,NMAX);
```

```
        free_vector(sbt,1,NMAX);
        free_vector(sb,1,NMAX);
        free_vector(sa,1,NMAX);
        free_lvector(ijc,1,NMAX);
        free_lvector(ijbt,1,NMAX);
        free_lvector(ijb,1,NMAX);
        free_lvector(ija,1,NMAX);
        return 0;
}
```

Routine sprstm operates in the same way as sprspm, except the output is all components of $\mathbf{A} \cdot \mathbf{B}^T$ greater than some threshold value, here chosen to be 0.99. The test program xsprstm is thus very similar to the previous test program.

```
/* Driver for routine sprstm */

#include <stdio.h>
#include "nr.h"
#include "nrutil.h"

#define NP 5
#define NMAX (2*NP*NP+1)
#define THRESH 0.99

int main(void)
{
    unsigned long i,j,k,*ija,*ijb,*ijbt,*ijc,msize;
    float *sa,*sb,*sbt,*sc,**a,**b,**c,**ab;
    static float ainit[NP][NP]={
        1.0,0.5,0.0,0.0,0.0,
        0.5,2.0,0.5,0.0,0.0,
        0.0,0.5,3.0,0.5,0.0,
        0.0,0.0,0.5,4.0,0.5,
        0.0,0.0,0.0,0.5,5.0};
    static float binit[NP][NP]={
        1.0,1.0,0.0,0.0,0.0,
        1.0,2.0,1.0,0.0,0.0,
        0.0,1.0,3.0,1.0,0.0,
        0.0,0.0,1.0,4.0,1.0,
        0.0,0.0,0.0,1.0,5.0};

    ija=lvector(1,NMAX);
    ijb=lvector(1,NMAX);
    ijbt=lvector(1,NMAX);
    ijc=lvector(1,NMAX);
    sa=vector(1,NMAX);
    sb=vector(1,NMAX);
    sbt=vector(1,NMAX);
    sc=vector(1,NMAX);
    c=matrix(1,NP,1,NP);
    ab=matrix(1,NP,1,NP);
    a=convert_matrix(&ainit[0][0],1,NP,1,NP);
    b=convert_matrix(&binit[0][0],1,NP,1,NP);
    sprsin(a,NP,0.5,NMAX,sa,ija);
    sprsin(b,NP,0.5,NMAX,sb,ijb);
    sprstp(sb,ijb,sbt,ijbt);
    msize=ija[ija[1]-1]-1;
```

```
sprstm(sa,ija,sbt,ijbt,THRESH,msize,sc,ijc);
for (i=1;i<=NP;i++) {
    for (j=1;j<=NP;j++) {
        ab[i][j]=0.0;
        for (k=1;k<=NP;k++) {
            ab[i][j]=ab[i][j]+a[i][k]*b[k][j];
        }
    }
}
printf("Reference matrix:\n");
for (i=1;i<=NP;i++) {
    for (j=1;j<=NP;j++) printf("%5.2f\t",ab[i][j]);
    printf("\n");
}
printf("sprstm matrix (off-diag. elements of mag >): %12.6f\n",THRESH);
for (i=1;i<=NP;i++) for (j=1;j<=NP;j++) c[i][j]=0.0;
for (i=1;i<=NP;i++) {
    c[i][i]=sc[i];
    for (j=ijc[i];j<=ijc[i+1]-1;j++) c[i][ijc[j]]=sc[j];
}
for (i=1;i<=NP;i++) {
    for (j=1;j<=NP;j++) printf("%5.2f\t",c[i][j]);
    printf("\n");
}
free_convert_matrix(b,1,NP,1,NP);
free_convert_matrix(a,1,NP,1,NP);
free_matrix(ab,1,NP,1,NP);
free_matrix(c,1,NP,1,NP);
free_vector(sc,1,NMAX);
free_vector(sbt,1,NMAX);
free_vector(sb,1,NMAX);
free_vector(sa,1,NMAX);
free_lvector(ijc,1,NMAX);
free_lvector(ijbt,1,NMAX);
free_lvector(ijb,1,NMAX);
free_lvector(ija,1,NMAX);
return 0;
}
```

Routine linbcg solves linear systems $\mathbf{A} \cdot \mathbf{x} = \mathbf{b}$ with a sparse matrix $\mathbf{A}$. The matrix $\mathbf{A}$ must be specified in sparse matrix format as described in *Numerical Recipes*. In our sample program xlinbcg we define $\mathbf{A}$ to be the 20×20 matrix

$$\begin{pmatrix} 1.0 & 2.0 & 0.0 & 0.0 & \dots \\ -2.0 & 1.0 & 2.0 & 0.0 & \dots \\ 0.0 & -2.0 & 1.0 & 2.0 & \dots \\ 0.0 & 0.0 & -2.0 & 1.0 & \dots \\ \vdots & \vdots & \vdots & \vdots & \ddots \end{pmatrix}$$

As a right-hand side vector $\mathbf{b}$ we have taken $(3.0, 1.0, 1.0, \dots, -1.0)$, and the solution is given as $\mathbf{x}$. Notice that the components of $\mathbf{x}$ are all initialized to zero. You will set them to some initial guess of the solution to your own problem, but a zero guess will usually suffice. The solution in xlinbcg is given the usual checks.

```c
/* Driver for routine linbcg */

#include <stdio.h>
#include <math.h>
#include "nr.h"
#include "nrutil.h"

#define NSIZE 59
#define NP 20
#define ITOL 1
#define TOL 1e-9
#define ITMAX 75

unsigned long ija[NSIZE+1] ={
    0,22,23,25,27,29,31,33,35,37,39,41,43,45,47,49,51,53,55,57,
    59,60,2,1,3,2,4,3,5,4,6,5,7,6,8,7,9,8,10,9,11,10,12,11,13,12,
    14,13,15,14,16,15,17,16,18,17,19,18,20,19};

double sa[NSIZE+1] ={
    0.0,3.0,3.0,3.0,3.0,3.0,3.0,3.0,3.0,3.0,3.0,3.0,3.0,3.0,3.0,3.0,3.0,
    3.0,3.0,3.0,3.0,0.0,2.0,-2.0,2.0,-2.0,2.0,-2.0,2.0,-2.0,2.0,-2.0,
    2.0,-2.0,2.0,-2.0,2.0,-2.0,2.0,-2.0,2.0,-2.0,2.0,-2.0,2.0,-2.0,2.0,
    -2.0,2.0,-2.0,2.0,-2.0,2.0,-2.0,2.0,-2.0,2.0,-2.0,2.0,-2.0};

int main(void)
{
    int i,ii,iter;
    double *b,*bcmp,*x,err;

    b=dvector(1,NP);
    bcmp=dvector(1,NP);
    x=dvector(1,NP);
    for (i=1;i<=NP;i++) {
        x[i]=0.0;
        b[i]=1.0;
    }
    b[1]=3.0;
    b[NP] = -1.0;
    linbcg(NP,b,x,ITOL,TOL,ITMAX,&iter,&err);
    printf("%s %15e\n","Estimated error:",err);
    printf("%s %6d\n","Iterations needed:",iter);
    printf("\nSolution vector:\n");
    for (ii=1;ii<=NP/5;ii++) {
        for (i=5*(ii-1)+1;i<=5*ii;i++) printf("%12.6f",x[i]);
        printf("\n");
    }
    for (i=1;i<=(NP % 5);i++)
        printf("%12.6f",x[5*(NP/5)+i]);
    printf("\n");
    dsprsax(sa,ija,x,bcmp,NP);
    /* this is a double precision version of sprsax */
    printf("\npress RETURN to continue...\n");
    (void) getchar();
    printf("test of solution vector:\n");
    printf("%9s %12s\n","a*x","b");
    for (i=1;i<=NP;i++)
        printf("%12.6f %12.6f\n",bcmp[i],b[i]);
```

```
        free_dvector(x,1,NP);
        free_dvector(bcmp,1,NP);
        free_dvector(b,1,NP);
        return 0;
}
```

Vandermonde matrices of dimension $N \times N$ have elements that are entirely integer powers of N arbitrary numbers $x_1 \ldots x_N$. (See *Numerical Recipes* for details). In the demonstration program xvander we provide five such numbers to specify a 5×5 matrix, and five elements of a right-hand side vector Q. Routine vander is used to find the solution vector W. This vector is tested by applying the matrix to W and comparing the result to Q.

```
/* Driver for routine vander */

#include <stdio.h>
#include "nr.h"
#include "nrutil.h"

#define N 5

int main(void)
{
    int i,j;
    double sum=0.0,*w,*term;
    static double x[]={0.0,1.0,1.5,2.0,2.5,3.0};
    static double q[]={0.0,1.0,1.5,2.0,2.5,3.0};

    w=dvector(1,N);
    term=dvector(1,N);
    vander(x,w,q,N);
    printf("\nSolution vector:\n");
    for (i=1;i<=N;i++)
        printf("%7s%1d%2s %12f \n","w[",i,"]=",w[i]);
    printf("\nTest of solution vector:\n");
    printf("%14s %11s\n","mtrx*sol'n","original");
    for (i=1;i<=N;i++) {
        term[i]=w[i];
        sum += w[i];
    }
    printf("%12.4f %12.4f\n",sum,q[1]);
    for (i=2;i<=N;i++) {
        sum=0.0;
        for (j=1;j<=N;j++) {
            term[j] *= x[j];
            sum += term[j];
        }
        printf("%12.4f %12.4f\n",sum,q[i]);
    }
    free_dvector(term,1,N);
    free_dvector(w,1,N);
    return 0;
}
```

A very similar test is applied to toeplz, which operates on Toeplitz matrices. The $N \times N$ Toeplitz matrix is specified by $2N - 1$ numbers r_i, in this case taken to be simply a linear progression of values. A right-hand side y_i is chosen likewise. toeplz finds the solution vector x_i, and checks it in the usual fashion.

```c
/* Driver for routine toeplz */

#include <stdio.h>
#include "nr.h"

#define N 5
#define TWON (2*N)

int main(void)
{
    int i,j;
    float sum,r[TWON+1],x[N+1],y[N+1];

    for (i=1;i<=N;i++) y[i]=0.1*i;
    for (i=1;i<TWON;i++) r[i]=1.0/i;
    toeplz(r,x,y,N);
    printf("Solution vector:\n");
    for (i=1;i<=N;i++)
        printf("%7s%1d%s %13f\n","x[",i,"] =",x[i]);
    printf("\nTest of solution:\n");
    printf("%13s %12s\n","mtrx*soln","original");
    for (i=1;i<=N;i++) {
        sum=0.0;
        for (j=1;j<=N;j++)
            sum += (r[N+i-j]*x[j]);
        printf("%12.4f %12.4f\n",sum,y[i]);
    }
    return 0;
}
```

The routines choldc and cholsl are tested together by xcholsl. First we form a 3×3 positive-definite, symmetric matrix. We compute its Cholesky decomposition, and check the answer by explicitly multiplying back the Cholesky factors. We then use the decomposition to solve a 3×3 system of equations, and check the answer by substituting back in the original system.

```c
/* Driver for routine cholsl */

#include <stdio.h>
#include "nr.h"
#include "nrutil.h"

#define N 3

int main(void)
{
    int i,j,k;
    float sum,**a,**atest,**chol,*p,*x;
    static float aorig[N+1][N+1]=
        {0.0,0.0,0.0,0.0,
         0.0,100.0,15.0,0.01,
         0.0,15.0,2.3,0.01,
         0.0,0.01,0.01,1.0};
    static float b[N+1]={0.0,0.4,0.02,99.0};
```

```
    a=matrix(1,N,1,N);
    atest=matrix(1,N,1,N);
    chol=matrix(1,N,1,N);
    p=vector(1,N);
    x=vector(1,N);
    for (i=1;i<=N;i++)
        for (j=1;j<=N;j++) a[i][j]=aorig[i][j];
    choldc(a,N,p);
    printf("Original matrix:\n");
    for (i=1;i<=N;i++) {
        for (j=1;j<=N;j++) {
            chol[i][j]=((i > j) ? a[i][j] : (i == j ? p[i] : 0.0));
            if (i > j) chol[i][j]=a[i][j];
            else chol[i][j]=(i == j ? p[i] : 0.0);
            printf("%16.6e",aorig[i][j]);
        }
        printf("\n");
    }
    printf("\n");
    printf("Product of Cholesky factors:\n");
    for (i=1;i<=N;i++) {
        for (j=1;j<=N;j++) {
            for (sum=0.0,k=1;k<=N;k++) sum += chol[i][k]*chol[j][k];
            atest[i][j]=sum;
            printf("%16.6e",atest[i][j]);
        }
        printf("\n");
    }
    printf("\n");
    printf("Check solution vector:\n");
    cholsl(a,N,p,b,x);
    for (i=1;i<=N;i++) {
        for (sum=0.0,j=1;j<=N;j++) sum += aorig[i][j]*x[j];
        p[i]=sum;
        printf("%16.6e%16.6e\n",p[i],b[i]);
    }
    free_vector(x,1,N);
    free_vector(p,1,N);
    free_matrix(chol,1,N,1,N);
    free_matrix(atest,1,N,1,N);
    free_matrix(a,1,N,1,N);
    return 0;
}
```

The QR decomposition routine qrdcmp is tested by a program xqrdcmp that is very similar to the program for testing the LU decomposition. It reads in the same set of test matrices from the file matrx1.dat, does the decomposition of each one, and then checks the result by multiplying the **Q** and **R** matrices.

```
/* Driver for routine qrdcmp */

#include <stdio.h>
#include <stdlib.h>
#include "nr.h"
#include "nrutil.h"
```

```c
#define NP 20
#define MAXSTR 80

int main(void)
{
    int i,j,k,l,m,n,sing;
    float con,**a,*c,*d,**q,**qt,**r,**x;
    char dummy[MAXSTR];
    FILE *fp;

    a=matrix(1,NP,1,NP);
    c=vector(1,NP);
    d=vector(1,NP);
    q=matrix(1,NP,1,NP);
    qt=matrix(1,NP,1,NP);
    r=matrix(1,NP,1,NP);
    x=matrix(1,NP,1,NP);
    if ((fp = fopen("matrx1.dat","r")) == NULL)
        nrerror("Data file matrx1.dat not found\n");
    while (!feof(fp)) {
        fgets(dummy,MAXSTR,fp);
        fgets(dummy,MAXSTR,fp);
        fscanf(fp,"%d %d ",&n,&m);
        fgets(dummy,MAXSTR,fp);
        for (k=1;k<=n;k++)
            for (l=1;l<=n;l++) fscanf(fp,"%f ",&a[k][l]);
        fgets(dummy,MAXSTR,fp);
        for (l=1;l<=m;l++)
            for (k=1;k<=n;k++) fscanf(fp,"%f ",&x[k][l]);
        /* Print out a-matrix for comparison with product of
           Q and R decomposition matrices */
        printf("Original matrix:\n");
        for (k=1;k<=n;k++) {
            for (l=1;l<=n;l++) printf("%12.6f",a[k][l]);
            printf("\n");
        }
        /* Perform the decomposition */
        qrdcmp(a,n,c,d,&sing);
        if (sing) fprintf(stderr,"Singularity in QR decomposition.\n");
        /* find the Q and R matrices */
        for (k=1;k<=n;k++) {
            for (l=1;l<=n;l++) {
                if (l > k) {
                    r[k][l]=a[k][l];
                    q[k][l]=0.0;
                } else if (l < k) {
                    r[k][l]=q[k][l]=0.0;
                } else {
                    r[k][l]=d[k];
                    q[k][l]=1.0;
                }
            }
        }
        for (i=n-1;i>=1;i--) {
            for (con=0.0,k=i;k<=n;k++) con += a[k][i]*a[k][i];
            con /= 2.0;
            for (k=i;k<=n;k++) {
```

```
            for (l=i;l<=n;l++) {
                qt[k][l]=0.0;
                for (j=i;j<=n;j++) {
                    qt[k][l] += q[j][l]*a[k][i]*a[j][i]/con;
                }
            }
        }
        for (k=i;k<=n;k++)
            for (l=i;l<=n;l++) q[k][l] -= qt[k][l];
    }
    /* compute product of Q and R matrices for comparison
       with original matrix. */
    for (k=1;k<=n;k++) {
        for (l=1;l<=n;l++) {
            x[k][l]=0.0;
            for (j=1;j<=n;j++)
                x[k][l] += q[k][j]*r[j][l];
        }
    }
    printf("\nProduct of Q and R matrices:\n");
    for (k=1;k<=n;k++) {
        for (l=1;l<=n;l++) printf("%12.6f",x[k][l]);
        printf("\n");
    }
    printf("\nQ matrix of the decomposition:\n");
    for (k=1;k<=n;k++) {
        for (l=1;l<=n;l++) printf("%12.6f",q[k][l]);
        printf("\n");
    }
    printf("\nR matrix of the decomposition:\n");
    for (k=1;k<=n;k++) {
        for (l=1;l<=n;l++) printf("%12.6f",r[k][l]);
        printf("\n");
    }
    printf("\n*********************************\n");
    printf("press return for next problem:\n");
    (void) getchar();
}
fclose(fp);
free_matrix(x,1,NP,1,NP);
free_matrix(r,1,NP,1,NP);
free_matrix(qt,1,NP,1,NP);
free_matrix(q,1,NP,1,NP);
free_vector(d,1,NP);
free_vector(c,1,NP);
free_matrix(a,1,NP,1,NP);
return 0;
}
```

Similarly routine qrsolv is tested by a program xqrsolv that is almost the same as the test program for LU backsubstitution, and uses the same set of test matrices in the file matrx1.dat.

```
/* Driver for routine qrdcmp */

#include <stdio.h>
#include <stdlib.h>
```

```c
#include "nr.h"
#include "nrutil.h"

#define NP 20
#define MAXSTR 80

int main(void)
{
    int j,k,l,m,n,sing;
    float *x,**a,**ai,**b,*c,*d;
    char dummy[MAXSTR];
    FILE *fp;

    x=vector(1,NP);
    a=matrix(1,NP,1,NP);
    b=matrix(1,NP,1,NP);
    ai=matrix(1,NP,1,NP);
    c=vector(1,NP);
    d=vector(1,NP);
    if ((fp = fopen("matrx1.dat","r")) == NULL)
        nrerror("Data file matrx1.dat not found\n");
    while (!feof(fp)) {
        fgets(dummy,MAXSTR,fp);
        fgets(dummy,MAXSTR,fp);
        fscanf(fp,"%d %d ",&n,&m);
        fgets(dummy,MAXSTR,fp);
        for (k=1;k<=n;k++)
            for (l=1;l<=n;l++) fscanf(fp,"%f ",&a[k][l]);
        fgets(dummy,MAXSTR,fp);
        for (l=1;l<=m;l++)
            for (k=1;k<=n;k++) fscanf(fp,"%f ",&b[k][l]);
        /* Save matrix a for later testing */
        for (l=1;l<=n;l++)
            for (k=1;k<=n;k++) ai[k][l]=a[k][l];
        /* Do qr decomposition */
        qrdcmp(a,n,c,d,&sing);
        if (sing) fprintf(stderr,"Singularity in QR decomposition.\n");
        /* Solve equations for each right-hand vector */
        for (k=1;k<=m;k++) {
            for (l=1;l<=n;l++) x[l]=b[l][k];
            qrsolv(a,n,c,d,x);
            /* Test results with original matrix */
            printf("right-hand side vector:\n");
            for (l=1;l<=n;l++)
                printf("%12.6f",b[l][k]);
            printf("\n%s%s\n","result of matrix applied",
                " to sol'n vector");
            for (l=1;l<=n;l++) {
                b[l][k]=0.0;
                for (j=1;j<=n;j++)
                    b[l][k] += (ai[l][j]*x[j]);
            }
            for (l=1;l<=n;l++)
                printf("%12.6f",b[l][k]);
            printf("\n*******************************\n");
        }
        printf("press RETURN for next problem:\n");
```

```
            (void) getchar();
        }
        fclose(fp);
        free_vector(d,1,NP);
        free_vector(c,1,NP);
        free_matrix(ai,1,NP,1,NP);
        free_matrix(b,1,NP,1,NP);
        free_matrix(a,1,NP,1,NP);
        free_vector(x,1,NP);
        return 0;
}
```

The test program for qrupdt once again reads in test matrices from the file matrx1.dat. However, it uses the solution vectors in the file as vectors defining the update to be performed. The solution is checked by explicitly updating the original matrices and then doing an *ab initio* QR decomposition.

```
/* Driver for routine qrdupd */

#include <stdio.h>
#include <stdlib.h>
#include "nr.h"
#include "nrutil.h"

#define NP 20
#define MAXSTR 80

int main(void)
{
    int i,j,k,l,m,n,sing;
    float con,**a,**au,*c,*d,**q,**qt,**r,**s,*u,*v,**x;
    char dummy[MAXSTR];
    FILE *fp;

    a=matrix(1,NP,1,NP);
    au=matrix(1,NP,1,NP);
    c=vector(1,NP);
    d=vector(1,NP);
    q=matrix(1,NP,1,NP);
    qt=matrix(1,NP,1,NP);
    r=matrix(1,NP,1,NP);
    s=matrix(1,NP,1,NP);
    u=vector(1,NP);
    v=vector(1,NP);
    x=matrix(1,NP,1,NP);
    if ((fp = fopen("matrx1.dat","r")) == NULL)
        nrerror("Data file matrx1.dat not found\n");
    while (!feof(fp)) {
        fgets(dummy,MAXSTR,fp);
        fgets(dummy,MAXSTR,fp);
        fscanf(fp,"%d %d ",&n,&m);
        fgets(dummy,MAXSTR,fp);
        for (k=1;k<=n;k++)
            for (l=1;l<=n;l++) fscanf(fp,"%f ",&a[k][l]);
        fgets(dummy,MAXSTR,fp);
        for (l=1;l<=m;l++)
            for (k=1;k<=n;k++) fscanf(fp,"%f ",&s[k][l]);
```

```
/* Print out a-matrix for comparison with product of
   Q and R decomposition matrices */
printf("Original matrix:\n");
for (k=1;k<=n;k++) {
    for (l=1;l<=n;l++) printf("%12.6f",a[k][l]);
    printf("\n");
}
/* updated matrix we'll use later */
for (k=1;k<=n;k++)
    for (l=1;l<=n;l++)
        au[k][l]=a[k][l]+s[k][l]*s[l][2];
/* Perform the decomposition */
qrdcmp(a,n,c,d,&sing);
if (sing) fprintf(stderr,"Singularity in QR decomposition.\n");
/* find the Q and R matrices */
for (k=1;k<=n;k++) {
    for (l=1;l<=n;l++) {
        if (l > k) {
            r[k][l]=a[k][l];
            q[k][l]=0.0;
        } else if (l < k) {
            r[k][l]=q[k][l]=0.0;
        } else {
            r[k][l]=d[k];
            q[k][l]=1.0;
        }
    }
}
for (i=n-1;i>=1;i--) {
    for (con=0.0,k=i;k<=n;k++) con += a[k][i]*a[k][i];
    con /= 2.0;
    for (k=i;k<=n;k++) {
        for (l=i;l<=n;l++) {
            qt[k][l]=0.0;
            for (j=i;j<=n;j++) {
                qt[k][l] += q[j][l]*a[k][i]*a[j][i]/con;
            }
        }
    }
    for (k=i;k<=n;k++)
        for (l=i;l<=n;l++) q[k][l] -= qt[k][l];
}
/* compute product of Q and R matrices for comparison
   with original matrix. */
for (k=1;k<=n;k++) {
    for (l=1;l<=n;l++) {
        x[k][l]=0.0;
        for (j=1;j<=n;j++)
            x[k][l] += q[k][j]*r[j][l];
    }
}
printf("\nProduct of Q and R matrices:\n");
for (k=1;k<=n;k++) {
    for (l=1;l<=n;l++) printf("%12.6f",x[k][l]);
    printf("\n");
}
printf("\nQ matrix of the decomposition:\n");
```

```
        for (k=1;k<=n;k++) {
            for (l=1;l<=n;l++) printf("%12.6f",q[k][l]);
            printf("\n");
        }
        printf("\nR matrix of the decomposition:\n");
        for (k=1;k<=n;k++) {
            for (l=1;l<=n;l++) printf("%12.6f",r[k][l]);
            printf("\n");
        }
        /* Q transpose */
        for (k=1;k<=n;k++)
            for (l=1;l<=n;l++)
                qt[k][l]=q[l][k];
        for (k=1;k<=n;k++) {
            v[k]=s[k][2];
            for (u[k]=0.0,l=1;l<=n;l++) u[k] += qt[k][l]*s[l][1];
        }
        qrupdt(r,qt,n,u,v);
        for (k=1;k<=n;k++)
            for (l=1;l<=n;l++)
                for (x[k][l]=0.0,j=1;j<=n;j++) x[k][l] += qt[j][k]*r[j][l];
        printf("Updated matrix:\n");
        for (k=1;k<=n;k++) {
            for (l=1;l<=n;l++) printf("%12.6f",au[k][l]);
            printf("\n");
        }
        printf("\nProduct of new Q and R matrices:\n");
        for (k=1;k<=n;k++) {
            for (l=1;l<=n;l++) printf("%12.6f",x[k][l]);
            printf("\n");
        }
        printf("\nNew Q matrix:\n");
        for (k=1;k<=n;k++) {
            for (l=1;l<=n;l++) printf("%12.6f",qt[l][k]);
            printf("\n");
        }
        printf("\nNew R matrix:\n");
        for (k=1;k<=n;k++) {
            for (l=1;l<=n;l++) printf("%12.6f",r[k][l]);
            printf("\n");
        }
        printf("\n**********************************\n");
        printf("press return for next problem:\n");
        (void) getchar();
    }
    fclose(fp);
    free_matrix(x,1,NP,1,NP);
    free_vector(v,1,NP);
    free_vector(u,1,NP);
    free_matrix(s,1,NP,1,NP);
    free_matrix(r,1,NP,1,NP);
    free_matrix(qt,1,NP,1,NP);
    free_matrix(q,1,NP,1,NP);
    free_vector(d,1,NP);
    free_vector(c,1,NP);
    free_matrix(au,1,NP,1,NP);
    free_matrix(a,1,NP,1,NP);
```

```
    return 0;
}
```

Appendix

File matrx1.dat:

```
MATRICES FOR INPUT TO TEST ROUTINES
Size of matrix (NxN), Number of solutions:
3 2
Matrix A:
1.0 0.0 0.0
0.0 2.0 0.0
0.0 0.0 3.0
Solution vectors:
1.0 0.0 0.0
1.0 1.0 1.0
NEXT PROBLEM
Size of matrix (NxN), Number of solutions:
3 2
Matrix A:
1.0 2.0 3.0
2.0 2.0 3.0
3.0 3.0 3.0
Solution vectors:
1.0 1.0 1.0
1.0 2.0 3.0
NEXT PROBLEM:
Size of matrix (NxN), Number of solutions:
5 2
Matrix A:
1.0 2.0 3.0 4.0 5.0
2.0 3.0 4.0 5.0 1.0
3.0 4.0 5.0 1.0 2.0
4.0 5.0 1.0 2.0 3.0
5.0 1.0 2.0 3.0 4.0
Solution vectors:
1.0 1.0 1.0 1.0 1.0
1.0 2.0 3.0 4.0 5.0
NEXT PROBLEM:
Size of matrix (NxN), Number of solutions:
5 2
Matrix A:
1.4 2.1 2.1 7.4 9.6
1.6 1.5 1.1 0.7 5.0
3.8 8.0 9.6 5.4 8.8
4.6 8.2 8.4 0.4 8.0
2.6 2.9 0.1 9.6 7.7
Solution vectors:
1.1 1.6 4.7 9.1 0.1
4.0 9.3 8.4 0.4 4.1
```

File matrx2.dat:

```
FILE OF TRIDIAGONAL MATRICES FOR PROGRAM 'TRIDAG'
Dimension of matrix
3
```

```
Diagonal elements (N)
1.0 2.0 3.0
Super-diagonal elements (N-1)
2.0 3.0
Sub-diagonal elements (N-1)
2.0 3.0
Right-hand side vector (N)
1.0 2.0 3.0
NEXT PROBLEM:
Dimension of matrix
5
Diagonal elements (N)
1.0 1.0 1.0 1.0 1.0
Super-diagonal elements (N-1)
1.0 2.0 3.0 4.0
Sub-diagonal elements (N-1)
2.0 3.0 4.0 5.0
Right-hand side vector (N)
1.0 2.0 3.0 4.0 5.0
NEXT PROBLEM:
Dimension of matrix
5
Diagonal elements (N)
1.0 2.0 3.0 4.0 5.0
Super-diagonal elements (N-1)
2.0 3.0 4.0 5.0
Sub-diagonal elements (N-1)
2.0 3.0 4.0 5.0
Right-hand side vector (N)
1.0 1.0 1.0 1.0 1.0
NEXT PROBLEM:
Dimension of matrix
6
Diagonal elements (N)
9.7 9.5 5.2 3.5 5.1 6.0
Super-diagonal elements (N-1)
6.0 1.2 0.7 3.0 1.5
Sub-diagonal elements (N-1)
2.1 9.4 3.3 7.5 8.8
Right-hand side vector (N)
2.0 7.5 0.6 7.4 9.8 8.8
```

File `matrx3.dat`:

```
FILE OF MATRICES FOR SVDCMP:
Number of Rows, Columns
5 3
Matrix
1.0 2.0 3.0
2.0 3.0 4.0
3.0 4.0 5.0
4.0 5.0 6.0
5.0 6.0 7.0
NEXT PROBLEM:
Number of Rows, Columns
5 5
Matrix
```

```
1.0 2.0 3.0 4.0 5.0
2.0 2.0 3.0 4.0 5.0
3.0 3.0 3.0 4.0 5.0
4.0 4.0 4.0 4.0 5.0
5.0 5.0 5.0 5.0 5.0
NEXT PROBLEM:
Number of Rows, Columns
6 6
Matrix
3.0 5.3 5.6 3.5 6.8 5.7
0.4 8.2 6.7 1.9 2.2 5.3
7.8 8.3 7.7 3.3 1.9 4.8
5.5 8.8 3.0 1.0 5.1 6.4
5.1 5.1 3.6 5.8 5.7 4.9
3.5 2.7 5.7 8.2 9.6 2.9
```

Chapter 3: Interpolation and Extrapolation

Chapter 3 of *Numerical Recipes* deals with interpolation and extrapolation (*the same routines are usable for both*). Three fundamental interpolation methods are first discussed,

1. *Polynomial interpolation (polint),*

2. *Rational function interpolation (ratint), and*

3. *Cubic spline interpolation (spline, splint).*

To find the place in an ordered table at which to perform an interpolation, two routines are given, locate and hunt. Also, for cases in which the actual coefficients of a polynomial interpolation are desired, the routines polcoe and polcof are provided (along with important warnings circumscribing their usefulness).

For higher-dimensional interpolations, *Numerical Recipes* treats only problems on a regularly spaced grid. Routine polin2 does a two-dimensional polynomial interpolation that aims at accuracy rather than smoothness. When smooth interpolation is desired, the methods shown in bcucof and bcuint for bicubic interpolation are recommended. In the case of two-dimensional spline interpolations, the routines splie2 and splin2 are offered.

★   ★   ★   ★

Program polint takes two arrays xa and ya of length N that express the known values of a function, and calculates the value, at a point x, of the unique polynomial of degree $N - 1$ passing through all the given values. For the purpose of illustration, in xpolint we have taken evenly spaced xa[i] and set ya[i] equal to simple functions (sines and exponentials) of these xa[i]. For the sine we use an interval of length π, and for the exponential an interval of length 1.0. You may choose the number N of reference points and observe the improvement of the results as N increases. The test points x are slightly shifted from the reference points so that you can compare the estimated error dy with the actual error. By removing the shift, you may check that the polynomial actually hits all reference points.

```
/* Driver for routine polint */

#include <stdio.h>
#include <math.h>
#include "nr.h"
#include "nrutil.h"

#define PI 3.1415926
```

```
int main(void)
{
    int i,n,nfunc;
    float dy,f,x,y,*xa,*ya;

    printf("generation of interpolation tables\n");
    printf(" ... sin(x)    0<x<PI\n");
    printf(" ... exp(x)    0<x<1 \n");
    printf("how many entries go in these tables?\n");
    if (scanf("%d",&n) == EOF) return 1;
    xa=vector(1,n);
    ya=vector(1,n);
    for (nfunc=1;nfunc<=2;nfunc++) {
        if (nfunc == 1) {
            printf("\nsine function from 0 to PI\n");
            for (i=1;i<=n;i++) {
                xa[i]=i*PI/n;
                ya[i]=sin(xa[i]);
            }
        } else if (nfunc == 2) {
            printf("\nexponential function from 0 to 1\n");
            for (i=1;i<=n;i++) {
                xa[i]=i*1.0/n;
                ya[i]=exp(xa[i]);
            }
        } else {
            break;
        }
        printf("\n%9s %13s %16s %13s\n",
            "x","f(x)","interpolated","error");
        for (i=1;i<=10;i++) {
            if (nfunc == 1) {
                x=(-0.05+i/10.0)*PI;
                f=sin(x);
            } else if (nfunc == 2) {
                x=(-0.05+i/10.0);
                f=exp(x);
            }
            polint(xa,ya,n,x,&y,&dy);
            printf("%12.6f %12.6f %12.6f %4s %11f\n",
                x,f,y," ",dy);
        }
        printf("\n***********************************\n");
        printf("press RETURN\n");
        (void) getchar();
    }
    free_vector(ya,1,n);
    free_vector(xa,1,n);
    return 0;
}
```

ratint is functionally similar to `polint` in that it also returns a value y for the function at point x, and an error estimate `dy` as well. In this case the values are determined from the unique diagonal rational function that passes through all the reference points. If you inspect the driver closely, you will find that two of the test points fall directly on top of reference points and should give exact results. The remainder do not. You can compare the estimated error **dyy** to the actual error |yy − yexp| for these cases.

```
/* Driver for routine ratint */

#include <stdio.h>
#include <math.h>
#include "nr.h"
#include "nrutil.h"

#define NPT 6
#define EPS 1.0

float f(float x,float eps)
{
    return x*exp(-x)/(SQR(x-1.0)+eps*eps);
}

int main(void)
{
    int i;
    float dyy,xx,yexp,yy,*x,*y;

    x=vector(1,NPT);
    y=vector(1,NPT);
    for (i=1;i<=NPT;i++) {
        x[i]=i*2.0/NPT;
        y[i]=f(x[i],EPS);
    }
    printf("\nDiagonal rational function interpolation\n");
    printf("\n%5s %13s %14s %12s\n","x","interp.","accuracy","actual");
    for (i=1;i<=10;i++) {
        xx=0.2*i;
        ratint(x,y,NPT,xx,&yy,&dyy);
        yexp=f(xx,EPS);
        printf("%6.2f %12.6f    %11f %13.6f\n",xx,yy,dyy,yexp);
    }
    free_vector(y,1,NPT);
    free_vector(x,1,NPT);
    return 0;
}
```

Procedure `spline` generates a cubic spline. Given an array of x_i and $f(x_i)$, and given values of the first derivative of function f at the two endpoints of the tabulated region, it returns the second derivative of f at each of the tabulation points. As an example we chose the function $\sin x$ and evaluated it at evenly spaced points `x[i]`. In this case the first derivatives at the end-points are $ypl = \cos x_1$ and $ypn = \cos x_N$. The output array of `spline` is `y2[i]` and this is listed along with $-\sin x_i$, the second derivative of $\sin x_i$, for comparison.

```
/* Driver for routine spline */

#include <stdio.h>
#include <math.h>
#include "nr.h"
```

```
#include "nrutil.h"

#define N 20
#define PI 3.1415926

int main(void)
{
    int i;
    float yp1,ypn,*x,*y,*y2;

    x=vector(1,N);
    y=vector(1,N);
    y2=vector(1,N);
    printf("\nsecond-derivatives for sin(x) from 0 to pi\n");
    /* Generate array for interpolation */
    for (i=1;i<=20;i++) {
        x[i]=i*PI/N;
        y[i]=sin(x[i]);
    }
    /* calculate 2nd derivative with spline */
    yp1=cos(x[1]);
    ypn=cos(x[N]);
    spline(x,y,N,yp1,ypn,y2);
    /* test result */
    printf("%23s %16s\n","spline","actual");
    printf("%11s %14s %16s\n","angle","2nd deriv","2nd deriv");
    for (i=1;i<=N;i++)
        printf("%10.2f %16.6f %16.6f\n",x[i],y2[i],-sin(x[i]));
    free_vector(y2,1,N);
    free_vector(y,1,N);
    free_vector(x,1,N);
    return 0;
}
```

Actual cubic-spline interpolations, however, are carried out by `splint`. This routine uses the output array from one call to `spline` to service any subsequent number of spline interpolations with different x's. The demonstration program `xsplint` tests this capability on both $\sin x$ and $\exp x$. The two are treated in succession according to whether `nfunc` is one or two. In each case the function is tabulated at equally spaced points, and the derivatives are found at the first and last point. A call to `spline` then produces an array of second derivatives `y2` which is fed to `splint`. The interpolated values `y` are compared with actual function values `f` at a different set of equally spaced points.

```
/* Driver for routine splint */

#include <stdio.h>
#include <math.h>
#include "nr.h"
#include "nrutil.h"

#define NP 10
#define PI 3.1415926

int main(void)
{
```

```
        int i,nfunc;
        float f,x,y,yp1,ypn,*xa,*ya,*y2;

        xa=vector(1,NP);
        ya=vector(1,NP);
        y2=vector(1,NP);
        for (nfunc=1;nfunc<=2;nfunc++) {
            if (nfunc == 1) {
                printf("\nsine function from 0 to pi\n");
                for (i=1;i<=NP;i++) {
                    xa[i]=i*PI/NP;
                    ya[i]=sin(xa[i]);
                }
                yp1=cos(xa[1]);
                ypn=cos(xa[NP]);
            } else if (nfunc == 2) {
                printf("\nexponential function from 0 to 1\n");
                for (i=1;i<=NP;i++) {
                    xa[i]=1.0*i/NP;
                    ya[i]=exp(xa[i]);
                }
                yp1=exp(xa[1]);
                ypn=exp(xa[NP]);
            } else {
                break;
            }
            /* Call spline to get second derivatives */
            spline(xa,ya,NP,yp1,ypn,y2);
            /* Call splint for interpolations */
            printf("\n%9s %13s %17s\n","x","f(x)","interpolation");
            for (i=1;i<=10;i++) {
                if (nfunc == 1) {
                    x=(-0.05+i/10.0)*PI;
                    f=sin(x);
                } else if (nfunc == 2) {
                    x = -0.05+i/10.0;
                    f=exp(x);
                }
                splint(xa,ya,y2,NP,x,&y);
                printf("%12.6f %12.6f %12.6f\n",x,f,y);
            }
            printf("\n***********************************\n");
            printf("Press RETURN\n");
            (void) getchar();
        }
        free_vector(y2,1,NP);
        free_vector(ya,1,NP);
        free_vector(xa,1,NP);
        return 0;
    }
```

The next program, `locate`, may be used in conjunction with any interpolation method to bracket the x-position for which $f(x)$ is sought by two adjacent tabulated positions. That is, given a monotonic array of x_i, and given a value of x, it finds the two values x_i, x_{i+1} that surround x. In `xlocate` we chose the array x_i to be non-uniform, varying exponentially with i. Then we took a uniform series of x-values and sought their position in the array using locate. For each x, locate finds the value j for which xx[j] is nearest below x. Then the driver shows j, and the two bracketing values xx[j] and xx[j+1]. If j is 0 or N, then x is not within the tabulated range. The program then flags 'lower lim' if x is below xx[1] or 'upper lim' if x is above xx[N].

```
/* Driver for routine locate */

#include <stdio.h>
#include <math.h>
#include "nr.h"
#include "nrutil.h"

#define N 100

int main(void)
{
    unsigned long i,j;
    float x,*xx;

    xx=vector(1,N);
    /* create array to be searched */
    for (i=1;i<=N;i++)
        xx[i]=exp(i/20.0)-74.0;
    printf("\nresult of:  j=0 indicates x too small\n");
    printf("%11s j=100 indicates x too large"," ");
    printf("\n%10s %6s %11s %12s \n","locate ","j","xx(j)","xx(j+1)");
    /* perform test */
    for (i=1;i<=19;i++) {
        x = -100.0+200.0*i/20.0;
        locate(xx,N,x,&j);
        if ((j < N) && (j > 0))
            printf("%10.4f %6lu %12.6f %12.6f\n",
                x,j,xx[j],xx[j+1]);
        else if (j == N)
            printf("%10.4f %6lu %12.6f %s\n",
                x,j,xx[j],"    upper lim");
        else
            printf("%10.4f %6lu %s %12.6f \n",
                x,j,"    lower lim",xx[j+1]);
    }
    free_vector(xx,1,N);
    return 0;
}
```

Routine hunt serves the same function as locate, but is used when the table is to be searched many times and the abscissa each time is close to its value on the previous search. xhunt sets up the array xx[i] and then a series x of points to locate. The hunt begins with a trial value ji (which is fed to hunt through variable j) and hunt returns solution j such that x lies between xx[j] and xx[j+1]. The two cases j=0 and j=N have the same meaning as in xlocate and are treated in the same way.

```
/* Driver for routine hunt */

#include <stdio.h>
#include <math.h>
```

```
#include "nr.h"
#include "nrutil.h"

#define N 100

int main(void)
{
    unsigned long i,j,ji;
    float x,*xx;

    xx=vector(1,N);
    /* create array to be searched */
    for (i=1;i<=N;i++)
        xx[i]=exp(i/20.0)-74.0;
    printf("\n  result of:    j=0 indicates x too small\n");
    printf("%14s j=100 indicates x too large"," ");
    printf("\n%12s %8s %4s %11s %13s \n",
        "locate:","guess","j","xx(j)","xx(j+1)");
    /* do test */
    for (i=1;i<=19;i++) {
        x = -100.0+10.0*i;
        /* trial parameter */
        j=(ji=5*i);
        /* begin search */
        hunt(xx,N,x,&j);
        if ((j < N) && (j > 0))
            printf("%12.5f %6lu %6lu %12.6f %12.6f \n",
                x,ji,j,xx[j],xx[j+1]);
        else if (j == N)
            printf("%12.5f %6lu %6lu %12.6f %s \n",
                x,ji,j,xx[j],"  upper lim");
        else
            printf("%12.5f %6lu %6lu %s %12.6f \n",
                x,ji,j,"  lower lim",xx[j+1]);
    }
    free_vector(xx,1,N);
    return 0;
}
```

The next two demonstration programs, xpolcoe and xpolcof, are so nearly identical that they may be discussed together. polcoe and polcof themselves both find coefficients of interpolating polynomials. In the present instance we have tried both a sine function and an exponential function for ya[i], each tabulated at uniformly spaced points xa[i]. The validity of the array of polynomial coefficients coeff is tested by calculating the value sum of the polynomials at a series of test points and listing these alongside the functions f that they represent.

```
/* Driver for routine polcoe */

#include <stdio.h>
#include <math.h>
#include "nr.h"
#include "nrutil.h"

#define NP 4
#define PI 3.1415926
```

```
int main(void)
{
    int i,j,nfunc;
    float f,sum,x,*coeff,*xa,*ya;

    coeff=vector(0,NP);
    xa=vector(0,NP);
    ya=vector(0,NP);
    for (nfunc=1;nfunc<=2;nfunc++) {
        if (nfunc == 1) {
            printf("sine function from 0 to PI\n\n");
            for (i=0;i<=NP;i++) {
                xa[i]=(i+1)*PI/(NP+1);
                ya[i]=sin(xa[i]);
            }
        } else if (nfunc == 2) {
            printf("exponential function from 0 to 1\n\n");
            for (i=0;i<=NP;i++) {
                xa[i]=1.0*(i+1)/(NP+1);
                ya[i]=exp(xa[i]);
            }
        } else {
            break;
        }
        polcoe(xa,ya,NP,coeff);
        printf("  coefficients\n");
        for (i=0;i<=NP;i++) printf("%12.6f",coeff[i]);
        printf("\n\n%9s %13s %15s\n","x","f(x)","polynomial");
        for (i=1;i<=10;i++) {
            if (nfunc == 1) {
                x=(-0.05+i/10.0)*PI;
                f=sin(x);
            } else if (nfunc == 2) {
                x = -0.05+i/10.0;
                f=exp(x);
            }
            sum=coeff[NP];
            for (j=NP-1;j>=0;j--)
                sum=coeff[j]+sum*x;
            printf("%12.6f %12.6f %12.6f\n",x,f,sum);
        }
        printf("\n***********************************\n");
        printf("press RETURN\n");
        (void) getchar();
    }
    free_vector(ya,0,NP);
    free_vector(xa,0,NP);
    free_vector(coeff,0,NP);
    return 0;
}

/* Driver for routine polcof */

#include <stdio.h>
#include <math.h>
#include "nr.h"
```

```
#include "nrutil.h"

#define NP 4
#define PI 3.1415926

int main(void)
{
    int i,j,nfunc;
    float f,sum,x,*coeff,*xa,*ya;

    coeff=vector(0,NP);
    xa=vector(0,NP);
    ya=vector(0,NP);
    for (nfunc=1;nfunc<=2;nfunc++) {
        if (nfunc == 1) {
            printf("sine function from 0 to PI\n\n");
            for (i=0;i<=NP;i++) {
                xa[i]=(i+1)*PI/(NP+1);
                ya[i]=sin(xa[i]);
            }
        } else if (nfunc == 2) {
            printf("exponential function from 0 to 1\n\n");
            for (i=0;i<=NP;i++) {
                xa[i]=1.0*(i+1)/(NP+1);
                ya[i]=exp(xa[i]);
            }
        } else {
            break;
        }
        polcof(xa,ya,NP,coeff);
        printf("  coefficients\n");
        for (i=0;i<=NP;i++) printf("%12.6f",coeff[i]);
        printf("\n\n%9s %13s %15s\n","x","f(x)","polynomial");
        for (i=1;i<=10;i++) {
            if (nfunc == 1) {
                x=(-0.05+i/10.0)*PI;
                f=sin(x);
            } else if (nfunc == 2) {
                x = -0.05+i/10.0;
                f=exp(x);
            }
            sum=coeff[NP];
            for (j=NP-1;j>=0;j--)
                sum=coeff[j]+sum*x;
            printf("%12.6f %12.6f %12.6f\n",x,f,sum);
        }
        printf("\n*********************************\n");
        printf("press RETURN\n");
        (void) getchar();
    }
    free_vector(ya,0,NP);
    free_vector(xa,0,NP);
    free_vector(coeff,0,NP);
    return 0;
}
```

For two-dimensional interpolation, polin2 implements a bilinear interpolation.

We feed it coordinates **x1a,x2a** for an $M \times N$ array of gridpoints as well as the function value at each gridpoint. In return it gives the value **y** of the interpolated function at a given point **x1,x2**, and the estimated accuracy **dy** of the interpolation. **xpolin2** runs the test on a uniform grid for the function $f(x, y) = \sin x \exp y$. Then, for an offset grid of test points, the interpolated value **y** is compared to the actual function value **f**, and the actual error is compared to the estimated error **dy**.

```
/* Driver for routine polin2 */

#include <stdio.h>
#include <math.h>
#include "nr.h"
#include "nrutil.h"

#define N 5
#define PI 3.1415926

int main(void)
{
    int i,j;
    float dy,f,x1,x2,y,*x1a,*x2a,**ya;

    x1a=vector(1,N);
    x2a=vector(1,N);
    ya=matrix(1,N,1,N);
    for (i=1;i<=N;i++) {
        x1a[i]=i*PI/N;
        for (j=1;j<=N;j++) {
            x2a[j]=1.0*j/N;
            ya[i][j]=sin(x1a[i])*exp(x2a[j]);
        }
    }
    /* test 2-dimensional interpolation */
    printf("\nTwo dimensional interpolation of sin(x1)exp(x2)\n");
    printf("%9s %12s %13s %16s %11s\n",
        "x1","x2","f(x)","interpolated","error");
    for (i=1;i<=4;i++) {
        x1=(-0.1+i/5.0)*PI;
        for (j=1;j<=4;j++) {
            x2 = -0.1+j/5.0;
            f=sin(x1)*exp(x2);
            polin2(x1a,x2a,ya,N,N,x1,x2,&y,&dy);
            printf("%12.6f %12.6f %12.6f %12.6f %15.6f\n",
                x1,x2,f,y,dy);
        }
        printf ("*********************************\n");
    }
    free_matrix(ya,1,N,1,N);
    free_vector(x2a,1,N);
    free_vector(x1a,1,N);
    return 0;
}
```

Bicubic interpolation in two dimensions is carried out with **bcucof** and **bcuint**. The first supplies interpolating coefficients within a grid square and the second calculates interpolated values. The calculation provides not only interpolated function values, but also interpolated values of two partial derivatives, all of which are guaranteed to be smooth. To get this, we are required to supply more information than we have needed in previous interpolation routines.

Demonstration program xbcucof works with the function $f(x, y) = xy\exp(-xy)$. The program computes the two first derivatives and the cross derivative of this function at the four corners of a rectangular grid cell, in this case a 2×2 unit square with one corner at the origin. The points are supplied counterclockwise around the cell. d1 and d2 are the dimensions of the cell. A call to bcucof provides the following sixteen coefficients, listed for comparison.

```
Coefficients for bicubic interpolation:
    0.000000     0.000000     0.000000     0.000000
    0.000000     4.000000     0.000000     0.000000
    0.000000     0.000000   -13.655600     6.095174
    0.000000     0.000000     6.095174    -2.461486
```

```c
/* Driver for routine bcucof */

#include <stdio.h>
#include <math.h>
#include "nr.h"
#include "nrutil.h"

int main(void)
{
    int i,j;
    float d1,d2,ee,x1x2;
    float y[5],y1[5],y2[5],y12[5],**c;
    static float x1[]={0.0,0.0,2.0,2.0,0.0};
    static float x2[]={0.0,0.0,0.0,2.0,2.0};

    c=matrix(1,4,1,4);
    d1=x1[2]-x1[1];
    d2=x2[4]-x2[1];
    for (i=1;i<=4;i++) {
        x1x2=x1[i]*x2[i];
        ee=exp(-x1x2);
        y[i]=x1x2*ee;
        y1[i]=x2[i]*(1.0-x1x2)*ee;
        y2[i]=x1[i]*(1.0-x1x2)*ee;
        y12[i]=(1.0-3.0*x1x2+x1x2*x1x2)*ee;
    }
    bcucof(y,y1,y2,y12,d1,d2,c);
    printf("\nCoefficients for bicubic interpolation:\n\n");
    for (i=1;i<=4;i++) {
        for (j=1;j<=4;j++) printf("%12.6f",c[i][j]);
        printf("\n");
    }
    free_matrix(c,1,4,1,4);
    return 0;
}
```

Program xbcuint works with the function $f(x, y) = (xy)^2$, which has derivatives $\partial f/\partial x = 2xy^2$, $\partial f/\partial y = 2yx^2$, and $\partial^2 f/\partial x\partial y = 4xy$. These are supplied to bcuint along with the locations of the grid points. bcuint calls bcucof internally to determine coefficients, and then calculates `ansy`, `ansy1`, `ansy2`, the interpolated values of f, $\partial f / \partial x$ and $\partial f / \partial y$ at the specified test point (`x1`,`x2`). These are compared by the demonstration program to expected values for the three quantities, which are called `ey`, `ey1`, and `ey2`. The test points run along the diagonal of the grid square.

```
/* Driver for routine bcuint */

#include <stdio.h>
#include "nr.h"

int main(void)
{
    int i;
    float ansy,ansy1,ansy2,ey,ey1,ey2;
    float x1,x11,x1u,x1x2,x2,x21,x2u,xxyy;
    float y[5],y1[5],y12[5],y2[5];
    static float xx[]={0.0,0.0,2.0,2.0,0.0};
    static float yy[]={0.0,0.0,0.0,2.0,2.0};

    x11=xx[1];
    x1u=xx[2];
    x21=yy[1];
    x2u=yy[4];
    for (i=1;i<=4;i++) {
        xxyy=xx[i]*yy[i];
        y[i]=xxyy*xxyy;
        y1[i]=2.0*yy[i]*xxyy;
        y2[i]=2.0*xx[i]*xxyy;
        y12[i]=4.0*xxyy;
    }
    printf("\n%6s %8s %7s %11s %6s %10s %6s %10s \n\n",
        "x1","x2","y","expect","y1","expect","y2","expect");
    for (i=1;i<=10;i++) {
        x2=(x1=0.2*i);
        bcuint(y,y1,y2,y12,x11,x1u,x21,x2u,x1,x2,&ansy,&ansy1,&ansy2);
        x1x2=x1*x2;
        ey=x1x2*x1x2;
        ey1=2.0*x2*x1x2;
        ey2=2.0*x1*x1x2;
        printf("%8.4f %8.4f %8.4f %8.4f %8.4f %8.4f %8.4f %8.4f\n",
            x1,x2,ansy,ey,ansy1,ey1,ansy2,ey2);
    }
    return 0;
}
```

Routines `splie2` and `splin2` work as a pair to perform bicubic spline interpolations. `splie2` takes a function tabulated on an $M \times N$ grid and performs one dimensional natural cubic splines along the rows of the grid to generate an array of second derivatives. These are fodder for `splin2` which takes the grid points, function values, and second derivative values and returns the interpolated function value for a desired point in the grid region.

Demonstration program `xsplie2` exercises `splie2` on a regular 10×10 grid of points with coordinates `x1` and `x2`, for the function $y = (x_1 x_2)^2$. The calculated second derivative array is compared with the actual second derivative $2x_1 x_2$ of the function. Keep in mind that a natural spline is assumed, so that agreement will not be so good near the boundaries of the grid. (This shows that you should *not* assume a natural spline if you have better derivative information at the endpoints.)

```c
/* Driver for routine splie2 */

#include <stdio.h>
#include "nr.h"
#include "nrutil.h"

#define M 10
#define N 10

int main(void)
{
    int i,j;
    float x1x2,*x1,*x2,**y,**y2;

    x1=vector(1,N);
    x2=vector(1,N);
    y=matrix(1,M,1,N);
    y2=matrix(1,M,1,N);
    for (i=1;i<=M;i++) x1[i]=0.2*i;
    for (i=1;i<=N;i++) x2[i]=0.2*i;
    for (i=1;i<=M;i++)
        for (j=1;j<=N;j++) {
            x1x2=x1[i]*x2[j];
            y[i][j]=x1x2*x1x2;
        }
    splie2(x1,x2,y,M,N,y2);
    printf("\nsecond derivatives from SPLIE2\n");
    printf("natural spline assumed\n");
    for (i=1;i<=5;i++) {
        for (j=1;j<=5;j++) printf("%12.6f",y2[i][j]);
        printf("\n");
    }
    printf("\nactual second derivatives\n");
    for (i=1;i<=5;i++) {
        for (j=1;j<=5;j++) printf("%12.6f",2.0*x1[i]*x1[i]);
        printf("\n");
    }
    free_matrix(y2,1,M,1,N);
    free_matrix(y,1,M,1,N);
    free_vector(x2,1,N);
    free_vector(x1,1,N);
    return 0;
}
```

The demonstration program `xsplin2` establishes a similar 10×10 grid for the function $y = x_1 x_2 \exp(-x_1 x_2)$. It makes a single call to `splie2` to produce second derivatives `y2`, and then finds function values `f` through calls to `splin2`, comparing them to actual function values `ff`. These values are determined, for no particular reason, along a quadratic path $x_2 = x_1^2$ through the grid region.

```c
/* Driver for routine splin2 */

#include <stdio.h>
#include <math.h>
```

```c
#include "nr.h"
#include "nrutil.h"

#define M 10
#define N 10

int main(void)
{
    int i,j;
    float f,ff,x1x2,xx1,xx2,*x1,*x2,**y,**y2;

    x1=vector(1,N);
    x2=vector(1,N);
    y=matrix(1,M,1,N);
    y2=matrix(1,M,1,N);
    for (i=1;i<=M;i++) x1[i]=0.2*i;
    for (i=1;i<=N;i++) x2[i]=0.2*i;
    for (i=1;i<=M;i++) {
        for (j=1;j<=N;j++) {
            x1x2=x1[i]*x2[j];
            y[i][j]=x1x2*exp(-x1x2);
        }
    }
    splie2(x1,x2,y,M,N,y2);
    printf("%9s %12s %14s %12s\n","x1","x2","splin2","actual");
    for (i=1;i<=10;i++) {
        xx1=0.1*i;
        xx2=xx1*xx1;
        splin2(x1,x2,y,y2,M,N,xx1,xx2,&f);
        x1x2=xx1*xx2;
        ff=x1x2*exp(-x1x2);
        printf("%12.6f %12.6f %12.6f %12.6f\n",xx1,xx2,f,ff);
    }
    free_matrix(y2,1,M,1,N);
    free_matrix(y,1,M,1,N);
    free_vector(x2,1,N);
    free_vector(x1,1,N);
    return 0;
}
```

Chapter 4: Integration of Functions

Numerical integration, or "quadrature", has been treated with some degree of detail in Numerical Recipes. Chapter 4 begins with `trapzd`*, a function for applying the extended trapezoidal rule. It can be used in successive calls for sequentially improving accuracy, and is used as a foundation for several other programs. For example* `qtrap` *is an integrating routine that makes repeated calls to* `trapzd` *until a certain fractional accuracy is achieved.* `qsimp` *also calls* `trapzd`*, and in this case performs integration by Simpson's rule. Romberg integration, a generalization of Simpson's rule to successively higher orders, is performed with* `qromb` *— this one also calls* `trapzd`*. For improper integrals a different "workhorse" is used, the procedure* `midpnt`*. This routine applies the extended midpoint rule to avoid function evaluations at an endpoint of the region of integration. It can be used in* `qtrap` *or* `qsimp` *in place of* `trapzd`*. Routine* `qromb` *can be generalized similarly, and we have implemented this idea in* `qromo`*, a Romberg integrator for open intervals. The chapter also offers a number of exact replacements for* `midpnt`*, to be used for various types of singularity in the integrand:*

1. `midinf` *– if one or the other of the limits of integration is infinite.*

2. `midsql` *– if there is an inverse square root singularity of the integrand at the lower limit of integration.*

3. `midsqu` *– if there is an inverse square root singularity of the integrand at the upper limit of integration.*

4. `midexp` *– when the upper limit of integration is infinite and the integrand decreases exponentially at infinity.*

The somewhat more subtle method of Gaussian quadrature uses unequally spaced abscissas, and weighting coefficients which can be read from tables. Routine `qgaus` *computes integrals with a ten-point Gauss-Legendre weighting using such coefficients.* `gauleg, gaulag, gauher` *and* `gaujac` *calculate the tables of abscissas and weights for N-point Gauss-Legendre, Gauss-Laguerre, Gauss-Hermite and Gauss-Jacobi quadratures, respectively. The routine* `orthog` *lets you construct nonclassical orthogonal polynomials, while* `gaucof` *lets you find the abscissas and weights for the corresponding Gaussian quadrature.*

$$\star \quad \star \quad \star \quad \star$$

`trapzd` applies the extended trapezoidal rule for integration. It is called sequentially for higher and higher stages of refinement of the integral. The sample program xtrapzd uses `trapzd` to perform a numerical integration of the function

$$\text{func} = x^2(x^2 - 2)\sin x$$

whose indefinite integral is

$$\text{fint} = 4x(x^2 - 7)\sin x - (x^4 - 14x^2 + 28)\cos x.$$

The integral is performed from $A = 0.0$ to $B = \pi/2$. To demonstrate the increasing accuracy on sequential calls, `trapzd` is called 14 times with the index i increasing by one each time. The improving values of the integral are listed for comparison to the actual value $\text{fint}(B) - \text{fint}(A)$.

```c
/* Driver for routine trapzd */

#include <stdio.h>
#include <math.h>
#include "nr.h"

#define NMAX 14
#define PIO2 1.5707963

/* Test function */
float func(float x)
{
    return (x*x)*(x*x-2.0)*sin(x);
}

/* Integral of test function */
float fint(float x)
{
    return 4.0*x*(x*x-7.0)*sin(x)-(pow(x,4.0)-14.0*(x*x)+28.0)*cos(x);
}

int main(void)
{
    int i;
    float a=0.0,b=PIO2,s;

    printf("\nIntegral of func with 2^(n-1) points\n");
    printf("Actual value of integral is %10.6f\n",fint(b)-fint(a));
    printf("%6s %24s\n","n","approx. integral");
    for (i=1;i<=NMAX;i++) {
        s=trapzd(func,a,b,i);
        printf("%6d %20.6f\n",i,s);
    }
    return 0;
}
```

qtrap carries out the same integration algorithm but allows us to specify the accuracy with which we wish the integration done. (It is specified within qtrap as EPS=1.0E-6.) qtrap itself makes the sequential calls to `trapzd` until the desired accuracy is reached. Then qtrap issues a single result. In sample program xqtrap we compare this result to the exact value of the integral.

```
/* Driver for routine qtrap */

#include <stdio.h>
#include <math.h>
#include "nr.h"

#define PIO2 1.5707963

/* Test function */
float func(float x)
{
    return x*x*(x*x-2.0)*sin(x);
}

/* Integral of test function */
float fint(float x)
{
    return 4.0*x*(x*x-7.0)*sin(x)-(pow(x,4.0)-14.0*x*x+28.0)*cos(x);
}

int main(void)
{
    float a=0.0,b=PIO2,s;

    printf("Integral of func computed with QTRAP\n\n");
    printf("Actual value of integral is %12.6f\n",fint(b)-fint(a));
    s=qtrap(func,a,b);
    printf("Result from routine QTRAP is %12.6f\n",s);
    return 0;
}
```

Alternatively, the integral may be handled by qsimp which applies Simpson's rule. Sample program xqsimp carries out the same integration as the previous program, and reports the result in the same way as well.

```
/* Driver for routine qsimp */

#include <stdio.h>
#include <math.h>
#include "nr.h"

#define PIO2 1.5707963

/* Test function */
float func(float x)
{
    return x*x*(x*x-2.0)*sin(x);
}

/* Integral of test function */
float fint(float x)
{
    return 4.0*x*(x*x-7.0)*sin(x)-(pow(x,4.0)-14.0*x*x+28.0)*cos(x);
}

int main(void)
{
```

```
    float a=0.0,b=PIO2,s;

    printf("Integral of func computed with QSIMP\n\n");
    printf("Actual value of integral is %12.6f\n",fint(b)-fint(a));
    s=qsimp(func,a,b);
    printf("Result from routine QSIMP is %11.6f\n",s);
    return 0;
}
```

qromb generalizes Simpson's rule to higher orders. It makes successive calls to trapzd and stores the results. Then it uses polint, the polynomial interpolater/extrapolator, to project the value of integral that would be obtained were we to continue indefinitely with trapzd. Sample program xqromb is essentially identical to the sample programs for qtrap and qsimp.

```
/* Driver for routine qromb */

#include <stdio.h>
#include <math.h>
#include "nr.h"

#define PIO2 1.5707963

/* Test function */
float func(float x)
{
    return x*x*(x*x-2.0)*sin(x);
}

/* Integral of test function func */
float fint(float x)
{
    return 4.0*x*(x*x-7.0)*sin(x)-(pow(x,4.0)-14.0*x*x+28.0)*cos(x);
}

int main(void)
{
    float a=0.0,b=PIO2,s;

    printf("Integral of func computed with QROMB\n\n");
    printf("Actual value of integral is %12.6f\n",fint(b)-fint(a));
    s=qromb(func,a,b);
    printf("Result from routine QROMB is %11.6f\n",s);
    return 0;
}
```

Sample program xmidpnt uses the function func $= 1/\sqrt{x}$ which is singular at the origin. Limits of integration are set at $A = 0.0$ and $B = 1.0$. midpnt, however, implements an open formula and does not evaluate the function exactly at $x = 0$. In this case the integral is compared to fint(B) − fint(A) where fint $= 2\sqrt{x}$, the integral of func.

```
/* Driver for routine midpnt */

#include <stdio.h>
#include <math.h>
```

```
#include "nr.h"

#define NMAX 10

/* Test function */
float func(float x)
{
    return 1.0/sqrt(x);
}

/* Integral of test function */
float fint(float x)
{
    return 2.0*sqrt(x);
}

int main(void)
{
    float a=0.0,b=1.0,s;
    int i;

    printf("\nIntegral of func computed with MIDPNT\n");
    printf("Actual value of integral is %7.4f\n",(fint(b)-fint(a)));
    printf("%6s %29s \n","n","Approx. integral");
    for (i=1;i<=NMAX;i++) {
        s=midpnt(func,a,b,i);
        printf("%6d %24.6f\n",i,s);
    }
    return 0;
}
```

Various special forms of `midpnt` (i.e., `midsql`, `midsqu`, `midinf`, `midexp`) are demonstrated by sample program `xqromo`. For those tests that integrate to infinity, we take infinity to be 1.0E20. The following integrations are performed:

1. Integral of $\sqrt{x}/\sin x$ from 0.0 to $\pi/2$. (This has a $1/\sqrt{x}$ singularity at $x = 0$, and uses `midsql`.)

2. Integral of $\sqrt{(\pi - x)}/\sin x$ from $\pi/2$ to π. (This has the $1/\sqrt{x}$ singularity at the upper limit $x = \pi$, and uses `midsqu`.)

3. Integral of $(\sin x)/x^2$ from $\pi/2$ to ∞. (This has a region of integration extending to ∞ and uses `midinf`. It is quite slowly convergent, as is the next integral.)

4. Integral of $(\sin x)/x^2$ from $-\infty$ to $-\pi/2$. (Region of integration goes to $-\infty$; uses `midinf`.)

5. Integral of $\exp(-x)/\sqrt{x}$ from 0.0 to ∞. (This has a singularity at $x = 0.0$ and also integrates up to ∞. It is performed in two pieces, (0.0 to $\pi/2$) and ($\pi/2$ to ∞) using `midsql` and `midinf`. The two calculations give results `res1` and `res2` respectively, which are added to give the entire integral.)

6. Same integral as in (5), but with the segment from ($\pi/2$ to ∞) done using `midexp`.

```
/* Driver for routine qromo */

#include <stdio.h>
```

```c
#include <math.h>
#include "nr.h"

#define X1 0.0
#define X2 1.5707963
#define X3 3.1415926
#define AINF 1.0E20
#define PI 3.1415926

static float funcl(float x)
{
    return (float) (sqrt(x)/sin(x));
}

static float funcu(float x)
{
    return (float) (sqrt(PI-x)/sin(x));
}

static float fncinf(float x)
{

    return (float) (sin(x)/(x*x));
}

static float fncend(float x)
{
    return (float) (exp(-x)/sqrt(x));
}

int main(void)
{
    float res1,res2,result;

    printf("\nImproper integrals:\n\n");
    result=qromo(funcl,X1,X2,midsql);
    printf("Function: sqrt(x)/sin(x)      Interval: (0,pi/2)\n");
    printf("Using: MIDSQL                 Result: %8.4f\n\n",result);
    result=qromo(funcu,X2,X3,midsqu);
    printf("Function: sqrt(pi-x)/sin(x)   Interval: (pi/2,pi)\n");
    printf("Using: MIDSQU                 Result: %8.4f\n\n",result);
    result=qromo(fncinf,X2,AINF,midinf);
    printf("Function: sin(x)/x**2         Interval: (pi/2,infty)\n");
    printf("Using: MIDINF                 Result: %8.4f\n\n",result);
    result=qromo(fncinf,-AINF,-X2,midinf);
    printf("Function: sin(x)/x**2         Interval: (-infty,-pi/2)\n");
    printf("Using: MIDINF                 Result: %8.4f\n\n",result);
    res1=qromo(fncend,X1,X2,midsql);
    res2=qromo(fncend,X2,AINF,midinf);
    printf("Function: exp(-x)/sqrt(x)     Interval: (0.0,infty)\n");
    printf("Using: MIDSQL,MIDINF          Result: %8.4f\n\n",res1+res2);
    res2=qromo(fncend,X2,AINF,midexp);
    printf("Function: exp(-x)/sqrt(x)     Interval: (0.0,infty)\n");
    printf("Using: MIDSQL,MIDEXP          Result: %8.4f\n\n",res1+res2);
    return 0;
}
```

Procedure `qgaus` performs a Gauss-Legendre integration, using only ten function evaluations. Sample program `xqgaus` applies it to the function $x \exp(-x)$ whose integral from x_1 to x is $(1 + x_1)\exp(-x_1) - (1 + x)\exp(-x)$. `qgaus` returns the value of this integral. The method is used for a series of intervals, as short as $(0.0 - 0.5)$ and as long as $(0.0 - 5.0)$. You may observe how the accuracy depends on the interval.

```
/* Driver for routine qgaus */

#include <stdio.h>
#include <math.h>
#include "nr.h"

#define X1 0.0
#define X2 5.0
#define NVAL 10

float func(float x)
{
    return x*exp(-x);
}

int main(void)
{
    float dx,ss,x;
    int i;

    dx=(X2-X1)/NVAL;
    printf("\n%s %10s %13s\n\n","0.0 to","qgaus","expected");
    for (i=1;i<=NVAL;i++) {
        x=X1+i*dx;
        ss=qgaus(func,X1,x);
        printf("%5.2f %12.6f %12.6f\n",x,ss,
            (-(1.0+x)*exp(-x)+(1.0+X1)*exp(-X1)));
    }
    return 0;
}
```

Sample program `xgauleg`, which drives `gauleg`, performs the same method of quadrature, and on the same function. However, it chooses its own abscissas and weights for the Gauss-Legendre calculation, and is not restricted to a ten-point formula; it can do an N-point calculation for any N. The N abscissas and weights appropriate to an interval $x = 0.0$ to 1.0 are found by sample program `xgauleg` for the case $N = 10$. The results you should find are listed below. Next the program applies these values to a quadrature and compares the result to that from a formal integration.

#	x[i]	w[i]
1	0.013047	0.033336
2	0.067468	0.074726
3	0.160295	0.109543
4	0.283302	0.134633
5	0.425563	0.147762
6	0.574437	0.147762
7	0.716698	0.134633
8	0.839705	0.109543
9	0.932532	0.074726

10   0.986953   0.033336

```
/* Driver for routine gauleg */

#include <stdio.h>
#include <math.h>
#include "nr.h"
#include "nrutil.h"

#define NPOINT 10
#define X1 0.0
#define X2 1.0
#define X3 10.0

float func(float x)
{
    return x*exp(-x);
}

int main(void)
{
    int i;
    float xx=0.0;
    float *x,*w;

    x=vector(1,NPOINT);
    w=vector(1,NPOINT);
    gauleg(X1,X2,x,w,NPOINT);
    printf("\n%2s %10s %12s\n","#","x[i]","w[i]");
    for (i=1;i<=NPOINT;i++)
        printf("%2d %12.6f %12.6f\n",i,x[i],w[i]);
    /* Demonstrate the use of gauleg for integration */
    gauleg(X1,X3,x,w,NPOINT);
    for (i=1;i<=NPOINT;i++)
        xx += (w[i]*func(x[i]));
    printf("\nIntegral from GAULEG: %12.6f\n",xx);
    printf("Actual value: %12.6f\n",
        (1.0+X1)*exp(-X1)-(1.0+X3)*exp(-X3));
    free_vector(w,1,NPOINT);
    free_vector(x,1,NPOINT);
    return 0;
}
```

The next three sample programs are all similar to the previous one. They compute abscissas and weights for some form of Gaussian quadrature and print them out. As a check, each routine sums the weights and compares the sum with an analytically known value. If the abscissas are not symmetric about the origin, their sum is also nonzero and can be compared with an analytic value. Next, each program calculates an appropriate integral using the coefficients just determined.

The first such sample program, xgaulag, exercises gaulag by computing coefficients for an N-point Gauss-Laguerre quadrature. The program fixes the parameter $\alpha = 1$. It then calculates the integral

$$\int_0^\infty e^{-x} J_0(x) x\, dx = 2^{-3/2}$$

and compares the answer with the analytic result. As you will find, N of order a dozen is enough to determine the value to full single precision.

```
/* Driver for routine gaulag */

#include <stdio.h>
#include <math.h>
#include "nr.h"
#include "nrutil.h"

#define NP 64

float func(float x)
{
    return bessj0(x);
}

int main(void)
{
    int i,n;
    float alf=1.0,checkw,checkx,xx,*x,*w;

    x=vector(1,NP);
    w=vector(1,NP);
    for (;;) {
        printf("Enter N\n");
        if (scanf("%d",&n) == EOF) break;
        gaulag(x,w,n,alf);
        printf("%3s %10s %14s\n","#","x(i)","w(i)");
        for (i=1;i<=n;i++) printf("%3d %14.6e %14.6e\n",i,x[i],w[i]);
        checkx=checkw=0.0;
        for (i=1;i<=n;i++) {
            checkx += x[i];
            checkw += w[i];
        }
        printf("\nCheck value: %15.7e  should be: %15.7e\n",checkx,n*(n+alf));
        printf("\nCheck value: %15.7e  should be: %15.7e\n",checkw,
            exp(gammln(1.0+alf)));
        /* demonstrate the use of GAULAG for an integral */
        for (xx=0.0,i=1;i<=n;i++) xx += w[i]*func(x[i]);
        printf("\nIntegral from gaulag: %12.6f\n",xx);
        printf("Actual value:         %12.6f\n",1.0/(2.0*sqrt(2.0)));
    }
    free_vector(w,1,NP);
    free_vector(x,1,NP);
    return 0;
}
```

The next demonstration program, xgauher, computes coefficients for Gauss-Hermite quadrature by calling gauher. The test integral in this case is

$$\int_{-\infty}^{\infty} e^{-x^2} \cos x \, dx = \sqrt{\pi} e^{-1/4}$$

```
/* Driver for routine gauher */

#include <stdio.h>
#include <math.h>
#include "nr.h"
#include "nrutil.h"

#define NP 64
#define SQRTPI 1.7724539

float func(float x)
{
    return cos(x);
}

int main(void)
{
    int i,n;
    float check,xx,*x,*w;

    x=vector(1,NP);
    w=vector(1,NP);
    for (;;) {
        printf("Enter N\n");
        if (scanf("%d",&n) == EOF) break;
        gauher(x,w,n);
        printf("%3s %10s %14s\n","#","x(i)","w(i)");
        for (i=1;i<=n;i++) printf("%3d %14.6e %14.6e\n",i,x[i],w[i]);
        for (check=0.0,i=1;i<=n;i++) check += w[i];
        printf("\nCheck value: %15.7e  should be: %15.7e\n",check,SQRTPI);
        /* demonstrate the use of GAUHER for an integral */
        for (xx=0.0,i=1;i<=n;i++) xx += w[i]*func(x[i]);
        printf("\nIntegral from gauher: %12.6f\n",xx);
        printf("Actual value:         %12.6f\n",SQRTPI*exp(-0.25));
    }
    free_vector(w,1,NP);
    free_vector(x,1,NP);
    return 0;
}
```

Program xgaujac exercises gaujac by computing the abscissas and weights for an N-point Gauss-Jacobi quadrature. A convenient test integral is

$$\int_{-1}^{1} \frac{dx}{\sqrt{1-x^2}} \frac{1}{\sqrt{1-k^2(1+x)/2}} = 2K(k)$$

so we set $\alpha = \beta = -1/2$ in the program. (The integration reduces to Gauss-Chebyshev quadrature for this choice of parameters.) We choose $k = 1/2$, but you can easily experiment with other values. A surprisingly small value of N will calculate the integral accurately.

```
/* Driver for routine gaujac */

#include <stdio.h>
#include <math.h>
#include "nr.h"
```

```
#include "nrutil.h"

#define NP 64
#define PIBY2 1.5707963

float func(float ak,float x)
{
    return 1.0/sqrt(1.0-ak*ak*(1.0+x)/2.0);
}

int main(void)
{
    int i,n;
    float ak,alf=(-0.5),bet=(-0.5),checkw,checkx,xx,*x,*w;

    x=vector(1,NP);
    w=vector(1,NP);
    for (;;) {
        printf("Enter N\n");
        if (scanf("%d",&n) == EOF) break;
        gaujac(x,w,n,alf,bet);
        printf("%3s %10s %14s\n","#","x(i)","w(i)");
        for (i=1;i<=n;i++) printf("%3d %14.6e %14.6e\n",i,x[i],w[i]);
        checkx=checkw=0.0;
        for (i=1;i<=n;i++) {
            checkx += x[i];
            checkw += w[i];
        }
        printf("\nCheck value: %15.7e  should be: %15.7e\n",
                checkx,n*(bet-alf)/(alf+bet+2*n));
        printf("\nCheck value: %15.7e  should be: %15.7e\n",
                checkw,exp(gammln(1.0+alf)+gammln(1.0+bet)-
                gammln(2.0+alf+bet))*pow(2.0,alf+bet+1.0));
        /* demonstrate the use of GAUJAC for an integral */
        ak=0.5;
        for (xx=0.0,i=1;i<=n;i++) xx += w[i]*func(ak,x[i]);
        printf("\nIntegral from gaujac: %12.6f\n",xx);
        printf("Actual value:         %12.6f\n",2.0*ellf(PIBY2,ak));
    }
    free_vector(w,1,NP);
    free_vector(x,1,NP);
    return 0;
}
```

Routine gaucof computes abscissas and weights given the coefficients in the recurrence relation for the corresponding orthogonal polynomials. We test it in xgaucof with Gauss-Hermite quadrature. The results should be the same as from xgauher.

```
/* Driver for routine gaucof */

#include <stdio.h>
#include "nr.h"
#include "nrutil.h"

#define NP 64
#define SQRTPI 1.7724539
```

```
int main(void)
{
    /* Test with Gauss-Hermite */
    int i,n;
    float amu0,check,*a,*b,*x,*w;

    a=vector(1,NP);
    b=vector(1,NP);
    x=vector(1,NP);
    w=vector(1,NP);
    for (;;) {
        printf("Enter N:\n");
        if (scanf("%d",&n) == EOF) break;
        for (i=1;i<n;i++) {
            a[i]=0.0;
            b[i+1]=i*0.5;
        }
        a[n]=0.0;
        /* b[1] is arbitrary for call to tqli */
        amu0=SQRTPI;
        gaucof(n,a,b,amu0,x,w);
        printf("%3s %10s %14s\n","#","x(i)","w(i)");
        for (i=1;i<=n;i++) printf("%3d %14.6e %14.6e\n",i,x[i],w[i]);
        for (check=0.0,i=1;i<=n;i++) check += w[i];
        printf("\nCheck value: %15.7e  should be: %15.7e\n",check,SQRTPI);
    }
    free_vector(w,1,NP);
    free_vector(x,1,NP);
    free_vector(b,1,NP);
    free_vector(a,1,NP);
    return 0;
}
```

Routine `orthog` computes abscissas and weights for a nonclassical weight function. We test it in `xorthog` with the weight function $-\log x$, as described in *Numerical Recipes*. The test integral here is

$$-\int_0^1 \frac{\log x}{(1+x)^2}\, dx = \log 2$$

Once again a rather small value of N suffices for good accuracy.

```
/* Driver for routine orthog */

#include <stdio.h>
#include <math.h>
#include "nr.h"
#include "nrutil.h"

#define NP 64

float func(float x)
{
    return 1.0/SQR(1.0+x);
}

int main(void)
```

```
{
    int i,n;
    float amu0,check,xx,*a,*b,*x,*w,*anu,*alpha,*beta;

    a=vector(1,NP);
    b=vector(1,NP);
    x=vector(1,NP);
    w=vector(1,NP);
    anu=vector(1,2*NP);
    alpha=vector(1,2*NP-1);
    beta=vector(1,2*NP-1);

    /* Test with w[x] = -log x */
    for (;;) {
        printf("Enter N\n");
        if (scanf("%d",&n) == EOF) break;
        alpha[1]=0.5;
        beta[1]=1.0;
        for (i=2;i<=2*n-1;i++) {
            alpha[i]=0.5;
            beta[i]=1.0/(4.0*(4.0-1.0/((i-1)*(i-1))));
        }
        anu[1]=1.0;
        anu[2] = -0.25;
        for (i=2;i<=2*n-1;i++) anu[i+1] = -anu[i]*i*(i-1)/(2.0*(i+1)*(2*i-1));
        orthog(n,anu,alpha,beta,a,b);
        amu0=1.0;
        gaucof(n,a,b,amu0,x,w);
        printf("%3s %10s %14s\n","#","x(i)","w(i)");
        for (i=1;i<=n;i++) printf("%3d %14.6e %14.6e\n",i,x[i],w[i]);
        for (check=0.0,i=1;i<=n;i++) check += w[i];
        printf("\nCheck value: %15.7e   should be: %15.7e\n",check,amu0);
        /* demonstrate the use of ORTHOG for an integral */
        for (xx=0.0,i=1;i<=n;i++) xx += w[i]*func(x[i]);
        printf("\nIntegral from orthog: %12.6f\n",xx);
        printf("Actual value:         %12.6f\n",log(2.0));
    }
    free_vector(beta,1,2*NP-1);
    free_vector(alpha,1,2*NP-1);
    free_vector(anu,1,2*NP);
    free_vector(w,1,NP);
    free_vector(x,1,NP);
    free_vector(b,1,NP);
    free_vector(a,1,NP);
    return 0;
}
```

Chapter 4 of *Numerical Recipes* ends with a short discussion of multidimensional integration, exemplified by routine quad3d which does a 3-dimensional integration by repeated 1-dimensional integration. The C version of this algorithm is quite simple because recursion can be used. Sample program xquad3d applies the method to the integration of func $= x^2 + y^2 + z^2$ over a spherical volume with a radius xmax which is taken successively as $0.1, 0.2, \ldots, 1.0$. The integral is done in Cartesian rather than spherical coordinates, but the result is compared to that easily found in spherical coordinates, $4\pi(\text{xmax})^5/5$. Procedure func generates the function. Procedures yy1 and yy2 supply the two limits of the y-integration for each value of x. Similarly z1 and z2 give the limits of z-integration for given x and y.

```c
/* Driver for routine quad3d */

#include <stdio.h>
#include <math.h>
#include "nr.h"

#define PI 3.1415927
#define NVAL 10

static float xmax;

float func(float x,float y,float z)
{
    return x*x+y*y+z*z;
}

float z1(float x,float y)
{
    return (float) -sqrt(xmax*xmax-x*x-y*y);
}

float z2(float x,float y)
{
    return (float) sqrt(xmax*xmax-x*x-y*y);
}

float yy1(float x)
{
    return (float) -sqrt(xmax*xmax-x*x);
}

float yy2(float x)
{
    return (float) sqrt(xmax*xmax-x*x);
}

int main(void)
{
    int i;
    float xmin,s;

    printf("Integral of r^2 over a spherical volume\n\n");
    printf("%13s %10s %11s\n","radius","QUAD3D","Actual");
    for (i=1;i<=NVAL;i++) {
        xmax=0.1*i;
        xmin = -xmax;
        s=quad3d(func,xmin,xmax);
        printf("%12.2f %12.6f %11.6f\n",
            xmax,s,4.0*PI*pow(xmax,5.0)/5.0);
    }
    return 0;
}
```

Chapter 5: Evaluation of Functions

Chapter 5 of *Numerical Recipes* treats the approximation and evaluation of functions. The methods, along with a few others, are applied in Chapter 6 to the calculation of a collection of "special" functions. Polynomial or power series expansions are perhaps the most often used approximations and a few tips are given for accelerating the convergence of some series. In the case of alternating series, Euler's transformation is popular, and is implemented in program `eulsum`. For general polynomials, `ddpoly` demonstrates the evaluation of both the polynomial and its derivatives from a list of its coefficients. The division of one polynomial into another, giving a quotient and remainder polynomial, is done by `poldiv`, while the evaluation of a rational function as the ratio of two polynomials is carried out by `ratval`. Ridders' method of finding numerical derivatives, `dfridr`, relies on polynomial extrapolation techniques from Chapter 3.

The approximation of functions by Chebyshev polynomial series is presented as a method of arriving at the approximation of nearly smallest deviation from the true function over a given region for a specified order of approximation. The coefficients for such polynomials are given by `chebft` and function approximations are subsequently carried out by `chebev`. To generate the derivative or integral of a function from its Chebyshev coefficients, use `chder` or `chint` respectively. To convert Chebyshev coefficients into coefficients of a polynomial for the same function (a dangerous procedure about which we offer due warning in the text) use `chebpc` and `pcshft` in succession. Finally, to economize power series, use `pccheb`, the inverse of `chebpc`.

The chapter concludes with a routine for finding the Padé approximant from power series coefficients, `pade`, and a routine `ratlsq` for finding a rational function fit by a least squares method.

Chapter 5 also treats several methods for which we supply no programs. These are continued fractions, complex arithmetic (see Appendix C), recurrence relations, and the solution of quadratic and cubic equations.

$$\star \quad \star \quad \star \quad \star$$

Procedure `eulsum` applies Euler's transformation to the summation of an alternating series. It is called successively for each term to be summed. Our sample program `xeulsum` evaluates the approximation

$$\ln(1 + x) = x - \frac{x^2}{2} + \frac{x^3}{3} - \frac{x^4}{4} + \cdots \qquad -1 < x < 1$$

It asks how many terms `mval` are to be included in the approximation and then makes `mval` calls to `eulsum`. Each time, index j increases and `term` takes the value $(-1)^{j+1}x^j/j$. Both this approximation and the function $\ln(1+x)$ itself are evaluated across the region -1 to 1 for comparison. If `mval` is set less than 1 or more than 40, the program terminates.

```
/* Driver for routine eulsum */

#include <stdio.h>
#include <math.h>
#include "nr.h"
#include "nrutil.h"

#define NVAL 40

int main(void)
{
    int i,j,mval;
    float sum,term,x,xpower,*wksp;

    wksp=vector(1,NVAL);
    /* evaluate ln(1+x)=x-x^2/2+x^3/3-x^4/4 ... for -1<x<1 */
    for (;;) {
        printf("\nHow many terms in polynomial?\n");
        printf("Enter n between 1 and %2d. (n=0 to end) ",NVAL);
        scanf("%d",&mval);
        printf("\n");
        if ((mval <= 0) || (mval > NVAL)) break;
        printf("%9s %14s %14s\n","x","actual","polynomial");
        for (i = -8;i<=8;i++) {
            x=i/10.0;
            sum=0.0;
            xpower = -1;
            for (j=1;j<=mval;j++) {
                xpower *= (-x);
                term=xpower/j;
                eulsum(&sum,term,j,wksp);
            }
            printf("%12.6f %12.6f %12.6f\n",x,log(1.0+x),sum);
        }
    }
    free_vector(wksp,1,NVAL);
    return 0;
}
```

`ddpoly` evaluates a polynomial and its derivatives, given the coefficients of the polynomial in the form of an input vector. Sample program `xddpoly` illustrates this for the polynomial:

$$(x-1)^5 = -1 + 5x - 10x^2 + 10x^3 - 5x^4 + x^5$$

(This is a foolish example, of course. No one would knowingly evaluate $(x-1)^5$ by multiplying it out and evaluating terms individually — but it gives us a convenient way to check the result!). Since this is a fifth order polynomial, we set NC, the degree of the polynomial, to 5, and initialize the array `c` of coefficients, with `c[0]` being the constant coefficient and `c[5]` the highest-order coefficient. There are two loops, one of which evaluates for x values from 0.0 to 2.0, and the other of which stores the value of the function and NC-1 derivatives. d[j][i] keeps the entire array of values for printing. In the second part of the program, the polynomial evaluations are compared with

$$f^{(n-1)}(x) = \frac{5!}{(6-n)!}(x-1.0)^{6-n} \qquad n = 1, \ldots, 5$$

```
/* Driver for routine ddpoly */

#include <stdio.h>
#include "nr.h"
#include "nrutil.h"

#define NC 5
#define ND NC-1
#define NP 20

int main(void)
{
    int i,j,k;
    float x,pwr,*pd,**d;
    static float c[NC+1]={-1.0,5.0,-10.0,10.0,-5.0,1.0};
    static char *a[ND+1]={"polynomial:", "first deriv:",
        "second deriv:","third deriv:","fourth deriv:"};

    pd=vector(0,ND);
    d=matrix(0,ND,1,NP);
    for (i=1;i<=NP;i++) {
        x=0.1*i;
        ddpoly(c,NC,x,pd,ND);
        for (j=0;j<=ND;j++) d[j][i]=pd[j];
    }
    for (i=0;i<=ND;i++) {
        printf("%6s %s \n"," ",a[i]);
        printf("%12s %17s %15s\n","x","DDPOLY","actual");
        for (j=1;j<=NP;j++) {
            x=0.1*j;
            pwr=1.0;
            for (k=1;k<=NC-i;k++) pwr *= x-1.0;
            printf("%15.6f %15.6f %15.6f\n",x,d[i][j],
                (factrl(NC)/factrl(NC-i))*pwr);
        }
        printf("press ENTER to continue...\n");
        (void) getchar();
    }
    free_matrix(d,0,ND,1,NP);
    free_vector(pd,0,ND);
    return 0;
}
```

poldiv divides polynomials. Given the coefficients of a numerator and denominator polynomial, poldiv returns the coefficients of a quotient and a remainder polynomial. Sample program xpoldiv takes

$$\text{Numerator} = u = -1 + 5x - 10x^2 + 10x^3 - 5x^4 + x^5 = (x-1)^5$$
$$\text{Denominator} = v = 1 + 3x + 3x^2 + x^3 = (x+1)^3$$

for which we expect

$$\text{Quotient} = q = 31 - 8x + x^2$$
$$\text{Remainder} = r = -32 - 80x - 80x^2$$

The program compares these with the output of `poldiv`.

```
/* Driver for routine poldiv */

#include <stdio.h>
#include "nr.h"
#include "nrutil.h"

#define N 5
#define NV 3

int main(void)
{
    int i;
    static float u[N+1]={-1.0,5.0,-10.0,10.0,-5.0,1.0};
    static float v[NV+1]={1.0,3.0,3.0,1.0};
    float *q,*r;

    q=vector(0,N);
    r=vector(0,N);
    poldiv(u,N,v,NV,q,r);
    printf("\n%10s %10s %10s %10s %10s %10s\n\n",
        "x^0","x^1","x^2","x^3","x^4","x^5");
    printf("quotient polynomial coefficients:\n");
    for (i=0;i<=5;i++) printf("%10.2f ",q[i]);
    printf("\nexpected quotient coefficients:\n");
    printf("%10.2f %10.2f %10.2f %10.2f %10.2f %10.2f\n\n",
        31.0,-8.0,1.0,0.0,0.0,0.0);
    printf("remainder polynomial coefficients:\n");
    for (i=0;i<=3;i++) printf("%10.2f ",r[i]);
    printf("\nexpected remainder coefficients:\n");
    printf("%10.2f %10.2f %10.2f %10.2f\n",-32.0,-80.0,-80.0,0.0);
    free_vector(r,0,N);
    free_vector(q,0,N);
    return 0;
}
```

The next routine in Chapter 5, `ratval`, does not have its own demonstration program. Instead, it is used in the demonstration programs for `pade` and `ratlsq` at the end of this chapter.

Ridders' method for numerical differentiation is carried out by routine `dfridr`. In sample program `xdfridr` we test it on the simple function $f(x) = \tan x$, for which $f'(x) = \sec^2 x$. You might try input values like $x = 1.0$, $h = 0.1$, to demonstrate that h need not (and should not) be very small.

```
/* Driver for routine dfridr */

#include <stdio.h>
#include <math.h>
#include "nr.h"
#include "nrutil.h"
```

```
float func(float x)
{
    return tan(x);
}

int main(void)
{
    float x,h,dx,err;

    printf("input x, h\n");
    while (scanf("%f %f",&x,&h) != EOF)
    {
        dx=dfridr(func,x,h,&err);
        printf("dfridr=%12.6f %12.6f %12.6f\n",dx,1.0/SQR(cos(x)),err);
    }
    return 0;
}
```

The next seven programs all deal with Chebyshev polynomials. chebft evaluates the coefficients for a Chebyshev polynomial approximation of a function on a specified interval and for a maximum degree N of polynomial. Demonstration program xchebft uses the function func $= x^2(x^2 - 2)\sin x$ on the interval $(-\pi/2, \pi/2)$ with the maximum degree of NVAL=40. Notice that chebft is called with this maximum degree specified, even though subsequent evaluations may truncate the Chebyshev series at much lower terms. After we choose the number mval of terms in the evaluation, the Chebyshev polynomial is evaluated term by term, for x values between -0.8π and 0.8π, and the result f is compared to the actual function value.

```
/* Driver for routine chebft */

#include <stdio.h>
#include <math.h>
#include "nr.h"

#define NVAL 40
#define PIO2 1.5707963

float func(float x)
{
    return x*x*(x*x-2.0)*sin(x);
}

int main(void)
{
    float a=(-PIO2),b=PIO2,dum,f;
    float t0,t1,term,x,y,c[NVAL];
    int i,j,mval;

    chebft(a,b,c,NVAL,func);
    /* test result */
    for (;;) {
        printf("\nHow many terms in Chebyshev evaluation?\n");
        printf("Enter n between 6 and %2d. (n=0 to end).\n",NVAL);
        scanf("%d",&mval);
        if ((mval <= 0) || (mval > NVAL)) break;
```

```
        printf("\n%9s %14s %16s\n","x","actual","chebyshev fit");
        for (i = -8;i<=8;i++) {
            x=i*PIO2/10.0;
            y=(x-0.5*(b+a))/(0.5*(b-a));
            /* Evaluate Chebyshev polynomial without CHEBEV */
            t0=1.0;
            t1=y;
            f=c[1]*t1+c[0]*0.5;
            for (j=2;j<mval;j++) {
                dum=t1;
                t1=2.0*y*t1-t0;
                t0=dum;
                term=c[j]*t1;
                f += term;
            }
            printf("%12.6f %12.6f %12.6f\n",x,func(x),f);
        }
    }
    return 0;
}
```

chebev is the Chebyshev polynomial evaluator and the next sample program
xchebev uses it for the same problem just discussed. In fact, the program is identical
except that it replaces the internal polynomial summation with chebev, which applies
Clenshaw's recurrence to find the polynomial values.

```
/* Driver for routine chebev */

#include <stdio.h>
#include <math.h>
#include "nr.h"

#define NVAL 40
#define PIO2 1.5707963

float func(float x)
{
    return x*x*(x*x-2.0)*sin(x);
}

int main(void)
{
    int i,mval;
    float a=(-PIO2),b=PIO2,x,c[NVAL];

    chebft(a,b,c,NVAL,func);
    /* Test Chebyshev evaluation routine */
    for (;;) {
        printf("\nHow many terms in Chebyshev evaluation?\n");
        printf("Enter n between 6 and %2d. (n=0 to end).\n",NVAL);
        scanf("%d",&mval);
        if ((mval <= 0) || (mval > NVAL)) break;
        printf("\n%9s %14s %16s \n","x","actual","chebyshev fit");
        for (i = -8;i<=8;i++) {
            x=i*PIO2/10.0;
            printf("%12.6f %12.6f %12.6f\n",
                x,func(x),chebev(a,b,c,mval,x));
```

```
      }
   }
   return 0;
}
```

By the same token, the tests for `chint` and `chder` needn't be much different. `chint` determines Chebyshev coefficients for the integral of the function, and `chder` for the derivative of the function, given the Chebyshev coefficients for the function itself (from `chebft`) and the interval (A, B) of evaluation. When applied to the function above, the true integral is

$$\texttt{fint} = 4x(x^2 - 7)\sin x - (x^4 - 14x^2 + 28)\cos x$$

and the true derivative is

$$\texttt{fder} = 4x(x^2 - 1)\sin x + x^2(x^2 - 2)\cos x$$

The code in sample programs `xchint` and `xchder` compares the true and Chebyshev-derived integral and derivative values for a range of x in the interval of evaluation. Since `chint` and `chder` return Chebyshev coefficients, and not the integral and derivative values themselves, calls to `chebev` are required for the comparison.

```
/* Driver for routine chint */

#include <stdio.h>
#include <math.h>
#include "nr.h"

#define NVAL 40
#define PIO2 1.5707963

float func(float x)
{
    return x*x*(x*x-2.0)*sin(x);
}

float fint(float x)
{
    return 4.0*x*(x*x-7.0)*sin(x)-(x*x*(x*x-14.0)+28.0)*cos(x);
}

int main(void)
{
    int i,mval;
    float a=(-PIO2),b=PIO2,x;
    float c[NVAL],cint[NVAL];

    chebft(a,b,c,NVAL,func);
    /* test integral */
    for (;;) {
        printf("\nHow many terms in Chebyshev evaluation?\n");
        printf("Enter n between 6 and %2d. (n=0 to end).\n",NVAL);
        scanf("%d",&mval);
        if ((mval <= 0) || (mval > NVAL)) break;
        chint(a,b,c,cint,mval);
        printf("\n%9s %14s %16s\n","x","actual","Cheby. integ.");
```

```
        for (i = -8;i<=8;i++) {
            x=i*PIO2/10.0;
            printf("%12.6f %12.6f %12.6f\n",
                x,fint(x)-fint(-PIO2),chebev(a,b,cint,mval,x));
        }
    }
    return 0;
}

/* Driver for routine chder */

#include <stdio.h>
#include <math.h>
#include "nr.h"

#define NVAL 40
#define PIO2 1.5707963

float func(float x)
{
    return x*x*(x*x-2.0)*sin(x);
}

float fder(float x)
{
    return 4.0*x*(x*x-1.0)*sin(x)+x*x*(x*x-2.0)*cos(x);
}

int main(void)
{
    int i,mval;
    float a=(-PIO2),b=PIO2,x;
    float c[NVAL],cder[NVAL];

    chebft(a,b,c,NVAL,func);
    /* Test derivative */
    for (;;) {
        printf("\nHow many terms in Chebyshev evaluation?\n");
        printf("Enter n between 6 and %2d. (n=0 to end).\n",NVAL);
        scanf("%d",&mval);
        if ((mval <= 0) || (mval > NVAL)) break;
        chder(a,b,c,cder,mval);
        printf("\n%9s %14s %16s\n","x","actual","Cheby. deriv.");
        for (i = -8;i<=8;i++) {
            x=i*PIO2/10.0;
            printf("%12.6f %12.6f %12.6f\n",
                x,fder(x),chebev(a,b,cder,mval,x));
        }
    }
    return 0;
}
```

The next two programs of this chapter turn the coefficients of a Chebyshev approximation into those of a polynomial approximation in the variable

$$y = \frac{x - \frac{1}{2}(B + A)}{\frac{1}{2}(B - A)}$$

(routine chebpc), or of a polynomial approximation in x itself (routine chebpc followed by pcshft). These procedures are discouraged for reasons discussed in *Numerical Recipes*, but should they serve some special purpose for you, we have at least warned that you will be sacrificing accuracy, particularly for polynomials above order 7 or 8. Sample program xchebpc calls chebft and chebpc to find polynomial coefficients in y for a truncated series. For a set of x values between $-\pi$ and π it calculates y and then the terms of the y-polynomial, which are summed in variable poly. Finally, poly is compared to the true function value. (The function func is the same used before.)

```c
/* Driver for routine chebpc */

#include <stdio.h>
#include <math.h>
#include "nr.h"

#define NVAL 40
#define PIO2 1.5707963

float func(float x)
{
    return x*x*(x*x-2.0)*sin(x);
}

int main(void)
{
    int i,j,mval;
    float a=(-PIO2),b=PIO2,poly,x,y;
    float c[NVAL],d[NVAL];

    chebft(a,b,c,NVAL,func);
    for (;;) {
        printf("\nHow many terms in Chebyshev evaluation?\n");
        printf("Enter n between 6 and %2d. (n=0 to end).\n",NVAL);
        scanf("%d",&mval);
        if ((mval <= 0) || (mval > NVAL)) break;
        chebpc(c,d,mval);
        /* Test polynomial */
        printf("\n%9s %14s %14s\n","x","actual","polynomial");
        for (i = -8;i<=8;i++) {
            x=i*PIO2/10.0;
            y=(x-0.5*(b+a))/(0.5*(b-a));
            poly=d[mval-1];
            for (j=mval-2;j>=0;j--) poly=poly*y+d[j];
            printf("%12.6f %12.6f %12.6f\n",x,func(x),poly);
        }
    }
    return 0;
}
```

pcshft shifts the polynomial to be one in variable x. Sample program xpcshft is like the previous program except that it follows the call to chebpc with a call to pcshft.

```
/* Driver for routine pcshft */

#include <stdio.h>
#include <math.h>
#include "nr.h"

#define NVAL 40
#define PIO2 1.5707963

float func(float x)
{
    return x*x*(x*x-2.0)*sin(x);
}

int main(void)
{
    int i,j,mval;
    float a=(-PIO2),b=PIO2,poly,x;
    float c[NVAL],d[NVAL];

    chebft(a,b,c,NVAL,func);
    for (;;) {
        printf("\nHow many terms in Chebyshev evaluation?\n");
        printf("Enter n between 6 and %2d. (n=0 to end).\n",NVAL);
        scanf("%d",&mval);
        if ((mval <= 0) || (mval > NVAL)) break;
        chebpc(c,d,mval);
        pcshft(a,b,d,mval);
        /* Test shifted polynomial */
        printf("\n%9s %14s %14s\n","x","actual","polynomial");
        for (i = -8;i<=8;i++) {
            x=i*PIO2/10.0;
            poly=d[mval-1];
            for (j=mval-2;j>=0;j--) poly=poly*x+d[j];
            printf("%12.6f %12.6f %12.6f\n",x,func(x),poly);
        }
    }
    return 0;
}
```

Program xpccheb shows how to use pccheb to economize a power series. The example chosen is the series on p. 294 of *Numerical Methods That Work*, by F.S. Acton, namely that of $\cos \pi y$ on the interval $[-1, 1]$. The coefficients of the series up to y^{17} are computed and converted to Chebyshev coefficients. These coefficients are truncated after the term in y^{13}, and the new power series coefficients are computed. The answer is checked at selected points in the interval.

```
/* Driver for routine pccheb */

#include <stdio.h>
#include <math.h>
#include "nr.h"
#include "nrutil.h"

#define NCHECK 15
#define NFEW 13
```

```
#define NMANY 17
#define NMAX 100
#define PI 3.14159265

int main(void)
{
    int i,j;
    float a=(-PI),b=PI,fac,f,sum,sume,py,py2,*c,*d,*e,*ee;

    c=vector(0,NMAX-1);
    d=vector(0,NMAX-1);
    e=vector(0,NMAX-1);
    ee=vector(0,NMAX-1);
    /* put power series of cos(PI*y) into e */
    fac=1.0;
    e[0]=ee[0]=0.0;
    for (j=0;j<NMANY;j++) {
        i=j & 3;    /* tricky way to perform j % 4 */
        if (i == 1 || i == 3) e[j]=0.0;
        else if (i == 0) e[j]=1.0/fac;
        else e[j] = -1.0/fac;
        fac *= (j+1);
        ee[j]=e[j];
    }
    pcshft((-2.0-b-a)/(b-a),(2.0-b-a)/(b-a),e,NMANY);
    /* i.e., inverse of pcshft(a,b,...) which we do below */
    pccheb(e,c,NMANY);
    printf("Index, series, Chebyshev coefficients\n");
    for (j=0;j<NMANY;j+=2)
        printf("%3d %15.6e %15.6e\n",j,e[j],c[j]);
    chebpc(c,d,NFEW);
    pcshft(a,b,d,NFEW);
    printf("Index, new series, coefficient ratios\n");
    for (j=0;j<NFEW;j+=2) {
        printf("%3d %15.6e %15.6e\n",
                j,d[j],d[j]/(ee[j]+1.0e-30));
    }
    printf("     Point tested, function value, error power series, error Cheb.\n");
    for (i=0;i<=NCHECK;i++) {
        py=a+i*(b-a)/(float)NCHECK;
        py2=py*py;
        sum=sume=0.0;
        fac=1.0;
        for (j=0;j<NFEW;j+=2) {
            sum += fac*d[j];
            sume += fac*ee[j];
            fac *= py2;
        }
        f=cos(py);
        printf("check: %15.6e %15.6e %15.6e %15.6e\n",py,f,sume-f,sum-f);
    }
    free_vector(ee,0,NMAX-1);
    free_vector(e,0,NMAX-1);
    free_vector(d,0,NMAX-1);
    free_vector(c,0,NMAX-1);
    return 0;
}
```

Test program `xpade` excercises `pade` by computing the Padé approximation to $\ln(1+x)/x$. The Padé approximation is accurate for a much larger value of x than the power series with the same number of terms from which it is derived. Note that this test program and the one following both use `ratval`, so we do not supply a separate sample program for it.

```
/* Driver for routine pade */

#include <stdio.h>
#include <math.h>
#include "nr.h"
#include "nrutil.h"

double fn(double x)
{
    return (x == 0.0 ? 1.0 : log(1.0+x)/x);
}

#define NMAX 100

int main(void)
{
    int j,k,n;
    float resid;
    double b,d,fac,x,*c,*cc;

    c=dvector(0,NMAX);
    cc=dvector(0,NMAX);
    for (;;) {
        printf("Enter n for PADE routine:\n");
        if (scanf("%d",&n) == EOF) break;
        fac=1;
        for (j=1;j<=2*n+1;j++) {
            c[j-1]=fac/((double) j);
            cc[j-1]=c[j-1];
            fac = -fac;
        }
        pade(c,n,&resid);
        printf("Norm of residual vector= %16.8e\n",resid);
        printf("point, func. value, pade series, power series\n");
        for (j=1;j<=21;j++) {
            x=(j-1)*0.25;
            for (b=0.0,k=2*n+1;k>=1;k--) {
                b *= x;
                b += cc[k-1];
            }
            d=ratval(x,c,n,n);
            printf("%16.8f %16.8f %16.8f %16.8f\n",x,fn(x),d,b);
        }
    }
    free_dvector(cc,0,NMAX);
    free_dvector(c,0,NMAX);
    return 0;
}
```

Sample program `xratlsq` finds a rational approximation to the function $\tan^{-1} x$

by the least squares method in `ratlsq`. You enter a desired interval and the orders
of the numerator and the denominator. The printed result shows the discrepancy
between the fit and the true function value.

```c
/* Driver for routine ratlsq */

#include <stdio.h>
#include <math.h>
#include "nr.h"
#include "nrutil.h"

#define NMAX 100

double fn(double t)
{
    return atan(t);
}

int main(void)
{
    int j,kk,mm;
    double a,b,*cof,dev,eee,fit,xs;

    cof=dvector(0,NMAX);
    for (;;) {
        printf("enter a,b,mm,kk\n");
        if (scanf("%lf %lf %d %d",&a,&b,&mm,&kk) == EOF) break;
        ratlsq(fn,a,b,mm,kk,cof,&dev);
        for (j=0;j<=mm+kk;j++) printf("cof(%3d)=%27.15e\n",j,cof[j]);
        printf("maximum absolute deviation= %12.6f\n",dev);
        printf("    x          error         exact\n");
        printf("--------- ------------ ---------\n");
        for (j=1;j<=50;j++) {
            xs=a+(b-a)*(j-1.0)/49.0;
            fit=ratval(xs,cof,mm,kk);
            eee=fn(xs);
            printf("%10.5f %15.7e %15.7e\n",xs,fit-eee,eee);
        }
    }
    free_dvector(cof,0,NMAX);
    return 0;
}
```

Chapter 6: Special Functions

This chapter on special functions provides illustrations of techniques developed in Chapter 5. At the same time, it offers routines for calculating many of the functions that arise frequently in analytical work, but which are not so common as to be included, for example, as a single keystroke on your pocket calculator. In terms of demonstration programs, they represent a simple collection. The test routines are all virtually identical, making reference to a single file of function values called fncval.dat *which is listed in the Appendix at the end of this chapter. In this file are accurate values for the individual functions for a variety of values for each argument. We have aimed to "stress" the routines a bit by throwing in some extreme values for the arguments.*

Many of the function values came from Abramowitz and Stegun's Handbook of Mathematical Functions. Some others, however, came from our library of dusty volumes from past masters. There is an implicit danger in a comparison test like this — namely, that our source has used the same algorithms as ours to construct the tables. In that case, we test only our mutual competence at computing, not the correctness of the result. Nevertheless, there is some assurance in knowing that the values we calculate are the ones that have been used and scrutinized for many years. Moreover, the expressions for the functions themselves can be worked out in certain special or limiting cases without computer aid, and in these instances the results have proven correct.

$$\star \quad \star \quad \star \quad \star$$

With few exceptions, the routines that follow work in this fashion:

1. Open file fncval.dat.

2. Find the appropriate data table according to its title.

3. Read the argument list for each table entry and pass it to the routine to be tested.

4. Print the arguments along with the expected and actual results.

For the routines in this list, therefore, we forego any further comment, but simply identify them by the special function that they evaluate. The exceptions are the test program for the utility procedure beschb and the test of the hypergeometric function routine hypgeo.

Natural logarithm of the gamma function for positive arguments:

```
/* Driver for routine gammln */

#include <stdio.h>
#include <stdlib.h>
#include <string.h>
#include <math.h>
#include "nr.h"
#include "nrutil.h"

#define MAXSTR 80

int main(void)
{
    char txt[MAXSTR];
    int i,nval;
    float actual,calc,x;
    FILE *fp;

    if ((fp = fopen("fncval.dat","r")) == NULL)
        nrerror("Data file fncval.dat not found\n");
    fgets(txt,MAXSTR,fp);
    while (strncmp(txt,"Gamma Function",14)) {
        fgets(txt,MAXSTR,fp);
        if (feof(fp)) nrerror("Data not found in fncval.dat\n");
    }
    fscanf(fp,"%d %*s",&nval);
    printf("\n%s\n",txt);
    printf("%10s %21s %21s\n","x","actual","gammln(x)");
    for (i=1;i<=nval;i++) {
        fscanf(fp,"%f %f",&x,&actual);
        if (x > 0.0) {
            calc=(x<1.0 ? gammln(x+1.0)-log(x) : gammln(x));
            printf("%12.2f %20.6f %20.6f\n",x,
                log(actual),calc);
        }
    }
    fclose(fp);
    return 0;
}
```

Factorial function $N!$:

```
/* Driver for routine factrl */

#include <stdio.h>
#include <stdlib.h>
#include <string.h>
#include <math.h>
#include "nr.h"
#include "nrutil.h"

#define MAXSTR 80

int main(void)
{
    char txt[MAXSTR];
    float actual;
```

```
        int i,n,nval;
        FILE *fp;

        if ((fp = fopen("fncval.dat","r")) == NULL)
            nrerror("Data file fncval.dat not found\n");
        fgets(txt,MAXSTR,fp);
        while (strncmp(txt,"N-factorial",11)) {
            fgets(txt,MAXSTR,fp);
            if (feof(fp)) nrerror("Data not found in fncval.dat\n");
        }
        fscanf(fp,"%d %*s",&nval);
        printf("\n%s\n",txt);
        printf("%6s %18s %20s \n","n","actual","factrl(n)");
        for (i=1;i<=nval;i++) {
            fscanf(fp,"%d %f ",&n,&actual);
            if (actual < 1.0e10)
                printf("%6d %20.0f %20.0f\n",n,actual,factrl(n));
            else
                printf("%6d %20e %20e \n",n,actual,factrl(n));
        }
        fclose(fp);
        return 0;
    }
```

Binomial coefficients:

```
/* Driver for routine bico */

#include <stdio.h>
#include <stdlib.h>
#include <string.h>
#include "nr.h"
#include "nrutil.h"

#define MAXSTR 80

int main(void)
{
    char txt[MAXSTR];
    int i,k,n,nval;
    float binco;
    FILE *fp;

    if ((fp = fopen("fncval.dat","r")) == NULL)
        nrerror("Data file fncval.dat not found\n");
    fgets(txt,MAXSTR,fp);
    while (strncmp(txt,"Binomial Coefficients",21)) {
        fgets(txt,MAXSTR,fp);
        if (feof(fp)) nrerror("Data not found in fncval.dat\n");
    }
    fscanf(fp,"%d %*s",&nval);
    printf("\n%s\n",txt);
    printf("%6s %6s %12s %12s \n","n","k","actual","bico(n,k)");
    for (i=1;i<=nval;i++) {
        fscanf(fp,"%d %d %f ",&n,&k,&binco);
        printf("%6d %6d %12.0f %12.0f \n",n,k,binco,bico(n,k));
    }
```

```
        fclose(fp);
        return 0;
}
```

Natural logarithm of $N!$:

```
/* Driver for routine factln */

#include <stdio.h>
#include <stdlib.h>
#include <string.h>
#include <math.h>
#include "nr.h"
#include "nrutil.h"

#define MAXSTR 80

int main(void)
{
    char txt[MAXSTR];
    int i,n,nval;
    float val;
    FILE *fp;

    if ((fp = fopen("fncval.dat","r")) == NULL)
        nrerror("Data file fncval.dat not found\n");
    fgets(txt,MAXSTR,fp);
    while (strncmp(txt,"N-factorial",11)) {
        fgets(txt,MAXSTR,fp);
        if (feof(fp)) nrerror("Data not found in fncval.dat\n");
    }
    fscanf(fp,"%d %*s",&nval);
    printf("\nlog of n_factorial\n");
    printf("\n%6s %19s %21s\n","n","actual","factln(n)");
    for (i=1;i<=nval;i++) {
        fscanf(fp,"%d %f",&n,&val);
        printf("%6d %20.7f %20.7f\n",n,log(val),factln(n));
    }
    fclose(fp);
    return 0;
}
```

Beta function:

```
/* Driver for routine beta */

#include <stdio.h>
#include <stdlib.h>
#include <string.h>
#include "nr.h"
#include "nrutil.h"

#define MAXSTR 80

int main(void)
{
    char txt[MAXSTR];
    int i,nval;
```

```
    float val,w,z;
    FILE *fp;

    if ((fp = fopen("fncval.dat","r")) == NULL)
        nrerror("Data file fncval.dat not found\n");
    fgets(txt,MAXSTR,fp);
    while (strncmp(txt,"Beta Function",13)) {
        fgets(txt,MAXSTR,fp);
        if (feof(fp)) nrerror("Data not found in fncval.dat\n");
    }
    fscanf(fp,"%d %*s",&nval);
    printf("\n%s\n",txt);
    printf("%5s %6s %16s %20s\n","w","z","actual","beta(w,z)");
    for (i=1;i<=nval;i++) {
        fscanf(fp,"%f %f %f",&w,&z,&val);
        printf("%6.2f %6.2f %18.6e %18.6e\n",w,z,val,beta(w,z));
    }
    fclose(fp);
    return 0;
}
```

Incomplete gamma function $P(a, x)$:

```
/* Driver for routine gammp */

#include <stdio.h>
#include <stdlib.h>
#include <string.h>
#include "nr.h"
#include "nrutil.h"

#define MAXSTR 80

int main(void)
{
    char txt[MAXSTR];
    int i,nval;
    float a,val,x;
    FILE *fp;

    if ((fp = fopen("fncval.dat","r")) == NULL)
        nrerror("Data file fncval.dat not found\n");
    fgets(txt,MAXSTR,fp);
    while (strncmp(txt,"Incomplete Gamma Function",25)) {
        fgets(txt,MAXSTR,fp);
        if (feof(fp)) nrerror("Data not found in fncval.dat\n");
    }
    fscanf(fp,"%d %*s",&nval);
    printf("\n%s\n",txt);
    printf("%4s %11s %14s %14s \n","a","x","actual","gammp(a,x)");
    for (i=1;i<=nval;i++) {
        fscanf(fp,"%f %f %f",&a,&x,&val);
        printf("%6.2f %12.6f %12.6f %12.6f \n",a,x,val,gammp(a,x));
    }
    fclose(fp);
    return 0;
}
```

Incomplete gamma function $Q(a, x) = 1 - P(a, x)$:

```c
/* Driver for routine gammq */

#include <stdio.h>
#include <stdlib.h>
#include <string.h>
#include "nr.h"
#include "nrutil.h"

#define MAXSTR 80

int main(void)
{
    char txt[MAXSTR];
    int i,nval;
    float a,val,x;
    FILE *fp;

    if ((fp = fopen("fncval.dat","r")) == NULL)
        nrerror("Data file fncval.dat not found\n");
    fgets(txt,MAXSTR,fp);
    while (strncmp(txt,"Incomplete Gamma Function",25)) {
        fgets(txt,MAXSTR,fp);
        if (feof(fp)) nrerror("Data not found in fncval.dat\n");
    }
    fscanf(fp,"%d %*s",&nval);
    printf("\n%s\n",txt);
    printf("%4s %11s %14s %14s \n","a","x","actual","gammq(a,x)");
    for (i=1;i<=nval;i++) {
        fscanf(fp,"%f %f %f",&a,&x,&val);
        printf("%6.2f %12.6f %12.6f %12.6f\n",a,x,(1.0-val),gammq(a,x));
    }
    fclose(fp);
    return 0;
}
```

Incomplete gamma function $P(a, x)$ evaluated from series representation:

```c
/* Driver for routine gser */

#include <stdio.h>
#include <stdlib.h>
#include <string.h>
#include "nr.h"
#include "nrutil.h"

#define MAXSTR 80

int main(void)
{
    char txt[MAXSTR];
    int i,nval;
    float a,gamser,gln,val,x;
    FILE *fp;

    if ((fp = fopen("fncval.dat","r")) == NULL)
        nrerror("Data file fncval.dat not found\n");
```

```
        fgets(txt,MAXSTR,fp);
        while (strncmp(txt,"Incomplete Gamma Function",25)) {
            fgets(txt,MAXSTR,fp);
            if (feof(fp)) nrerror("Data not found in fncval.dat\n");
        }
        fscanf(fp,"%d %*s",&nval);
        printf("\n%s\n",txt);
        printf("%4s %11s %14s %14s %12s %8s\n","a","x",
            "actual","gser(a,x)","gammln(a)","gln");
        for (i=1;i<=nval;i++) {
            fscanf(fp,"%f %f %f",&a,&x,&val);
            gser(&gamser,a,x,&gln);
            printf("%6.2f %12.6f %12.6f %12.6f %12.6f %12.6f\n",
                a,x,val,gamser,gammln(a),gln);
        }
        fclose(fp);
        return 0;
}
```

Incomplete gamma function $Q(a, x)$ evaluated by continued fraction representation:

```
/* Driver for routine gcf */

#include <stdio.h>
#include <stdlib.h>
#include <string.h>
#include "nr.h"
#include "nrutil.h"

#define MAXSTR 80

int main(void)
{
    char txt[MAXSTR];
    int i,nval;
    float a,val,x,gammcf,gln;
    FILE *fp;

    if ((fp = fopen("fncval.dat","r")) == NULL)
        nrerror("Data file fncval.dat not found\n");
    fgets(txt,MAXSTR,fp);
    while (strncmp(txt,"Incomplete Gamma Function",25)) {
        fgets(txt,MAXSTR,fp);
        if (feof(fp)) nrerror("Data not found in fncval.dat\n");
    }
    fscanf(fp,"%d %*s",&nval);
    printf("\n%s\n",txt);
    printf("%4s %11s %14s %13s %13s %8s\n","a","x",
        "actual","gcf(a,x)","gammln(a)","gln");
    for (i=1;i<=nval;i++) {
        fscanf(fp,"%f %f %f",&a,&x,&val);
        if (x >= (a+1.0)) {
            gcf(&gammcf,a,x,&gln);
            printf("%6.2f%13.6f%13.6f%13.6f%12.6f%13.6f\n",
                a,x,(1.0-val),gammcf,gammln(a),gln);
        }
    }
```

```
    fclose(fp);
    return 0;
}
```

Error function:

```
/* Driver for routine erff */

#include <stdio.h>
#include <stdlib.h>
#include <string.h>
#include "nr.h"
#include "nrutil.h"

#define MAXSTR 80

int main(void)
{
    char txt[MAXSTR];
    int i,nval;
    float val,x;
    FILE *fp;

    if ((fp = fopen("fncval.dat","r")) == NULL)
        nrerror("Data file fncval.dat not found\n");
    fgets(txt,MAXSTR,fp);
    while (strncmp(txt,"Error Function",14)) {
        fgets(txt,MAXSTR,fp);
        if (feof(fp)) nrerror("Data not found in fncval.dat\n");
    }
    fscanf(fp,"%d %*s",&nval);
    printf("\n%s\n",txt);
    printf("%4s %12s %12s\n","x","actual","erf(x)");
    for (i=1;i<=nval;i++) {
        fscanf(fp,"%f %f",&x,&val);
        printf("%6.2f %12.7f %12.7f\n",x,val,erff(x));
    }
    fclose(fp);
    return 0;
}
```

Complementary error function:

```
/* Driver for routine erffc */

#include <stdio.h>
#include <stdlib.h>
#include <string.h>
#include "nr.h"
#include "nrutil.h"

#define MAXSTR 80

int main(void)
{
    char txt[MAXSTR];
    int i,nval;
    float x,val;
```

```
    FILE *fp;

    if ((fp = fopen("fncval.dat","r")) == NULL)
        nrerror("Data file fncval.dat not found\n");
    fgets(txt,MAXSTR,fp);
    while (strncmp(txt,"Error Function",14)) {
        fgets(txt,MAXSTR,fp);
        if (feof(fp)) nrerror("Data not found in fncval.dat\n");
    }
    fscanf(fp,"%d %*s",&nval);
    printf("\ncomplementary error function\n");
    printf("%5s %12s %13s\n","x","actual","erfc(x)");
    for (i=1;i<=nval;i++) {
        fscanf(fp,"%f %f",&x,&val);
        val=1.0-val;
        printf("%6.2f %12.7f %12.7f\n",x,val,erffc(x));
    }
    fclose(fp);
    return 0;
}
```

Complementary error function from a Chebyshev fit to a guessed functional form:

```
/* Driver for routine erfcc */

#include <stdio.h>
#include <stdlib.h>
#include <string.h>
#include "nr.h"
#include "nrutil.h"

#define MAXSTR 80

int main(void)
{
    char txt[MAXSTR];
    int i,nval;
    float x,val;
    FILE *fp;

    if ((fp = fopen("fncval.dat","r")) == NULL)
        nrerror("Data file fncval.dat not found\n");
    fgets(txt,MAXSTR,fp);
    while (strncmp(txt,"Error Function",14)) {
        fgets(txt,MAXSTR,fp);
        if (feof(fp)) nrerror("Data not found in fncval.dat\n");
    }
    fscanf(fp,"%d %*s",&nval);
    printf("\ncomplementary error function\n");
    printf("%5s %12s %13s\n","x","actual","erfcc(x)");
    for (i=1;i<=nval;i++) {
        fscanf(fp,"%f %f",&x,&val);
        val=1.0-val;
        printf("%6.2f %12.7f %12.7f\n",x,val,erfcc(x));
    }
    fclose(fp);
    return 0;
```

```
}
```

Exponential integral $E_n(x)$:

```
/* Driver for routine expint */

#include <stdio.h>
#include <stdlib.h>
#include <string.h>
#include "nr.h"
#include "nrutil.h"

#define MAXSTR 80

int main(void)
{
    char txt[MAXSTR];
    int i,nval,n;
    float val,x;
    FILE *fp;

    if ((fp = fopen("fncval.dat","r")) == NULL)
        nrerror("Data file fncval.dat not found\n");
    fgets(txt,MAXSTR,fp);
    while (strncmp(txt,"Exponential Integral En",23)) {
        fgets(txt,MAXSTR,fp);
        if (feof(fp)) nrerror("Data not found in fncval.dat\n");
    }
    fscanf(fp,"%d %*s",&nval);
    printf("\n%s\n",txt);
    printf("%4s %7s %15s %21s \n","n","x","actual","expint(n,x)");
    for (i=1;i<=nval;i++) {
        fscanf(fp,"%d %f %f",&n,&x,&val);
        printf("%4d %8.2f %18.6e %18.6e\n",n,x,val,expint(n,x));
    }
    fclose(fp);
    return 0;
}
```

Exponential integral $Ei(x)$:

```
/* Driver for routine ei */

#include <stdio.h>
#include <stdlib.h>
#include <string.h>
#include "nr.h"
#include "nrutil.h"

#define MAXSTR 80

int main(void)
{
    char txt[MAXSTR];
    int i,nval;
    float val,x;
    FILE *fp;
```

```
    if ((fp = fopen("fncval.dat","r")) == NULL)
        nrerror("Data file fncval.dat not found\n");
    fgets(txt,MAXSTR,fp);
    while (strncmp(txt,"Exponential Integral Ei",23)) {
        fgets(txt,MAXSTR,fp);
        if (feof(fp)) nrerror("Data not found in fncval.dat\n");
    }
    fscanf(fp,"%d %*s",&nval);
    printf("\n%s\n",txt);
    printf("%5s %12s %11s \n","x","actual","ei(x)");
    for (i=1;i<=nval;i++) {
        fscanf(fp,"%f %f",&x,&val);
        printf("%6.2f %12.6f %12.6f\n",x,val,ei(x));
    }
    fclose(fp);
    return 0;
}
```

Incomplete Beta function:

```
/* Driver for routine betai */

#include <stdio.h>
#include <stdlib.h>
#include <string.h>
#include "nr.h"
#include "nrutil.h"

#define MAXSTR 80

int main(void)
{
    char txt[MAXSTR];
    int i,nval;
    float a,b,x,val;
    FILE *fp;

    if ((fp = fopen("fncval.dat","r")) == NULL)
        nrerror("Data file fncval.dat not found\n");
    fgets(txt,MAXSTR,fp);
    while (strncmp(txt,"Incomplete Beta Function",24)) {
        fgets(txt,MAXSTR,fp);
        if (feof(fp)) nrerror("Data not found in fncval.dat\n");
    }
    fscanf(fp,"%d %*s",&nval);
    printf("\n%s\n",txt);
    printf("%5s %10s %12s %14s %13s \n",
        "a","b","x","actual","betai(x)");
    for (i=1;i<=nval;i++) {
        fscanf(fp,"%f %f %f %f",&a,&b,&x,&val);
        printf("%6.2f %12.6f %12.6f %12.6f %12.6f\n",
            a,b,x,val,betai(a,b,x));
    }
    fclose(fp);
    return 0;
}
```

Bessel function J_0:

```
/* Driver for routine bessj0 */

#include <stdio.h>
#include <stdlib.h>
#include <string.h>
#include "nr.h"
#include "nrutil.h"

#define MAXSTR 80

int main(void)
{
    char txt[MAXSTR];
    int i,nval;
    float val,x;
    FILE *fp;

    if ((fp = fopen("fncval.dat","r")) == NULL)
        nrerror("Data file fncval.dat not found\n");
    fgets(txt,MAXSTR,fp);
    while (strncmp(txt,"Bessel Function J0",18)) {
        fgets(txt,MAXSTR,fp);
        if (feof(fp)) nrerror("Data not found in fncval.dat\n");
    }
    fscanf(fp,"%d %*s",&nval);
    printf("\n%s\n",txt);
    printf("%5s %12s %13s \n","x","actual","bessj0(x)");
    for (i=1;i<=nval;i++) {
        fscanf(fp,"%f %f",&x,&val);
        printf("%6.2f %12.7f %12.7f \n",x,val,bessj0(x));
    }
    fclose(fp);
    return 0;
}
```

Bessel function Y_0:

```
/* Driver for routine bessy0 */

#include <stdio.h>
#include <stdlib.h>
#include <string.h>
#include "nr.h"
#include "nrutil.h"

#define MAXSTR 80

int main(void)
{
    char txt[MAXSTR];
    int i,nval;
    float val,x;
    FILE *fp;

    if ((fp = fopen("fncval.dat","r")) == NULL)
        nrerror("Data file fncval.dat not found\n");
```

```
        fgets(txt,MAXSTR,fp);
        while (strncmp(txt,"Bessel Function Y0",18)) {
            fgets(txt,MAXSTR,fp);
            if (feof(fp)) nrerror("Data not found in fncval.dat\n");
        }
        fscanf(fp,"%d %*s",&nval);
        printf("\n%s\n",txt);
        printf("%5s %12s %13s \n","x","actual","bessy0(x)");
        for (i=1;i<=nval;i++) {
            fscanf(fp,"%f %f",&x,&val);
            printf("%6.2f %12.7f %12.7f\n",x,val,bessy0(x));
        }
        fclose(fp);
        return 0;
}
```

Bessel function J_1:

```
/* Driver for routine bessj1 */

#include <stdio.h>
#include <stdlib.h>
#include <string.h>
#include "nr.h"
#include "nrutil.h"

#define MAXSTR 80

int main(void)
{
    char txt[MAXSTR];
    int i,nval;
    float val,x;
    FILE *fp;

    if ((fp = fopen("fncval.dat","r")) == NULL)
        nrerror("Data file fncval.dat not found\n");
    fgets(txt,MAXSTR,fp);
    while (strncmp(txt,"Bessel Function J1",18)) {
        fgets(txt,MAXSTR,fp);
        if (feof(fp)) nrerror("Data not found in fncval.dat\n");
    }
    fscanf(fp,"%d %*s",&nval);
    printf("\n%s\n",txt);
    printf("%5s %12s %13s \n","x","actual","bessj1(x)");
    for (i=1;i<=nval;i++) {
        fscanf(fp,"%f %f",&x,&val);
        printf("%6.2f %12.7f %12.7f\n",x,val,bessj1(x));
    }
    fclose(fp);
    return 0;
}
```

Bessel function Y_1:

```
/* Driver for routine bessy1 */

#include <stdio.h>
```

```
#include <stdlib.h>
#include <string.h>
#include "nr.h"
#include "nrutil.h"

#define MAXSTR 80

int main(void)
{
    char txt[MAXSTR];
    int i,nval;
    float val,x;
    FILE *fp;

    if ((fp = fopen("fncval.dat","r")) == NULL)
        nrerror("Data file fncval.dat not found\n");
    fgets(txt,MAXSTR,fp);
    while (strncmp(txt,"Bessel Function Y1",18)) {
        fgets(txt,MAXSTR,fp);
        if (feof(fp)) nrerror("Data not found in fncval.dat\n");
    }
    fscanf(fp,"%d %*s",&nval);
    printf("\n%s\n",txt);
    printf("%5s %12s %13s \n","x","actual","bessy1(x)");
    for (i=1;i<=nval;i++) {
        fscanf(fp,"%f %f",&x,&val);
        printf("%6.2f %12.7f %12.7f\n",x,val,bessy1(x));
    }
    fclose(fp);
    return 0;
}
```

Bessel function Y_n for $n > 1$:

```
/* Driver for routine bessy */

#include <stdio.h>
#include <stdlib.h>
#include <string.h>
#include "nr.h"
#include "nrutil.h"

#define MAXSTR 80

int main(void)
{
    char txt[MAXSTR];
    int i,nval,n;
    float val,x;
    FILE *fp;

    if ((fp = fopen("fncval.dat","r")) == NULL)
        nrerror("Data file fncval.dat not found\n");
    fgets(txt,MAXSTR,fp);
    while (strncmp(txt,"Bessel Function Yn",18)) {
        fgets(txt,MAXSTR,fp);
        if (feof(fp)) nrerror("Data not found in fncval.dat\n");
```

```
    }
    fscanf(fp,"%d %*s",&nval);
    printf("\n%s\n",txt);
    printf("%4s %7s %15s %20s \n","n","x","actual","bessy(n,x)");
    for (i=1;i<=nval;i++) {
        fscanf(fp,"%d %f %f",&n,&x,&val);
        printf("%4d %8.2f %18.6e %18.6e\n",n,x,val,bessy(n,x));
    }
    fclose(fp);
    return 0;
}
```

Bessel function J_n for $n > 1$:

```
/* Driver for routine bessj */

#include <stdio.h>
#include <stdlib.h>
#include <string.h>
#include "nr.h"
#include "nrutil.h"

#define MAXSTR 80

int main(void)
{
    char txt[MAXSTR];
    int i,nval,n;
    float val,x;
    FILE *fp;

    if ((fp = fopen("fncval.dat","r")) == NULL)
        nrerror("Data file fncval.dat not found\n");
    fgets(txt,MAXSTR,fp);
    while (strncmp(txt,"Bessel Function Jn",18)) {
        fgets(txt,MAXSTR,fp);
        if (feof(fp)) nrerror("Data not found in fncval.dat\n");
    }
    fscanf(fp,"%d %*s",&nval);
    printf("\n%s\n",txt);
    printf("%4s %7s %15s %20s \n","n","x","actual","bessj(n,x)");
    for (i=1;i<=nval;i++) {
        fscanf(fp,"%d %f %f",&n,&x,&val);
        printf("%4d %8.2f %18.6e %18.6e\n",n,x,val,bessj(n,x));
    }
    fclose(fp);
    return 0;
}
```

Bessel function I_0:

```
/* Driver for routine bessi0 */

#include <stdio.h>
#include <stdlib.h>
#include <string.h>
#include "nr.h"
#include "nrutil.h"
```

```
#define MAXSTR 80

int main(void)
{
    char txt[MAXSTR];
    int i,nval;
    float val,x;
    FILE *fp;

    if ((fp = fopen("fncval.dat","r")) == NULL)
        nrerror("Data file fncval.dat not found\n");
    fgets(txt,MAXSTR,fp);
    while (strncmp(txt,"Modified Bessel Function I0",27)) {
        fgets(txt,MAXSTR,fp);
        if (feof(fp)) nrerror("Data not found in fncval.dat\n");
    }
    fscanf(fp,"%d %*s",&nval);
    printf("\n%s\n",txt);
    printf("%5s %12s %13s \n","x","actual","bessi0(x)");
    for (i=1;i<=nval;i++) {
        fscanf(fp,"%f %f",&x,&val);
        printf("%6.2f %12.7f %12.7f\n",x,val,bessi0(x));
    }
    fclose(fp);
    return 0;
}
```

Bessel function K_0:

```
/* Driver for routine bessk0 */

#include <stdio.h>
#include <stdlib.h>
#include <string.h>
#include "nr.h"
#include "nrutil.h"

#define MAXSTR 80

int main(void)
{
    char txt[MAXSTR];
    int i,nval;
    float val,x;
    FILE *fp;

    if ((fp = fopen("fncval.dat","r")) == NULL)
        nrerror("Data file fncval.dat not found\n");
    fgets(txt,MAXSTR,fp);
    while (strncmp(txt,"Modified Bessel Function K0",27)) {
        fgets(txt,MAXSTR,fp);
        if (feof(fp)) nrerror("Data not found in fncval.dat\n");
    }
    fscanf(fp,"%d %*s",&nval);
    printf("\n%s\n",txt);
    printf("%5s %13s %18s \n","x","actual","bessk0(x)");
```

```
    for (i=1;i<=nval;i++) {
        fscanf(fp,"%f %f",&x,&val);
        printf("%6.2f %16.7e %16.7e\n",x,val,bessk0(x));
    }
    fclose(fp);
    return 0;
}
```

Bessel function I_1:

```
/* Driver for routine bessi1 */

#include <stdio.h>
#include <stdlib.h>
#include <string.h>
#include "nr.h"
#include "nrutil.h"

#define MAXSTR 80

int main(void)
{
    char txt[MAXSTR];
    int i,nval;
    float val,x;
    FILE *fp;

    if ((fp = fopen("fncval.dat","r")) == NULL)
        nrerror("Data file fncval.dat not found\n");
    fgets(txt,MAXSTR,fp);
    while (strncmp(txt,"Modified Bessel Function I1",27)) {
        fgets(txt,MAXSTR,fp);
        if (feof(fp)) nrerror("Data not found in fncval.dat\n");
    }
    fscanf(fp,"%d %*s",&nval);
    printf("\n%s\n",txt);
    printf("%5s %12s %13s \n","x","actual","bessi1(x)");
    for (i=1;i<=nval;i++) {
        fscanf(fp,"%f %f",&x,&val);
        printf("%6.2f %12.7f %12.7f\n",x,val,bessi1(x));
    }
    fclose(fp);
    return 0;
}
```

Bessel function K_1:

```
/* Driver for routine bessk1 */

#include <stdio.h>
#include <stdlib.h>
#include <string.h>
#include "nr.h"
#include "nrutil.h"

#define MAXSTR 80

int main(void)
```

```
{
    char txt[MAXSTR];
    int i,nval;
    float val,x;
    FILE *fp;

    if ((fp = fopen("fncval.dat","r")) == NULL)
        nrerror("Data file fncval.dat not found\n");
    fgets(txt,MAXSTR,fp);
    while (strncmp(txt,"Modified Bessel Function K1",27)) {
        fgets(txt,MAXSTR,fp);
        if (feof(fp)) nrerror("Data not found in fncval.dat\n");
    }
    fscanf(fp,"%d %*s",&nval);
    printf("\n%s\n",txt);
    printf("%5s %13s %17s \n","x","actual","bessk1(x)");
    for (i=1;i<=nval;i++) {
        fscanf(fp,"%f %f",&x,&val);
        printf("%6.2f %16.7e %16.7e\n",x,val,bessk1(x));
    }
    fclose(fp);
    return 0;
}
```

Bessel function K_n for $n > 1$:

```
/* Driver for routine bessk */

#include <stdio.h>
#include <stdlib.h>
#include <string.h>
#include "nr.h"
#include "nrutil.h"

#define MAXSTR 80

int main(void)
{
    char txt[MAXSTR];
    int i,nval,n;
    float val,x;
    FILE *fp;

    if ((fp = fopen("fncval.dat","r")) == NULL)
        nrerror("Data file fncval.dat not found\n");
    fgets(txt,MAXSTR,fp);
    while (strncmp(txt,"Modified Bessel Function Kn",27)) {
        fgets(txt,MAXSTR,fp);
        if (feof(fp)) nrerror("Data not found in fncval.dat\n");
    }
    fscanf(fp,"%d %*s",&nval);
    printf("\n%s\n",txt);
    printf("%4s %7s %14s %19s\n","n","x","actual","bessk(n,x)");
    for (i=1;i<=nval;i++) {
        fscanf(fp,"%d %f %f",&n,&x,&val);
        printf("%4d %8.2f %18.7e %16.7e \n",n,x,val,bessk(n,x));
    }
```

```
        fclose(fp);
        return 0;
}
```

Bessel function I_n for $n > 1$:

```
/* Driver for routine bessi */

#include <stdio.h>
#include <stdlib.h>
#include <string.h>
#include "nr.h"
#include "nrutil.h"

#define MAXSTR 80

int main(void)
{
    char txt[MAXSTR];
    int i,nval,n;
    float val,x;
    FILE *fp;

    if ((fp = fopen("fncval.dat","r")) == NULL)
        nrerror("Data file fncval.dat not found\n");
    fgets(txt,MAXSTR,fp);
    while (strncmp(txt,"Modified Bessel Function In",27)) {
        fgets(txt,MAXSTR,fp);
        if (feof(fp)) nrerror("Data not found in fncval.dat\n");
    }
    fscanf(fp,"%d %*s",&nval);
    printf("\n%s\n",txt);
    printf("%4s %7s %15s %20s\n","n","x","actual","bessi(n,x)");
    for (i=1;i<=nval;i++) {
        fscanf(fp,"%d %f %f",&n,&x,&val);
        printf("%4d %8.2f %18.7e %18.7e\n",n,x,val,bessi(n,x));
    }
    fclose(fp);
    return 0;
}
```

Bessel functions of fractional order, J_ν and Y_ν:

```
/* Driver for routine bessjy */

#include <stdio.h>
#include <stdlib.h>
#include <string.h>
#include "nr.h"
#include "nrutil.h"

#define MAXSTR 80

int main(void)
{
    char txt[MAXSTR];
    int i,nval;
    float rj,ry,rjp,ryp,x,xnu,xrj,xry,xrjp,xryp;
```

```
    FILE *fp;

    if ((fp = fopen("fncval.dat","r")) == NULL)
        nrerror("Data file fncval.dat not found\n");
    fgets(txt,MAXSTR,fp);
    while (strncmp(txt,"Ordinary Bessel Functions",25)) {
        fgets(txt,MAXSTR,fp);
        if (feof(fp)) nrerror("Data not found in fncval.dat\n");
    }
    fscanf(fp,"%d %*s",&nval);
    printf("\n%s\n",txt);
    printf("%5s %7s\n%14s %16s %17s %16s\n%14s %16s %17s %16s\n",
            "xnu","x","rj","ry","rjp","ryp","xrj","xry","xrjp","xryp");
    for (i=1;i<=nval;i++) {
        fscanf(fp,"%f %f %f %f %f %f",&xnu,&x,&rj,&ry,&rjp,&ryp);
        bessjy(x,xnu,&xrj,&xry,&xrjp,&xryp);
        printf("%5.2f %5.2f\n\t%16.6e %16.6e %16.6e %16.6e\n",xnu,x,rj,ry,rjp,ryp);
        printf("\t%16.6e %16.6e %16.6e %16.6e\n",xrj,xry,xrjp,xryp);
    }
    fclose(fp);
    return 0;
}
```

The routine `bessjy` calls the utility procedure `beschb` to calculate the following quantities:

$$\Gamma_+ \equiv \frac{1}{\Gamma(1+x)}, \qquad \Gamma_- \equiv \frac{1}{\Gamma(1-x)}$$

$$\Gamma_1 \equiv \frac{1}{2x}(\Gamma_- - \Gamma_+), \qquad \Gamma_2 \equiv \frac{1}{2}(\Gamma_- + \Gamma_+)$$

Test program `xbeschb` compares the result of calling `beschb` with the results obtained by calling `gammln`.

```
/* Driver for routine beschb */

#include <stdio.h>
#include <math.h>
#include "nr.h"
#include "nrutil.h"

int main(void)
{
    double gam1,gam2,gampl,gammi,x,xgam1,xgam2,xgampl,xgammi;

    for (;;) {
        printf("Enter x:\n");
        if (scanf("%lf",&x) == EOF) break;
        beschb(x,&xgam1,&xgam2,&xgampl,&xgammi);
        printf("%5s\n%17s %16s %17s %15s\n%17s %16s %17s %15s\n",
                "x","gam1","gam2","gampl","gammi","xgam1","xgam2","xgampl","xgammi");
        gampl=1/exp(gammln((float)(1+x)));
        gammi=1/exp(gammln((float)(1-x)));
        gam1=(gammi-gampl)/(2*x);
        gam2=(gammi+gampl)/2;
        printf("%5.2f\n\t%16.6e %16.6e %16.6e %16.6e\n",x,gam1,gam2,gampl,gammi);
        printf("\t%16.6e %16.6e %16.6e %16.6e\n",xgam1,xgam2,xgampl,xgammi);
    }
```

```
        return 0;
}
```

Modified Bessel functions of fractional order, I_ν and K_ν:

```c
/* Driver for routine bessik */

#include <stdio.h>
#include <stdlib.h>
#include <string.h>
#include "nr.h"
#include "nrutil.h"

#define MAXSTR 80

int main(void)
{
    char txt[MAXSTR];
    int i,nval;
    float ri,rk,rip,rkp,x,xnu,xri,xrk,xrip,xrkp;
    FILE *fp;

    if ((fp = fopen("fncval.dat","r")) == NULL)
        nrerror("Data file fncval.dat not found\n");
    fgets(txt,MAXSTR,fp);
    while (strncmp(txt,"Modified Bessel Functions",25)) {
        fgets(txt,MAXSTR,fp);
        if (feof(fp)) nrerror("Data not found in fncval.dat\n");
    }
    fscanf(fp,"%d %*s",&nval);
    printf("\n%s\n",txt);
    printf("%5s %7s\n%14s %16s %17s %16s\n%14s %16s %17s %16s\n",
            "xnu","x","ri","rk","rip","rkp","xri","xrk","xrip","xrkp");
    for (i=1;i<=nval;i++) {
        fscanf(fp,"%f %f %f %f %f %f",&xnu,&x,&ri,&rk,&rip,&rkp);
        bessik(x,xnu,&xri,&xrk,&xrip,&xrkp);
        printf("%5.2f %5.2f\n\t%16.6e %16.6e %16.6e %16.6e\n",xnu,x,ri,rk,rip,rkp);
        printf("\t%16.6e %16.6e %16.6e %16.6e\n",xri,xrk,xrip,xrkp);
    }
    fclose(fp);
    return 0;
}
```

Airy functions:

```c
/* Driver for routine airy */

#include <stdio.h>
#include <stdlib.h>
#include <string.h>
#include "nr.h"
#include "nrutil.h"

#define MAXSTR 80

int main(void)
{
    char txt[MAXSTR];
```

```
        int i,nval;
        float ai,bi,aip,bip,x,xai,xbi,xaip,xbip;
        FILE *fp;

        if ((fp = fopen("fncval.dat","r")) == NULL)
            nrerror("Data file fncval.dat not found\n");
        fgets(txt,MAXSTR,fp);
        while (strncmp(txt,"Airy Functions",14)) {
            fgets(txt,MAXSTR,fp);
            if (feof(fp)) nrerror("Data not found in fncval.dat\n");
        }
        fscanf(fp,"%d %*s",&nval);
        printf("\n%s\n",txt);
        printf("%5s\n%14s %16s %17s %16s\n%14s %16s %17s %16s\n",
               "x","ai","bi","aip","bip","xai","xbi","xaip","xbip");
        for (i=1;i<=nval;i++) {
            fscanf(fp,"%f %f %f %f %f",&x,&ai,&bi,&aip,&bip);
            airy(x,&xai,&xbi,&xaip,&xbip);
            printf("%5.2f\n\t%16.6e %16.6e %16.6e %16.6e\n",x,ai,bi,aip,bip);
            printf("\t%16.6e %16.6e %16.6e %16.6e\n",xai,xbi,xaip,xbip);
        }
        fclose(fp);
        return 0;
}
```

Spherical Bessel functions:

```
/* Driver for routine rj */

#include <stdio.h>
#include <stdlib.h>
#include <string.h>
#include "nr.h"
#include "nrutil.h"

#define MAXSTR 80

int main(void)
{
        char txt[MAXSTR];
        int i,n,nval;
        float sj,sy,sjp,syp,x,xsj,xsy,xsjp,xsyp;
        FILE *fp;

        if ((fp = fopen("fncval.dat","r")) == NULL)
            nrerror("Data file fncval.dat not found\n");
        fgets(txt,MAXSTR,fp);
        while (strncmp(txt,"Spherical Bessel Functions",26)) {
            fgets(txt,MAXSTR,fp);
            if (feof(fp)) nrerror("Data not found in fncval.dat\n");
        }
        fscanf(fp,"%d %*s",&nval);
        printf("\n%s\n",txt);
        for (i=1;i<=nval;i++) {
            fscanf(fp,"%d %f %f %f %f %f",&n,&x,&sj,&sy,&sjp,&syp);
            printf("%5s %4s\n%14s %16s %17s %16s\n%14s %16s %17s %16s\n",
                   "n","x","sj","sy","sjp","syp","xsj","xsy","xsjp","xsyp");
```

```
        sphbes(n,x,&xsj,&xsy,&xsjp,&xsyp);
        printf("%5d %5.2f\n\t%16.6e %16.6e %16.6e %16.6e\n",n,x,sj,sy,sjp,syp);
        printf("\t%16.6e %16.6e %16.6e %16.6e\n",xsj,xsy,xsjp,xsyp);
    }
    fclose(fp);
    return 0;
}
```

Legendre polynomials:

```
/* Driver for routine plgndr */

#include <stdio.h>
#include <stdlib.h>
#include <string.h>
#include <math.h>
#include "nr.h"
#include "nrutil.h"

#define MAXSTR 80

int main(void)
{
    char txt[MAXSTR];
    int i,j,m,n,nval;
    float fac,val,x;
    FILE *fp;

    if ((fp = fopen("fncval.dat","r")) == NULL)
        nrerror("Data file fncval.dat not found\n");
    fgets(txt,MAXSTR,fp);
    while (strncmp(txt,"Legendre Polynomials",20)) {
        fgets(txt,MAXSTR,fp);
        if (feof(fp)) nrerror("Data not found in fncval.dat\n");
    }
    fscanf(fp,"%d %*s",&nval);
    printf("\n%s\n",txt);
    printf("%4s %4s %10s %17s %24s\n","n",
        "m","x","actual","plgndr(n,m,x)");
    for (i=1;i<=nval;i++) {
        fscanf(fp,"%d %d %f %f",&n,&m,&x,&val);
        fac=1.0;
        if (m > 0)
            for (j=n-m+1;j<=n+m;j++) fac *= j;
        fac *= 2.0/(2.0*n+1.0);
        val *= sqrt(fac);
        printf("%4d %4d %13.6f %19.6e %19.6e\n",
            n,m,x,val,plgndr(n,m,x));
    }
    fclose(fp);
    return 0;
}
```

Fresnel integrals:

```
/* Driver for routine fresnel */

#include <stdio.h>
```

```
#include <stdlib.h>
#include <string.h>
#include "nr.h"
#include "nrutil.h"

#define MAXSTR 80

int main(void)
{
    char txt[MAXSTR];
    int i,nval;
    float c,s,x,xc,xs;
    FILE *fp;

    if ((fp = fopen("fncval.dat","r")) == NULL)
        nrerror("Data file fncval.dat not found\n");
    fgets(txt,MAXSTR,fp);
    while (strncmp(txt,"Fresnel Integrals",17)) {
        fgets(txt,MAXSTR,fp);
        if (feof(fp)) nrerror("Data not found in fncval.dat\n");
    }
    fscanf(fp,"%d %*s",&nval);
    printf("\n%s\n",txt);
    printf("%5s %12s %14s %16s %14s \n",
        "x","actual","c(x)","actual","s(x)");
    for (i=1;i<=nval;i++) {
        fscanf(fp,"%f %f %f",&x,&xs,&xc);
        frenel(x,&s,&c);
        printf("%6.2f %15.6e %15.6e %15.6e %15.6e\n",
            x,xs,s,xc,c);
    }
    fclose(fp);
    return 0;
}
```

Cosine and sine integrals:

```
/* Driver for routine cisi */

#include <stdio.h>
#include <stdlib.h>
#include <string.h>
#include "nr.h"
#include "nrutil.h"

#define MAXSTR 80

int main(void)
{
    char txt[MAXSTR];
    int i,nval;
    float ci,si,x,xci,xsi;
    FILE *fp;

    if ((fp = fopen("fncval.dat","r")) == NULL)
        nrerror("Data file fncval.dat not found\n");
    fgets(txt,MAXSTR,fp);
```

```
        while (strncmp(txt,"Cosine and Sine Integrals",25)) {
            fgets(txt,MAXSTR,fp);
            if (feof(fp)) nrerror("Data not found in fncval.dat\n");
        }
        fscanf(fp,"%d %*s",&nval);
        printf("\n%s\n",txt);
        printf("%5s %12s %12s %12s \n",
            "x","actual","ci(x)","actual","si(x)");
        for (i=1;i<=nval;i++) {
            fscanf(fp,"%f %f %f",&x,&xci,&xsi);
            cisi(x,&ci,&si);
            printf("%6.2f %12.6f %12.6f %12.6f %12.6f\n",
                x,xci,ci,xsi,si);
        }
        fclose(fp);
        return 0;
}
```

Dawson's integral:

```
/* Driver for routine dawson */

#include <stdio.h>
#include <stdlib.h>
#include <string.h>
#include "nr.h"
#include "nrutil.h"

#define MAXSTR 80

int main(void)
{
    char txt[MAXSTR];
    int i,nval;
    float val,x;
    FILE *fp;

    if ((fp = fopen("fncval.dat","r")) == NULL)
        nrerror("Data file fncval.dat not found\n");
    fgets(txt,MAXSTR,fp);
    while (strncmp(txt,"Dawson integral",15)) {
        fgets(txt,MAXSTR,fp);
        if (feof(fp)) nrerror("Data not found in fncval.dat\n");
    }
    fscanf(fp,"%d %*s",&nval);
    printf("\n%s\n",txt);
    printf("%5s %12s %11s \n","x","actual","dawson(x)");
    for (i=1;i<=nval;i++) {
        fscanf(fp,"%f %f",&x,&val);
        printf("%6.2f %12.6f %12.6f\n",x,val,dawson(x));
    }
    fclose(fp);
    return 0;
}
```

Carlson's elliptic integral of the first kind:

```
/* Driver for routine rf */

#include <stdio.h>
#include <stdlib.h>
#include <string.h>
#include "nr.h"
#include "nrutil.h"

#define MAXSTR 80

int main(void)
{
    char txt[MAXSTR];
    int i,nval;
    float val,x,y,z;
    FILE *fp;

    if ((fp = fopen("fncval.dat","r")) == NULL)
        nrerror("Data file fncval.dat not found\n");
    fgets(txt,MAXSTR,fp);
    while (strncmp(txt,"Elliptic Integral First Kind RF",31)) {
        fgets(txt,MAXSTR,fp);
        if (feof(fp)) nrerror("Data not found in fncval.dat\n");
    }
    fscanf(fp,"%d %*s",&nval);
    printf("\n%s\n",txt);
    printf("%7s %8s %8s %16s %20s\n","x","y","z","actual","rf(x,y,z)");
    for (i=1;i<=nval;i++) {
        fscanf(fp,"%f %f %f %f",&x,&y,&z,&val);
        printf("%8.2f %8.2f %8.2f %18.6e %18.6e\n",x,y,z,val,rf(x,y,z));
    }
    fclose(fp);
    return 0;
}
```

Carlson's elliptic integral of the second kind:

```
/* Driver for routine rd */

#include <stdio.h>
#include <stdlib.h>
#include <string.h>
#include "nr.h"
#include "nrutil.h"

#define MAXSTR 80

int main(void)
{
    char txt[MAXSTR];
    int i,nval;
    float val,x,y,z;
    FILE *fp;

    if ((fp = fopen("fncval.dat","r")) == NULL)
        nrerror("Data file fncval.dat not found\n");
    fgets(txt,MAXSTR,fp);
```

```
        while (strncmp(txt,"Elliptic Integral Second Kind RD",32)) {
            fgets(txt,MAXSTR,fp);
            if (feof(fp)) nrerror("Data not found in fncval.dat\n");
        }
        fscanf(fp,"%d %*s",&nval);
        printf("\n%s\n",txt);
        printf("%7s %8s %8s %16s %20s\n","x","y","z","actual","rd(x,y,z)");
        for (i=1;i<=nval;i++) {
            fscanf(fp,"%f %f %f %f",&x,&y,&z,&val);
            printf("%8.2f %8.2f %8.2f %18.6e %18.6e\n",x,y,z,val,rd(x,y,z));
        }
        fclose(fp);
        return 0;
}
```

Carlson's elliptic integral of the third kind:

```
/* Driver for routine rj */

#include <stdio.h>
#include <stdlib.h>
#include <string.h>
#include "nr.h"
#include "nrutil.h"

#define MAXSTR 80

int main(void)
{
    char txt[MAXSTR];
    int i,nval;
    float p,val,x,y,z;
    FILE *fp;

    if ((fp = fopen("fncval.dat","r")) == NULL)
        nrerror("Data file fncval.dat not found\n");
    fgets(txt,MAXSTR,fp);
    while (strncmp(txt,"Elliptic Integral Third Kind RJ",31)) {
        fgets(txt,MAXSTR,fp);
        if (feof(fp)) nrerror("Data not found in fncval.dat\n");
    }
    fscanf(fp,"%d %*s",&nval);
    printf("\n%s\n",txt);
    printf("%7s %8s %8s %8s %16s %20s\n","x","y","z","p","actual","rj(x,y,z,p)");
    for (i=1;i<=nval;i++) {
        fscanf(fp,"%f %f %f %f %f",&x,&y,&z,&p,&val);
        printf("%8.2f %8.2f %8.2f %8.2f %18.6e %18.6e\n",x,y,z,p,val,rj(x,y,z,p));
    }
    fclose(fp);
    return 0;
}
```

Carlson's degenerate elliptic integral:

```
/* Driver for routine rc */

#include <stdio.h>
#include <stdlib.h>
```

```
#include <string.h>
#include "nr.h"
#include "nrutil.h"

#define MAXSTR 80

int main(void)
{
    char txt[MAXSTR];
    int i,nval;
    float val,x,y;
    FILE *fp;

    if ((fp = fopen("fncval.dat","r")) == NULL)
        nrerror("Data file fncval.dat not found\n");
    fgets(txt,MAXSTR,fp);
    while (strncmp(txt,"Elliptic Integral Degenerate RC",31)) {
        fgets(txt,MAXSTR,fp);
        if (feof(fp)) nrerror("Data not found in fncval.dat\n");
    }
    fscanf(fp,"%d %*s",&nval);
    printf("\n%s\n",txt);
    printf("%7s %8s %16s %18s\n","x","y","actual","rc(x,y)");
    for (i=1;i<=nval;i++) {
        fscanf(fp,"%f %f %f",&x,&y,&val);
        printf("%8.2f %8.2f %18.6e %18.6e\n",x,y,val,rc(x,y));
    }
    fclose(fp);
    return 0;
}
```

Legendre elliptic integral of the first kind:

```
/* Driver for routine ellf */

#include <stdio.h>
#include <stdlib.h>
#include <string.h>
#include <math.h>
#include "nr.h"
#include "nrutil.h"

#define MAXSTR 80

#define FAC (3.1415926535/180.0)

int main(void)
{
    char txt[MAXSTR];
    int i,nval;
    float ak,alpha,phi,val;
    FILE *fp;

    if ((fp = fopen("fncval.dat","r")) == NULL)
        nrerror("Data file fncval.dat not found\n");
    fgets(txt,MAXSTR,fp);
    while (strncmp(txt,"Legendre Elliptic Integral First Kind",37)) {
```

```
        fgets(txt,MAXSTR,fp);
        if (feof(fp)) nrerror("Data not found in fncval.dat\n");
    }
    fscanf(fp,"%d %*s",&nval);
    printf("\n%s\n",txt);
    printf("%5s %10s %11s %22s\n","phi","sin(alpha)","actual","ellf(phi,ak)");
    for (i=1;i<=nval;i++) {
        fscanf(fp,"%f %f %f",&phi,&alpha,&val);
        alpha=alpha*FAC;
        ak=sin(alpha);
        phi=phi*FAC;
        printf("%6.2f %6.2f %18.6e %18.6e\n",phi,ak,val,ellf(phi,ak));
    }
    fclose(fp);
    return 0;
}
```

Legendre elliptic integral of the second kind:

```
/* Driver for routine elle */

#include <stdio.h>
#include <stdlib.h>
#include <string.h>
#include <math.h>
#include "nr.h"
#include "nrutil.h"

#define MAXSTR 80

#define FAC (3.1415926535/180.0)

int main(void)
{
    char txt[MAXSTR];
    int i,nval;
    float ak,alpha,phi,val;
    FILE *fp;

    if ((fp = fopen("fncval.dat","r")) == NULL)
        nrerror("Data file fncval.dat not found\n");
    fgets(txt,MAXSTR,fp);
    while (strncmp(txt,"Legendre Elliptic Integral Second Kind",38)) {
        fgets(txt,MAXSTR,fp);
        if (feof(fp)) nrerror("Data not found in fncval.dat\n");
    }
    fscanf(fp,"%d %*s",&nval);
    printf("\n%s\n",txt);
    printf("%5s %10s %11s %22s\n","phi","sin(alpha)","actual","elle(phi,ak)");
    for (i=1;i<=nval;i++) {
        fscanf(fp,"%f %f %f",&phi,&alpha,&val);
        alpha=alpha*FAC;
        ak=sin(alpha);
        phi=phi*FAC;
        printf("%6.2f %6.2f %18.6e %18.6e\n",phi,ak,val,elle(phi,ak));
    }
    fclose(fp);
```

```
    return 0;
}
```

Legendre elliptic integral of the third kind:

```
/* Driver for routine ellpi */

#include <stdio.h>
#include <stdlib.h>
#include <string.h>
#include <math.h>
#include "nr.h"
#include "nrutil.h"

#define MAXSTR 80

#define FAC (3.1415926535/180.0)

int main(void)
{
    char txt[MAXSTR];
    int i,nval;
    float ak,alpha,en,phi,val;
    FILE *fp;

    if ((fp = fopen("fncval.dat","r")) == NULL)
        nrerror("Data file fncval.dat not found\n");
    fgets(txt,MAXSTR,fp);
    while (strncmp(txt,"Legendre Elliptic Integral Third Kind",37)) {
        fgets(txt,MAXSTR,fp);
        if (feof(fp)) nrerror("Data not found in fncval.dat\n");
    }
    fscanf(fp,"%d %*s",&nval);
    printf("\n%s\n",txt);
    printf("%6s %6s %6s %10s %23s\n",
            "phi","-en","sin(alpha)","actual","ellpi(phi,ak)");
    for (i=1;i<=nval;i++) {
        fscanf(fp,"%f %f %f %f",&phi,&en,&alpha,&val);
        alpha=alpha*FAC;
        ak=sin(alpha);
        en = -en;
        phi=phi*FAC;
        printf("%6.2f %6.2f %6.2f %18.6e %18.6e\n",
                phi,en,ak,val,ellpi(phi,en,ak));
    }
    fclose(fp);
    return 0;
}
```

Routine sncndn returns Jacobian elliptic functions. The file fncval.dat contains information only about function sn. However, the values of cn and dn satisfy the relationships

$$sn^2 + cn^2 = 1, \qquad k^2 sn^2 + dn^2 = 1.$$

The program xsncndn works exactly as the others in terms of testing sn, but for verifying cn and dn it lists the values of the left sides of the two equations above.

Each of them should have the value 1.0 for all choices of arguments.

```c
/* Driver for routine sncndn */

#include <stdio.h>
#include <stdlib.h>
#include <string.h>
#include "nr.h"
#include "nrutil.h"

#define MAXSTR 80

int main(void)
{
    char txt[MAXSTR];
    int i,nval;
    float em,emmc,uu,val,sn,cn,dn;
    FILE *fp;

    if ((fp = fopen("fncval.dat","r")) == NULL)
        nrerror("Data file fncval.dat not found\n");
    fgets(txt,MAXSTR,fp);
    while (strncmp(txt,"Jacobian Elliptic Function",26)) {
        fgets(txt,MAXSTR,fp);
        if (feof(fp)) nrerror("Data not found in fncval.dat\n");
    }
    fscanf(fp,"%d %*s",&nval);
    printf("\n%s\n",txt);
    printf("%4s %8s %16s %13s %15s %18s\n","mc","u","actual",
        "sn","sn^2+cn^2","(mc)*(sn^2)+dn^2");
    for (i=1;i<=nval;i++) {
        fscanf(fp,"%f %f %f",&em,&uu,&val);
        emmc=1.0-em;
        sncndn(uu,emmc,&sn,&cn,&dn);
        printf("%5.2f %8.2f %15.5f %15.5f %12.5f %14.5f\n",
            emmc,uu,val,sn,(sn*sn+cn*cn),(em*sn*sn+dn*dn));
    }
    fclose(fp);
    return 0;
}
```

The test program **xhypgeo** tests out the suite **hypgeo**, **hypdrv** and **hypser** by calculating a hypergeometric function that can be expressed in closed form:

$$_2F_1\left(\tfrac{1}{2},1,\tfrac{3}{2},z^2\right) = \frac{1}{2z}\log\frac{1+z}{1-z}$$

$$_2F_1\left(\tfrac{1}{2},1,\tfrac{3}{2},-z^2\right) = \frac{1}{2iz}\log\frac{1+iz}{1-iz}$$

The program prompts you to enter various values of x and y and compares the results for $z = x + iy$.

```c
/* Driver for routine hypgeo */

#include <stdio.h>
#include <math.h>
#include "nr.h"
#include "nrutil.h"
```

```c
#include "complex.h"

fcomplex Clog(fcomplex a)
{
    fcomplex b;

    b.r=log(Cabs(a));
    b.i=atan2(a.i,a.r);
    return b;
}

int main(void)
{
    fcomplex a,b,c,z,zi,q1,q2,q3,q4;
    float x,y;

    a=Complex(0.5,0.0);
    b=Complex(1.0,0.0);
    c=Complex(1.5,0.0);
    for (;;) {
        printf("INPUT X,Y OF COMPLEX ARGUMENT:\n");
        if (scanf("%f %f",&x,&y) == EOF) break;
        z=Complex(x,y);
        q1=hypgeo(a,b,c,Cmul(z,z));
        q2=RCmul(0.5,Cdiv(Clog(Cdiv(Cadd(b,z),Csub(b,z))),z));
        q3=hypgeo(a,b,c,RCmul(-1.0,Cmul(z,z)));
        zi=Complex(-y,x);
        q4=RCmul(0.5,Cdiv(Clog(Cdiv(Cadd(b,zi),Csub(b,zi))),zi));
        printf("2F1(0.5,1.0,1.5;z^2) =%12.6f %12.6f\n",q1.r,q1.i);
        printf("check using log form: %12.6f %12.6f\n",q2.r,q2.i);
        printf("2F1(0.5,1.0,1.5;-z^2)=%12.6f %12.6f\n",q3.r,q3.i);
        printf("check using log form: %12.6f %12.6f\n",q4.r,q4.i);
    }
    return 0;
}
```

Appendix

File fncval.dat:

Values of Special Functions in format x,F(x) or x,y,F(x,y)
Dawson integral
4 Values

0.04	0.0399573606
0.16	0.1572970920
1.6	0.3999398943
10.0	0.0502538471

Ordinary Bessel Functions   nu,x,jnu(x),ynu(x),jnup(x),ynup(x)
20 Values

nu	x	jnu(x)	ynu(x)	jnup(x)	ynup(x)
2.0	1.0	1.149034849E-01	-1.650682607E+00	2.102436159E-01	2.520152392E+00
2.0	2.0	3.528340286E-01	-6.174081042E-01	2.238907791E-01	5.103756726E-01
2.0	5.0	4.656511628E-02	3.676628826E-01	-3.462051841E-01	7.979903490E-04
2.0	10.0	2.546303137E-01	-5.868082442E-03	-7.453316568E-03	2.501890407E-01
2.0	50.0	-5.971280079E-02	9.579316873E-02	-9.512331609E-02	-6.062739531E-02
5.0	1.0	2.497577302E-04	-2.604058666E+02	1.227850313E-03	1.268750910E+03
5.0	2.0	7.039629756E-03	-9.935989128E+00	1.639664542E-02	2.207402959E+01
5.0	5.0	2.611405461E-01	-4.536948225E-01	1.300918143E-01	2.615525351E-01

```
5.0   10.0   -2.340615282E-01   1.354030477E-01  -1.025719220E-01  -2.126510357E-01
5.0   50.0   -8.140024770E-02  -7.854841391E-02   7.898100205E-02  -8.020323269E-02
10.0   1.0    2.630615124E-10  -1.216180143E+08   2.618635056E-09   1.209399938E+09
10.0   2.0    2.515386283E-07  -1.291845422E+05   1.234650294E-06   6.313628817E+05
10.0   5.0    1.467802647E-03  -2.512911010E+01   2.584677845E-03   4.249433700E+01
10.0  10.0    2.074861066E-01  -3.598141522E-01   8.436957863E-02   1.605148864E-01
10.0  50.0   -1.138478491E-01   5.723897182E-03  -4.422891214E-03  -1.116145748E-01
20.0   1.0    3.873503009E-25  -4.113970315E+22   7.737778395E-24   8.217106466E+23
20.0   2.0    3.918972805E-19  -4.081651389E+16   3.900270468E-18   4.060105807E+17
20.0   5.0    2.770330052E-11  -5.933965297E+08   1.074693821E-10   2.294022549E+09
20.0  10.0    1.151336925E-05  -1.597483848E+03   2.011953903E-05   2.737803151E+03
20.0  50.0   -1.167043528E-01   1.644263395E-02  -1.368329398E-02  -1.071717186E-01
Modified Bessel Functions      nu,x,inu(x),knu(x),inup(x),knup(x)
28 Values
2.0    0.2    5.016687514E-03   4.951242929E+01   5.033395889E-02  -4.999002654E+02
2.0    1.0    1.357476698E-01   1.624838899E+00   2.936637645E-01  -3.851585027E+00
2.0    2.0    6.889484477E-01   2.537597546E-01   9.016884069E-01  -3.936256364E-01
2.0    2.5    1.276466148E+00   1.214602063E-01   1.495543327E+00  -1.710589814E-01
2.0    3.0    2.245212441E+00   6.151045847E-02   2.456561923E+00  -8.116340344E-02
2.0    5.0    1.750561497E+01   5.308943712E-03   1.733339616E+01  -6.168190930E-03
2.0   10.0    2.281518968E+03   2.150981701E-05   2.214684510E+03  -2.295073686E-05
2.0   20.0    3.931278522E+07   6.329543612E-10   3.852369486E+07  -6.516012331E-10
3.0    1.0    2.216842492E-02   7.101262825E+00   6.924239499E-02  -2.292862737E+01
3.0    2.0    2.127399592E-01   6.473853909E-01   3.698385088E-01  -1.224837841E+00
3.0    5.0    1.033115017E+01   8.291768415E-03   1.130692487E+01  -1.028400476E-02
3.0   10.0    1.758380717E+03   2.725270026E-05   1.754004753E+03  -2.968562708E-05
3.0   50.0    2.677764139E+20   3.727936774E-23   2.655764792E+20  -3.771608045E-23
5.0    1.0    2.714631560E-04   3.609605896E+02   1.379804441E-03  -1.849035364E+03
5.0    2.0    9.825679323E-03   9.431049101E+00   2.616437167E-02  -2.577353868E+01
5.0    5.0    2.157974547E+00   3.270627371E-02   2.950260216E+00  -4.796533952E-02
5.0   10.0    7.771882864E+02   5.754184999E-05   8.378963946E+02  -6.663236215E-05
5.0   50.0    2.278548308E+20   4.367182254E-23   2.267244113E+20  -4.432002477E-23
10.0   1.0    2.752948040E-10   1.807132899E+08   2.765437823E-09  -1.817137940E+09
10.0   2.0    3.016963879E-07   1.624824040E+05   1.535703963E-06  -8.302224962E+05
10.0   5.0    4.580044419E-03   9.758562829E+00   1.015562998E-02  -2.202940355E+01
10.0  10.0    2.189170616E+01   1.614255300E-03   3.042758634E+01  -2.324264134E-03
10.0  50.0    1.071597159E+20   9.150988210E-23   1.082473779E+20  -9.419859989E-23
20.0   1.0    3.966835986E-25   6.294369360E+22   7.943111714E-24  -1.260529076E+24
20.0   2.0    4.310560576E-19   5.770856853E+16   4.331042808E-18  -5.801141519E+17
20.0   5.0    5.024239358E-11   4.827000521E+08   2.068719274E-10  -1.993195442E+09
20.0  10.0    1.250799736E-04   1.787442782E+02   2.784778118E-04  -4.015325804E+02
20.0  50.0    5.442008403E+18   1.706148380E-21   5.814242572E+18  -1.852264589E-21
Spherical Bessel Functions      n,x,sjn(x),syn(x),sjnp(x),synp(x)
11 Values
0    0.1    9.9833417E-01   -9.9500417E+00   -3.3300012E-02    1.0049875E+02
0    1.0    8.4147098E-01   -5.4030231E-01   -3.0116868E-01    1.3817733E+00
0    5.0   -1.9178485E-01   -5.6732437E-02    9.5089408E-02   -1.8043837E-01
0   50.0   -5.2474971E-03   -1.9299321E-02    1.9404271E-02   -4.8615107E-03
1    0.1    3.3300012E-02   -1.0049875E+02    3.3233393E-01    2.0000250E+03
1    1.0    3.0116868E-01   -1.3817733E+00    2.3913563E-01    2.2232443E+00
1    5.0   -9.5089408E-02    1.8043837E-01   -1.5374909E-01   -1.2890778E-01
1   50.0   -1.9404271E-02    4.8615107E-03   -4.4713263E-03   -1.9493781E-02
20   1.0    7.5377957E-26   -3.2395922E+23    1.5058053E-24    6.7948312E+24
20   5.0    5.4277268E-12   -9.2679514E+08    2.1071414E-11    3.7715819E+09
20  50.0   -1.5785030E-02    1.3759531E-02   -1.2204181E-02   -1.4702296E-02
Airy Functions
8 Values
```

```
-3.5   -0.37553382   0.16893984  -0.34344343  -0.69311628
-2.0    0.22740743  -0.41230259   0.61825902   0.27879517
-1.0    0.53556088   0.10399739  -0.01016057   0.59237563
-0.01   0.35761619   0.61044364  -0.25880157   0.44831896
 0.00   0.35502805   0.61492663  -0.25881940   0.44828836
 0.01   0.35243992   0.61940962  -0.25880174   0.44831926
 0.5    0.23169361   0.85427704  -0.22491053   0.54457256
 1.00   0.13529242   1.20742359  -0.15914744   0.93243593
Elliptic Integral First Kind RF
3 Values
   0.1      0.1      0.1     3.16227766
   0.0      1.0      2.0     1.31102878
 100.0    100.0    100.0     0.1
Elliptic Integral Second Kind RD
3 Values
   0.1      0.1      0.1    31.6227766
   0.0      2.0      1.0     1.79721036
 100.0    100.0    100.0     0.001
Elliptic Integral Third Kind RJ
3 Values
   0.1      0.1      0.1      0.1    31.6227766
   2.0      3.0      4.0      5.0    0.142975797
 100.0    100.0    100.0    100.0    0.001
Elliptic Integral Degenerate RC
5 Values
   0.1      0.1     3.16227766
   0.0      0.25    3.14159265
   0.0625   0.125   3.14159265
   2.25     2.0     0.69314718
 100.0    100.0     0.1
Legendre Elliptic Integral First Kind
9 Values
   5.0     2.0     0.08726660
   5.0    30.0     0.08729413
   5.0    88.0     0.08737730
  30.0     2.0     0.52362636
  30.0    30.0     0.52942863
  30.0    88.0     0.54927042
  90.0     2.0     1.57127495
  90.0    30.0     1.68575035
  90.0    88.0     4.74271727
Legendre Elliptic Integral Second Kind
9 Values
   5.0     2.0     0.08726633
   5.0    30.0     0.08723881
   5.0    88.0     0.08715588
  30.0     2.0     0.52357119
  30.0    30.0     0.51788193
  30.0    88.0     0.50003003
  90.0     2.0     1.57031792
  90.0    30.0     1.46746221
  90.0    88.0     1.00258409
Legendre Elliptic Integral Third Kind
12 Values
  15.0     0.4    30.0     0.264953
  60.0     0.6    90.0     1.71951
  90.0     0.0    30.0     1.68575
```

```
90.0    0.9    75.0    12.46407
15.0    1.0    15.0     0.268156
45.0    0.0625 30.0     0.813845
45.0    0.625  30.0     0.921130
45.0    1.25   30.0     1.132136
45.0   -0.25   30.0     0.769872
90.0    0.0625 30.0     1.743055
90.0    0.625  30.0     2.800989
90.0   -0.25   30.0     1.501762
```

Fresnel Integrals
6 Values

```
 0.1    0.0005236    0.0999975
 1.0    0.4382591    0.7798934
 1.5    0.6975050    0.4452612
 2.0    0.3434157    0.4882534
 5.0    0.4991914    0.5636312
20.0    0.4840845    0.4999873
```

Exponential Integral En   n,x,En(x)
13 Values

```
0     1.0     0.3678794
2     0.0     1.0000000
3     0.0     0.5000000
4     0.0     0.3333333
2     0.5     0.3266439
3     0.5     0.2216044
4     0.5     0.1652428
10    0.5     0.0634583
20    0.5     0.0310612
2     5.0     0.9964690E-03
20    5.0     0.2782746E-03
2    50.0     0.3711783E-23
20   50.0     0.2766423E-23
```

Cosine and Sine Integrals
17 Values

```
 0.1   -1.727868E+00    9.994446E-02
 0.2   -1.042206E+00    1.995561E-01
 0.6   -2.227071E-02    5.881288E-01
 0.7    1.005147E-01    6.812222E-01
 1.8    4.568111E-01    1.505817E+00
 1.9    4.419403E-01    1.557775E+00
 2.0    4.229808E-01    1.605413E+00
 2.1    4.005120E-01    1.648699E+00
 2.2    3.750746E-01    1.687625E+00
 3.3    2.467829E-02    1.848081E+00
 6.3   -1.988822E-02    1.418174E+00
 6.4   -4.181411E-03    1.419223E+00
 6.5    1.110152E-02    1.421794E+00
 6.6    2.582314E-02    1.425816E+00
10.0   -4.545644E-02    1.658348E+00
12.5   -1.140835E-02    1.492337E+00
15.0    4.627868E-02    1.618194E+00
```

Exponential Integral Ei
21 Values

```
 0.1   -1.62281
 0.2   -0.821761
 0.3   -0.302669
 0.4    0.104768
```

```
 0.5    0.454220
 0.6    0.769881
 0.7    1.06491
 0.8    1.34740
 0.9    1.62281
 2.2    5.73261
 2.4    6.60067
 3.5    13.9254
 3.9    18.3157
 5.0    40.1853
 5.4    54.1935
12.5    23565.1
13.0    37197.7
13.5    58827.0
14.0    93193.0
14.5    147866.
15.0    234955.
Gamma Function
17 Values
 1.0    1.000000
 1.2    0.918169
 1.4    0.887264
 1.6    0.893515
 1.8    0.931384
 2.0    1.000000
 0.2    4.590845
 0.4    2.218160
 0.6    1.489192
 0.8    1.164230
-0.2   -5.821149
-0.4   -3.722981
-0.6   -3.696933
-0.8   -5.738555
10.0    3.6288000E05
20.0    1.2164510E17
30.0    8.8417620E30
N-factorial
18 Values
1       1
2       2
3       6
4       24
5       120
6       720
7       5040
8       40320
9       362880
10      3628800
11      39916800
12      479001600
13      6227020800
14      87178291200
15      1.3076755E12
20      2.4329042E18
25      1.5511222E25
30      2.6525281E32
Binomial Coefficients
```

```
20 Values
1      0      1
6      1      6
6      3      20
6      5      6
15     1      15
15     3      455
15     5      3003
15     7      6435
15     9      5005
15     11     1365
15     13     105
25     1      25
25     3      2300
25     5      53130
25     7      480700
25     9      2042975
25     11     4457400
25     13     5200300
25     15     3268760
25     17     1081575
```

Beta Function

```
15 Values
1.0    1.0    1.000000
0.2    1.0    5.000000
1.0    0.2    5.000000
0.4    1.0    2.500000
1.0    0.4    2.500000
0.6    1.0    1.666667
0.8    1.0    1.250000
6.0    6.0    3.607504E-04
6.0    5.0    7.936508E-04
6.0    4.0    1.984127E-03
6.0    3.0    5.952381E-03
6.0    2.0    0.238095E-01
7.0    7.0    8.325008E-05
5.0    5.0    1.587302E-03
4.0    4.0    7.142857E-03
3.0    3.0    0.333333E-01
2.0    2.0    1.666667E-01
```

Incomplete Gamma Function

```
20 Values
0.1    3.1622777E-02    0.7420263
0.1    3.1622777E-01    0.9119753
0.1    1.5811388        0.9898955
0.5    7.0710678E-02    0.2931279
0.5    7.0710678E-01    0.7656418
0.5    3.5355339        0.9921661
1.0    0.1000000        0.0951626
1.0    1.0000000        0.6321206
1.0    5.0000000        0.9932621
1.1    1.0488088E-01    0.0757471
1.1    1.0488088        0.6076457
1.1    5.2440442        0.9933425
2.0    1.4142136E-01    0.0091054
2.0    1.4142136        0.4130643
2.0    7.0710678        0.9931450
```

```
6.0     2.4494897      0.0387318
6.0    12.247449       0.9825937
11.0   16.583124       0.9404267
26.0   25.495098       0.4863866
41.0   44.821870       0.7359709
```
Error Function
20 Values
```
0.0    0.000000
0.1    0.1124629
0.2    0.2227026
0.3    0.3286268
0.4    0.4283924
0.5    0.5204999
0.6    0.6038561
0.7    0.6778012
0.8    0.7421010
0.9    0.7969082
1.0    0.8427008
1.1    0.8802051
1.2    0.9103140
1.3    0.9340079
1.4    0.9522851
1.5    0.9661051
1.6    0.9763484
1.7    0.9837905
1.8    0.9890905
1.9    0.9927904
```
Incomplete Beta Function
20 Values
```
0.5     0.5     0.01    0.0637686
0.5     0.5     0.10    0.2048328
0.5     0.5     1.00    1.0000000
1.0     0.5     0.01    0.0050126
1.0     0.5     0.10    0.0513167
1.0     0.5     1.00    1.0000000
1.0     1.0     0.5     0.5000000
5.0     5.0     0.5     0.5000000
10.0    0.5     0.9     0.1516409
10.0    5.0     0.5     0.0897827
10.0    5.0     1.0     1.0000000
10.0   10.0     0.5     0.5000000
20.0    5.0     0.8     0.4598773
20.0   10.0     0.6     0.2146816
20.0   10.0     0.8     0.9507365
20.0   20.0     0.5     0.5000000
20.0   20.0     0.6     0.8979414
30.0   10.0     0.7     0.2241297
30.0   10.0     0.8     0.7586405
40.0   20.0     0.7     0.7001783
```
Bessel Function J0
20 Values
```
-5.0   -0.1775968
-4.0   -0.3971498
-3.0   -0.2600520
-2.0    0.2238908
-1.0    0.7651976
 0.0    1.0000000
```

```
 1.0    0.7651977
 2.0    0.2238908
 3.0   -0.2600520
 4.0   -0.3971498
 5.0   -0.1775968
 6.0    0.1506453
 7.0    0.3000793
 8.0    0.1716508
 9.0   -0.0903336
10.0   -0.2459358
11.0   -0.1711903
12.0    0.0476893
13.0    0.2069261
14.0    0.1710735
15.0   -0.0142245
Bessel Function Y0
15 Values
 0.1   -1.5342387
 1.0    0.0882570
 2.0    0.51037567
 3.0    0.37685001
 4.0   -0.0169407
 5.0   -0.3085176
 6.0   -0.2881947
 7.0   -0.0259497
 8.0    0.2235215
 9.0    0.2499367
10.0    0.0556712
11.0   -0.1688473
12.0   -0.2252373
13.0   -0.0782079
14.0    0.1271926
15.0    0.2054743
Bessel Function J1
20 Values
-5.0    0.3275791
-4.0    0.0660433
-3.0   -0.3390590
-2.0   -0.5767248
-1.0   -0.4400506
 0.0    0.0000000
 1.0    0.4400506
 2.0    0.5767248
 3.0    0.3390590
 4.0   -0.0660433
 5.0   -0.3275791
 6.0   -0.2766839
 7.0   -0.0046828
 8.0    0.2346364
 9.0    0.2453118
10.0    0.0434728
11.0   -0.1767853
12.0   -0.2234471
13.0   -0.0703181
14.0    0.1333752
15.0    0.2051040
Bessel Function Y1
```

```
15 Values
 0.1  -6.4589511
 1.0  -0.7812128
 2.0  -0.1070324
 3.0   0.3246744
 4.0   0.3979257
 5.0   0.1478631
 6.0  -0.1750103
 7.0  -0.3026672
 8.0  -0.1580605
 9.0   0.1043146
10.0   0.2490154
11.0   0.1637055
12.0  -0.0570992
13.0  -0.2100814
14.0  -0.1666448
15.0   0.0210736
Bessel Function Jn, n>=2
20 Values
 2    1.0     1.149034849E-01
 2    2.0     3.528340286E-01
 2    5.0     4.656511628E-02
 2   10.0     2.546303137E-01
 2   50.0    -5.971280079E-02
 5    1.0     2.497577302E-04
 5    2.0     7.039629756E-03
 5    5.0     2.611405461E-01
 5   10.0    -2.340615282E-01
 5   50.0    -8.140024770E-02
10    1.0     2.630615124E-10
10    2.0     2.515386283E-07
10    5.0     1.467802647E-03
10   10.0     2.074861066E-01
10   50.0    -1.138478491E-01
20    1.0     3.873503009E-25
20    2.0     3.918972805E-19
20    5.0     2.770330052E-11
20   10.0     1.151336925E-05
20   50.0    -1.167043528E-01
Bessel Function Yn, n>=2
20 Values
 2    1.0    -1.650682607
 2    2.0    -6.174081042E-01
 2    5.0     3.676628826E-01
 2   10.0    -5.868082460E-03
 2   50.0     9.579316873E-02
 5    1.0    -2.604058666E02
 5    2.0    -9.935989128
 5    5.0    -4.536948225E-01
 5   10.0     1.354030477E-01
 5   50.0    -7.854841391E-02
10    1.0    -1.216180143E08
10    2.0    -1.291845422E05
10    5.0    -2.512911010E01
10   10.0    -3.598141522E-01
10   50.0     5.723897182E-03
20    1.0    -4.113970315E22
```

```
20     2.0      -4.081651389E16
20     5.0      -5.933965297E08
20    10.0      -1.597483848E03
20    50.0       1.644263395E-02
```
Modified Bessel Function I0
20 Values
```
0.0    1.0000000
0.2    1.0100250
0.4    1.0404018
0.6    1.0920453
0.8    1.1665149
1.0    1.2660658
1.2    1.3937256
1.4    1.5533951
1.6    1.7499807
1.8    1.9895593
2.0    2.2795852
2.5    3.2898391
3.0    4.8807925
3.5    7.3782035
4.0   11.301922
4.5   17.481172
5.0   27.239871
6.0   67.234406
8.0  427.56411
10.0  2815.7167
```
Modified Bessel Function K0
20 Values
```
0.1    2.4270690
0.2    1.7527038
0.4    1.1145291
0.6    0.77752208
0.8    0.56534710
1.0    0.42102445
1.2    0.31850821
1.4    0.24365506
1.6    0.18795475
1.8    0.14593140
2.0    0.11389387
2.5    6.2347553E-02
3.0    3.4739500E-02
3.5    1.9598897E-02
4.0    1.1159676E-02
4.5    6.3998572E-03
5.0    3.6910983E-03
6.0    1.2439943E-03
8.0    1.4647071E-04
10.0   1.7780062E-05
```
Modified Bessel Function I1
20 Values
```
0.0    0.00000000
0.2    0.10050083
0.4    0.20402675
0.6    0.31370403
0.8    0.43286480
1.0    0.56515912
1.2    0.71467794
```

```
1.4    0.88609197
1.6    1.0848107
1.8    1.3171674
2.0    1.5906369
2.5    2.5167163
3.0    3.9533700
3.5    6.2058350
4.0    9.7594652
4.5    15.389221
5.0    24.335643
6.0    61.341937
8.0    399.87313
10.0   2670.9883
```

Modified Bessel Function K1
20 Values

```
0.1    9.8538451
0.2    4.7759725
0.4    2.1843544
0.6    1.3028349
0.8    0.86178163
1.0    0.60190724
1.2    0.43459241
1.4    0.32083589
1.6    0.24063392
1.8    0.18262309
2.0    0.13986588
2.5    7.3890816E-02
3.0    4.0156431E-02
3.5    2.2239393E-02
4.0    1.2483499E-02
4.5    7.0780949E-03
5.0    4.0446134E-03
6.0    1.3439197E-03
8.0    1.5536921E-04
10.0   1.8648773E-05
```

Modified Bessel Function Kn, n>=2
28 Values

```
2     0.2     49.512430
2     1.0     1.6248389
2     2.0     2.5375975E-01
2     2.5     1.2146021E-01
2     3.0     6.1510459E-02
2     5.0     5.3089437E-03
2     10.0    2.1509817E-05
2     20.0    6.3295437E-10
3     1.0     7.101262825
3     2.0     6.473853909E-01
3     5.0     8.291768415E-03
3     10.0    2.725270026E-05
3     50.0    3.72793677E-23
5     1.0     3.609605896E02
5     2.0     9.431049101
5     5.0     3.270627371E-02
5     10.0    5.754184999E-05
5     50.0    4.36718224E-23
10    1.0     1.807132899E08
10    2.0     1.624824040E05
```

10	5.0	9.758562829
10	10.0	1.614255300E-03
10	50.0	9.15098819E-23
20	1.0	6.294369369E22
20	2.0	5.770856853E16
20	5.0	4.827000521E08
20	10.0	1.787442782E02
20	50.0	1.70614838E-21

Modified Bessel Function In, n>=2
28 Values

2	0.2	5.0166876E-03
2	1.0	1.3574767E-01
2	2.0	6.8894844E-01
2	2.5	1.2764661
2	3.0	2.2452125
2	5.0	17.505615
2	10.0	2281.5189
2	20.0	3.9312785E07
3	1.0	2.216842492E-02
3	2.0	2.127399592E-01
3	5.0	1.033115017E01
3	10.0	1.758380717E01
3	50.0	2.67776414E20
5	1.0	2.714631560E-04
5	2.0	9.825679323E-03
5	5.0	2.157974547
5	10.0	7.771882864E02
5	50.0	2.27854831E20
10	1.0	2.752948040E-10
10	2.0	3.016963879E-07
10	5.0	4.580044419E-03
10	10.0	2.189170616E01
10	50.0	1.07159716E20
20	1.0	3.966835986E-25
20	2.0	4.310560576E-19
20	5.0	5.024239358E-11
20	10.0	1.250799736E-04
20	50.0	5.44200840E18

Legendre Polynomials
19 Values

1	0	1.0	1.224745
10	0	1.0	3.240370
20	0	1.0	4.527693
1	0	0.7071067	0.866025
10	0	0.7071067	0.373006
20	0	0.7071067	-0.874140
1	0	0.0	0.000000
10	0	0.0	-0.797435
20	0	0.0	0.797766
2	2	0.7071067	0.484123
10	2	0.7071067	-0.204789
20	2	0.7071067	0.910208
2	2	0.0	0.968246
10	2	0.0	0.804785
20	2	0.0	-0.799672
10	10	0.7071067	0.042505
20	10	0.7071067	-0.707252

10	10	0.0	1.360172
20	10	0.0	-0.853705

Jacobian Elliptic Function
20 Values

0.0	0.1	0.099833
0.0	0.2	0.19867
0.0	0.5	0.47943
0.0	1.0	0.84147
0.0	2.0	0.90930
0.5	0.1	0.099751
0.5	0.2	0.19802
0.5	0.5	0.47075
0.5	1.0	0.80300
0.5	2.0	0.99466
1.0	0.1	0.099668
1.0	0.2	0.19738
1.0	0.5	0.46212
1.0	1.0	0.76159
1.0	2.0	0.96403
1.0	4.0	0.99933
1.0	-0.2	-0.19738
1.0	-0.5	-0.46212
1.0	-1.0	-0.76159
1.0	-2.0	-0.96403

Chapter 7: Random Numbers

Chapter 7 of Numerical Recipes deals with the generation of random num-bers drawn from various distributions. The first four procedures produce uniform deviates with a range of 0.0 to 1.0. ran0 *is a procedure that imple-ments the Park and Miller "minimal standard" random number generator.* ran1 *adds a shuffle to the minimal standard.* ran2 *obtains deviates by the L'Ecuyer long period algorithm plus a shuffle.* ran3 *is Knuth's portable gen-erator, based on a subtractive rather than a congruential method.*

The transformation method is used to generate some non-uniform distri-butions. Resulting from this are routines expdev, *which gives exponentially distributed deviates, and* gasdev *for Gaussian deviates. The rejection method of producing non-uniform deviates is also discussed, and is used in* gamdev *(deviate with a gamma function distribution),* poidev *(deviate with a Poisson distribution), and* bnldev *(deviate with a binomial distribution). For gener-ating random sequences of zeros and ones, there are two procedures,* irbit1 *and* irbit2, *both based on the 18 lowest significant bits in the seed* iseed, *but each using a different recurrence to proceed from step to step.*

The "pseudo-DES" (Data Encryption Standard) hashing of 64-bit words is carried out by psdes. *It is used to generate random deviates by* ran4. sobseq *produces Sobol's quasi-random sequence. The routines* vegas *and* miser *are used for multidimensional Monte-Carlo integration, with different strategies for picking points in n-dimensional space.*

$$\star \quad \star \quad \star \quad \star$$

Sample program xran tests each random number generator, ran0 to ran3, in turn. For each one, it first draws four consecutive random numbers $X_1, \ldots, X_4$ from the generator in question. Then it treats the numbers as coordinates of a point. For example, it takes (X_1, X_2) as a point in two dimensions, (X_1, X_2, X_3) as a point in three dimensions, etc. These points are inside boxes of unit dimension in their respective n-space. They may, however, be either inside or outside of the unit sphere in that space. For $n = 2, 3, 4$ we seek the probability that a point is inside the unit n-sphere. This number is easily calculated theoretically. For $n = 2$ it is $\pi/4$, for $n = 3$ it is $\pi/6$, and for $n = 4$ it is $\pi^2/32$. If the random number generator is not faulty, the points will fall within the unit n-sphere this fraction of the time, and the result should become increasingly accurate as the number of points increases. In the program xran we have taken out a factor of 2^n for convenience, and used the random number generators as a statistical means of determining the value of π, $4\pi/3$, and $\pi^2/2$.

```
/* Driver for routines ran0, ran1, ran2, ran3 */

#include <stdio.h>
#include <math.h>
#include "nr.h"

#define PI 3.1415926

static unsigned long twotoj[16]={0x1L,0x2L,0x4L,0x8L,0x10L,0x20L,0x40L,0x80L,
    0x100L,0x200L,0x400L,0x800L,0x1000L,0x2000L,0x4000L,0x8000L};

float fnc(float x1,float x2,float x3,float x4)
{
    return (float) sqrt(x1*x1+x2*x2+x3*x3+x4*x4);
}

void integ(float (*func)(long *))
{
    long idum=(-1),iy[4],jpower,k;
    int i,j;
    float x1,x2,x3,x4,yprob[4];

    /* Calculates pi statistically using volume of unit n-sphere */
    for (i=1;i<=3;i++) iy[i]=0;
    printf("volume of unit n-sphere, n = 2, 3, 4\n");
    printf("# points      pi      (4/3)*pi    (1/2)*pi^2\n\n");
    for (j=1;j<=15;j++) {
        for (k=twotoj[j-1];k>=0;k--) {
            x1=(*func)(&idum);
            x2=(*func)(&idum);
            x3=(*func)(&idum);
            x4=(*func)(&idum);
            if (fnc(x1,x2,0.0,0.0) < 1.0) ++iy[1];
            if (fnc(x1,x2,x3,0.0) < 1.0) ++iy[2];
            if (fnc(x1,x2,x3,x4) < 1.0) ++iy[3];
        }
        jpower=twotoj[j];
        for (i=1;i<=3;i++)
            yprob[i]=(float) twotoj[i+1]*iy[i]/jpower;
        printf("%6ld %12.6f %12.6f %12.6f\n",
            jpower,yprob[1],yprob[2],yprob[3]);
    }
    printf("\nactual %12.6f %12.6f %12.6f\n",
        PI,4.0*PI/3.0,0.5*PI*PI);
}

int main(void)
{
    printf("\nTesting ran0:\n"); integ(ran0);
    printf("\nTesting ran1:\n"); integ(ran1);
    printf("\nTesting ran2:\n"); integ(ran2);
    printf("\nTesting ran3:\n"); integ(ran3);
    return 0;
}
```

Routine `expdev` generates random numbers drawn from an exponential deviate. Sample program `xexpdev` makes ten thousand calls to `expdev` and bins the results into 21 bins, the contents of which are tallied in array `x[i]`. Then the sum `total` of all bins is taken, since some of the numbers will be too large to have fallen in any of the bins. The `x[i]` are scaled to `total`, and then compared to a similarly normalized exponential which is called `expect`.

```
/* Driver for routine expdev */

#include <stdio.h>
#include <math.h>
#include "nr.h"

#define NPTS 10000
#define EE 2.718281828

int main(void)
{
    long idum=(1);
    int i,j,total=0,x[21];
    float expect,xx,y,trig[21];

    for (i=0;i<=20;i++) {
        trig[i]=i/20.0;
        x[i]=0;
    }
    for (i=1;i<=NPTS;i++) {
        y=expdev(&idum);
        for (j=1;j<=20;j++)
            if ((y < trig[j]) && (y > trig[j-1])) ++x[j];
    }
    for (i=1;i<=20;i++) total += x[i];
    printf("\nexponential distribution with %7d points\n",NPTS);
    printf("    interval       observed       expected\n\n");
    for (i=1;i<=20;i++) {
        xx=(float) x[i]/total;
        expect=exp(-(trig[i-1]+trig[i])/2.0);
        expect *= (0.05*EE/(EE-1));
        printf("%6.2f %6.2f %12.6f %12.6f \n",
            trig[i-1],trig[i],xx,expect);
    }
    return 0;
}
```

`gasdev` generates random numbers from a Gaussian deviate. Example `xgasdev` takes ten thousand of these and puts them into 21 bins. For the purpose of binning, the center of the Gaussian is shifted over by `NOVER2=10` bins, to put it in the middle bin. The remainder of the program simply plots the contents of the bins, to illustrate that they have the characteristic Gaussian bell-shape. This allows a quick, though superficial, check of the integrity of the routine.

```
/* Driver for routine gasdev */

#include <stdio.h>
#include "nr.h"

#define N 20
#define NOVER2 (N/2)
```

```
#define NPTS 10000
#define ISCAL 400
#define LLEN 50

int main(void)
{
    char words[LLEN+1];
    int i,j,k,klim,dist[N+1];
    long idum=(-13);
    float dd,x;

    for (j=0;j<=N;j++) dist[j]=0;
    for (i=1;i<=NPTS;i++) {
        x=0.25*N*gasdev(&idum);
        j=(int)(x > 0 ? x+0.5 : x-0.5);
        if ((j >= -NOVER2) && (j <= NOVER2)) ++dist[j+NOVER2];
    }
    printf("Normally distributed deviate of %6d points\n",NPTS);
    printf ("%5s %10s %9s\n","x","p(x)","graph:");
    for (j=0;j<=N;j++) {
        dd=(float) dist[j]/NPTS;
        for (k=1;k<=LLEN;k++) words[k]=' ';
        klim=(int) (ISCAL*dd);
        if (klim > LLEN)  klim=LLEN;
        for (k=1;k<=klim;k++) words[k]='*';
        printf("%8.4f %8.4f  ",j/(0.25*N),dd);
        for (k=1;k<=LLEN;k++) printf("%c",words[k]);
        printf("\n");
    }
    return 0;
}
```

The next three sample programs, xgamdev, xpoidev and xbnldev, are identical to the previous one, but each drives a different random number generator and produces a different graph. xgamdev drives gamdev and displays a gamma distribution of order ia specified by the user. xpoidev drives poidev and produces a Poisson distribution with mean xm specified by the user. xbnldev drives bnldev and produces a binomial distribution, also with specified xm.

```
/* Driver for routine gamdev */

#include <stdio.h>
#include "nr.h"

#define N 20
#define NPTS 10000
#define ISCAL 200
#define LLEN 50

int main(void)
{
    char words[LLEN+1];
    long idum=(-13);
    int i,ia,j,k,klim,dist[N+1];
    float dd;
```

```
    for (;;) {
        for (j=0;j<=N;j++) dist[j]=0;
        do {
            printf("Select order of Gamma distribution (n=1..%d), -1 to end\n",N);
            scanf("%d",&ia);
        } while (ia > N);
        if (ia < 0) break;
        for (i=1;i<=NPTS;i++) {
            j=(int) gamdev(ia,&idum);
            if ((j >= 0) && (j <= N)) ++dist[j];
        }
        printf("\ngamma-distribution deviate, order %2d of %6d points\n",
            ia,NPTS);
        printf("%6s %7s %9s \n","x","p(x)","graph:");
        for (j=0;j<N;j++) {
            dd=(float) dist[j]/NPTS;
            for (k=1;k<=50;k++) words[k]=' ';
            klim=(int) (ISCAL*dd);
            if (klim > LLEN) klim=LLEN;
            for (k=1;k<=klim;k++) words[k]='*';
            printf("%6d %8.4f  ",j,dd);
            for (k=1;k<=klim;k++) printf("%c",words[k]);
            printf("\n");
        }
    }
    return 0;
}


/* Driver for routine poidev */

#include <stdio.h>
#include "nr.h"

#define N 20
#define NPTS 10000
#define ISCAL 200
#define LLEN 50

int main(void)
{
    char txt[LLEN+1];
    long idum=(-13);
    int i,j,k,klim,dist[N+1];
    float xm,dd;

    for (;;) {
        for (j=0;j<=N;j++) dist[j]=0;
        do {
            printf("Mean of Poisson distribution (0.0<x<%d.0) ",N);
            printf("- Negative to end:\n");
            scanf("%f",&xm);
        } while (xm > N);
        if (xm < 0.0) break;
        for (i=1;i<=NPTS;i++) {
            j=(int) (0.5+poidev(xm,&idum));
            if ((j >= 0) && (j <= N)) ++dist[j];
        }
```

```
        printf("Poisson-distributed deviate, mean %5.2f of %6d points\n",
            xm,NPTS);
        printf("%5s %8s %10s\n","x","p(x)","graph:");
        for (j=0;j<=N;j++) {
            dd=(float) dist[j]/NPTS;
            for (k=0;k<=LLEN;k++) txt[k]=' ';
            klim=(int) (ISCAL*dd);
            if (klim > LLEN) klim=LLEN;
            for (k=1;k<=klim;k++) txt[k]='*';
            txt[LLEN]='\0';
            printf("%6d %8.4f   %s\n",j,dd,txt);
        }
    }
    return 0;
}


/* Driver for routine bnldev */

#include <stdio.h>
#include "nr.h"

#define N 20
#define NPTS 1000
#define ISCAL 200
#define NN 100
#define LLEN 50

int main(void)
{
    char txt[LLEN+1];
    int i,j,k,klim,dist[N+1];
    long idum=(-133);
    float pp,xm,dd;

    for (;;) {
        for (j=0;j<=N;j++) dist[j]=0;
        do {
            printf("Mean of binomial distribution (0.0 to %d.0)",N);
            printf(" - Negative to end:\n");
            scanf("%f",&xm);
        } while (xm > 20.0);
        if (xm < 0.0) break;
        pp=xm/NN;
        for (i=1;i<=NPTS;i++) {
            j=bnldev(pp,NN,&idum);
            if (j >= 0 && j <= N) ++dist[j];
        }
        printf("Binomial-distributed deviate, mean %5.2f of %6d points\n",
            xm,NPTS);
        printf("%4s %8s %10s\n","x","p(x)","graph:");
        for (j=0;j<N;j++) {
            for (k=0;k<=LLEN;k++) txt[k]=' ';
            dd=(float) dist[j]/NPTS;
            klim=(int) (ISCAL*dd+1);
            if (klim > LLEN) klim=LLEN;
            for (k=1;k<=klim;k++) txt[k]='*';
            txt[LLEN]='\0';
```

```
            printf("%4d %9.4f    %s\n",j,dd,txt);
        }
    }
    return 0;
}
```

Functions `irbit1` and `irbit2` both generate random series of ones and zeros.
The sample programs `xirbit1` and `xirbit2` for the two are the same, and they check
that the series have correct statistical properties (or more exactly, that they have at
least one correct property). They look for a 1 in the series and count how many zeros
follow it before the next 1 appears. The result is stored as a distribution. There
should be, for example, a 50% chance of no zeros, a 25% chance of exactly one zero,
and so on.

```
/* Driver for routine irbit1 */

#include <stdio.h>
#include "nr.h"

#define NBIN 15
#define NTRIES 10000

int main(void)
{
    int i,iflg,ipts=0,j,n;
    unsigned long iseed=12345;
    float twoinv,delay[NBIN+1];

    /* Calculate distribution of runs of zeros */
    for (i=1;i<=NBIN;i++) delay[i]=0.0;
    printf("distribution of runs of n zeros\n");
    printf("%6s %22s %18s\n","n","probability","expected");
    for (i=1;i<=NTRIES;i++) {
        if (irbit1(&iseed) == 1) {
            ++ipts;
            iflg=0;
            for (j=1;j<=NBIN;j++) {
                if ((irbit1(&iseed) == 1) && (iflg == 0)) {
                    iflg=1;
                    ++delay[j];
                }
            }
        }
    }
    twoinv=0.5;
    for (n=1;n<=NBIN;n++) {
        printf("%6d %19.4f %20.4f\n",
            (n-1),delay[n]/ipts,twoinv);
        twoinv /= 2.0;
    }
    return 0;
}

/* Driver for routine irbit2 */

#include <stdio.h>
```

```
#include "nr.h"

#define NBIN 15
#define NTRIES 10000

static unsigned long twoton[16]={0x1L,0x2L,0x4L,0x8L,0x10L,0x20L,0x40L,0x80L,
    0x100L,0x200L,0x400L,0x800L,0x1000L,0x2000L,0x4000L,0x8000L};

int main(void)
{
    int i,iflg,ipts=0,j,n;
    unsigned long iseed=111;
    float delay[NBIN+1];

    /* Calculate distribution of runs of zeros */
    for (i=1;i<=NBIN;i++) delay[i]=0.0;
    for (i=1;i<=NTRIES;i++) {
        if (irbit2(&iseed) == 1) {
            ++ipts;
            iflg=0;
            for (j=1;j<=NBIN;j++) {
                if ((irbit2(&iseed) == 1) && (iflg == 0)) {
                    iflg=1;
                    ++delay[j];
                }
            }
        }
    }
    printf("distribution of runs of n zeros\n");
    printf("%6s %22s %18s \n","n","probability","expected");
    for (n=1;n<=NBIN;n++)
        printf("%6d %19.4f %20.4f\n",
            (n-1),delay[n]/ipts,1.0/(double)twoton[n]);
    return 0;
}
```

The routine `psdes` uses the logic of the Data Encryption Standard, though with a simplified nonlinear function, to fully "hash" both of its 4-byte arguments. Since the routine assumes that `unsigned long` integers are 32 bits, it is not strictly portable. The following test routine simply verifies that the test values suggested in the main book are in fact obtained.

```
/* Driver for routine psdes */

#include <stdio.h>
#include "nr.h"

int main(void)
{
    static unsigned long lword[5]={0,1,1,99,99};
    static unsigned long irword[5]={0,1,99,1,99};
    static char *ans[5]={
        "",
        "0x604d1dce 0x509c0c23",
        "0xd97f8571 0xa66cb41a",
        "0x7822309d 0x64300984",
        "0xd7f376f0 0x59ba89eb"};
```

```
        int i;

        for (i=1;i<=4;i++) {
            psdes(&lword[i],&irword[i]);
            printf("PSDES now calculates:      0x%x 0x%x\n",lword[i],irword[i]);
            printf("Known correct answers are:  %s\n",ans[i]);
        }
        return 0;
}
```

Once `psdes` is known to work, then the only trick in making the encryption-based random number generator `ran4` functional is to be sure that its hexadecimal masking values are compatible with your floating point format. The following program attempts to reproduce the test values given in the main book. If your computer produces *either* the IEEE values, *or* the VAX values, then you very likely have things set up correctly. If not, go back and read the discussion in the main book.

Note that the test program, in resetting the value `idum` between calls, forces `ran4` to access certain values (the 1st and 99th) in two different random sequences (sequence -1 and sequence -99). While this random access is an interesting, and sometimes useful, property, most uses of `ran4` would leave `idum` unchanged between calls after a single initializing call with a negative value.

```
/* Driver for routine ran4 */

#include <stdio.h>
#include "nr.h"

int main(void)
{
    int i;
    static long idum[5]={0,-1,99,-99,99};
    static char *ansvax[5]={"","0.275898","0.208204","0.034307","0.838676"};
    static char *ansiee[5]={"","0.219120","0.849246","0.375290","0.457334"};
    float random[5];

    for (i=1;i<=4;i++) random[i]=ran4(&idum[i]);
    printf("ran4 gets values: ");
    for (i=1;i<=4;i++) printf("%15.6f",random[i]);
    printf("\n    IEEE answers: ");
    for (i=1;i<=4;i++) printf("%15s",ansiee[i]);
    printf("\n     VAX answers: ");
    for (i=1;i<=4;i++) printf("%15s",ansvax[i]);
    printf("\n");
    return 0;
}
```

The routine `sobseq` returns successive values of a multi-dimensional, quasi-random Sobol' sequence. The following example shows how the routine is initialized via a negative argument, and then used to obtain the first 32 values of a 3-dimensional sequence such as might be used to sample points in a 3-dimensional cube. In this example, we do nothing with the values other than to print them out. You can check your output against the following:

```
    0.50000    0.50000    0.50000    1
    0.25000    0.75000    0.25000    2
```

0.75000	0.25000	0.75000	3
0.37500	0.62500	0.12500	4
0.87500	0.12500	0.62500	5
0.12500	0.37500	0.37500	6
0.62500	0.87500	0.87500	7
0.31250	0.31250	0.68750	8
0.81250	0.81250	0.18750	9
0.06250	0.56250	0.93750	10
0.56250	0.06250	0.43750	11
0.18750	0.93750	0.56250	12
0.68750	0.43750	0.06250	13
0.43750	0.18750	0.81250	14
0.93750	0.68750	0.31250	15
0.46875	0.84375	0.40625	16
0.96875	0.34375	0.90625	17
0.21875	0.09375	0.15625	18
0.71875	0.59375	0.65625	19
0.09375	0.46875	0.28125	20
0.59375	0.96875	0.78125	21
0.34375	0.71875	0.03125	22
0.84375	0.21875	0.53125	23
0.15625	0.53125	0.84375	24
0.65625	0.03125	0.34375	25
0.40625	0.28125	0.59375	26
0.90625	0.78125	0.09375	27
0.28125	0.15625	0.96875	28
0.78125	0.65625	0.46875	29
0.03125	0.90625	0.71875	30
0.53125	0.40625	0.21875	31
0.26562	0.60938	0.57812	32

```
/* Driver for routine sobseq */

#include <stdio.h>
#include "nr.h"
#include "nrutil.h"

int main(void)
{
    int i,n1=(-1),n2=3;
    float *x;

    x=vector(1,n2);
    sobseq(&n1,x);
    for (i=1;i<=32;i++) {
        sobseq(&n2,x);
        printf(" %10.5f %10.5f %10.5f %5d\n",x[1],x[2],x[3],i);
    }
    free_vector(x,1,n2);
    return 0;
}
```

We demonstrate the adaptive Monte Carlo integration routine vegas on a narrowly peaked, n-dimensional Gaussian function,

$$f(\mathbf{x}) = \left(\frac{10}{\sqrt{\pi}}\right) e^{-100(\mathbf{x}-\mathbf{x}_0)^2}$$

The function is normalized so that its integral over all space is unity, for any number of dimensions n.

After prompting for a (negative) seed for the random number generator, the demonstration program prompts for several further quantities: First, the number of dimensions. Second, a scalar offset x_0 which is used to construct the vector $\mathbf{x}_0$ as

$$\mathbf{x}_0 = (x_0, x_0, \ldots, x_0)$$

Third, the maximum number of evaluations of (calls to) the function that are allowed on each adaptive iteration. Fourth, the number of adaptive iterations. Fifth, a code that controls how much information is printed out by **vegas**. The program calls **vegas** twice, first from a "cold start", and then a second time, inheriting the adapted grid from the first call.

If the five input parameters are entered as

$$4, \quad 0.0, \quad 1000, \quad 4, \quad 0$$

the answer for the integral ought to come out $(1/2)^4 = 0.0625$. Why not 1? Because the offset of 0.0 bisects the Gaussian in each dimension with the boundary of the region.

The following is typical of the actual output produced. Notice that, with this random seed, the first pass gets a very unlucky value at iteration no. 2 (a low value with an unfortunately small error estimate), and the average of all iterations never fully recovers. You would know that something is wrong, however, from the large value (6.1) of the χ^2 per iteration. That is why you might then do a second pass with the now-adapted grid. By the end of the 4th iteration on the second pass, you have a good value, a reliable error estimate, and a credible value for the χ^2.

```
Input parameters for vegas:  ndim=   4  ncall=      768
                             it=    0   itmx=    4
                             nprn=  0   ALPH= 1.50
                             mds=  1   nd=  50
                             xl[ 1]=          0  xu[ 1]=        1
                             xl[ 2]=          0  xu[ 2]=        1
                             xl[ 3]=          0  xu[ 3]=        1
                             xl[ 4]=          0  xu[ 4]=        1
iteration no.   1 : integral =     0.07706086 +/-      0.076
all iterations:   integral =    0.07706086+/-    0.076 chi**2/IT n =    0.00018
iteration no.   2 : integral =     0.02334202 +/-      0.0075
all iterations:   integral =    0.02386367+/-    0.0075 chi**2/IT n =       0.5
iteration no.   3 : integral =     0.04258567 +/-      0.0066
all iterations:   integral =    0.0343241+/-     0.005 chi**2/IT n =         2
iteration no.   4 : integral =     0.06005578 +/-      0.0046
all iterations:   integral =    0.04810574+/-    0.0034 chi**2/IT n =       6.1
Number of iterations performed: 4
Integral, Standard Dev., Chi-sq. =      0.048106      0.003385    6.122937
 Input parameters for vegas:  ndim=   4  ncall=      768
                             it=    5   itmx=    4
                             nprn=  0   ALPH= 1.50
                             mds=  1   nd=  50
                             xl[ 1]=          0  xu[ 1]=        1
                             xl[ 2]=          0  xu[ 2]=        1
                             xl[ 3]=          0  xu[ 3]=        1
```

```
                          xl[ 4]=         0 xu[ 4]=           1
iteration no.   1 : integral =     0.06294661 +/-     0.0026
all iterations:    integral =     0.06294661+/-   0.0026 chi**2/IT n =        0
iteration no.   2 : integral =     0.06291871 +/-     0.002
all iterations:    integral =     0.06292916+/-   0.0016 chi**2/IT n =   0.00016
iteration no.   3 : integral =     0.06065985 +/-     0.0014
all iterations:    integral =     0.06163905+/-    0.001 chi**2/IT n =     0.58
iteration no.   4 : integral =     0.06289233 +/-     0.0013
all iterations:    integral =     0.06214152+/-  0.00081 chi**2/IT n =     0.58
Additional iterations performed: 4
Integral, Standard Dev., Chi-sq. =     0.062142     0.000810   0.576426
```

```c
/* Driver for routine vegas */

#include <stdio.h>
#include <math.h>
#include "nr.h"
#include "nrutil.h"

long idum;        /* for ranno */

int ndim;         /* for fxn */
float xoff;

float fxn(float pt[],float wgt)
{
    int j;
    float ans,sum;

    for (sum=0.0,j=1;j<=ndim;j++) sum += (100.0*SQR(pt[j]-xoff));
    ans=(sum < 80.0 ? exp(-sum) : 0.0);
    ans *= pow(5.64189,(double)ndim);
    return ans;
}

int main(void)
{
    int init,itmax,j,ncall,nprn;
    float avgi,chi2a,sd,xoff;
    float *regn;

    regn=vector(1,20);
    printf("IDUM=\n");
    scanf("%ld",&idum);
    if (idum > 0) idum = -idum;
    for (;;) {
        printf("ENTER N,NDIM,XOFF,NCALL,ITMAX,NPRN\n");
        if (scanf("%d %f %d %d %d",&ndim,&xoff,&ncall,&itmax,&nprn) == EOF) break;
        avgi=sd=chi2a=0.0;
        for (j=1;j<=ndim;j++) {
            regn[j]=0.0;
            regn[j+ndim]=1.0;
        }
        init = -1;
        vegas(regn,ndim,fxn,init,ncall,itmax,nprn,&avgi,&sd,&chi2a);
        printf("Number of iterations performed: %d\n",itmax);
        printf("Integral, Standard Dev., Chi-sq. = %12.6f %12.6f% 12.6f\n"
```

```
            ,avgi,sd,chi2a);
        init = 1;
        vegas(regn,ndim,fxn,init,ncall,itmax,nprn,&avgi,&sd,&chi2a);
        printf("Additional iterations performed: %d \n",itmax);
        printf("Integral, Standard Dev., Chi-sq. = %12.6f %12.6f% 12.6f\n",
            avgi,sd,chi2a);
    }
    free_vector(regn,1,20);
    printf("Normal completion\n");
    return 0;
}
```

Our demonstration of `miser`, the routine for multidimensional Monte Carlo integration with recursive stratified sampling, performs the same integral repeatedly, so as to test whether `miser`'s claimed standard deviation in fact corresponds (at least roughly) to the actual spread of its answers around the known correct value. The function being integrated is the same narrow Gaussian as was used for `vegas`. Now, however, we assume an offset x_0 that puts the peak of the Gaussian well into the region of integration, say $x_0 = 0.33333$.

The program prompts for the number of points to try in each integration, the number of dimensions, the value for x_0, a dithering parameter which can generally be set to zero (see discussion in main book), and the number of separate times to repeat the `miser` integration. A typical run with the input values

$$10000, \quad 4, \quad 0.333333, \quad 0.0, \quad 5$$

produces output like this:

```
Fractional error: actual,indicated=      0.015751    0.022333
```

The first number is the average of the actual errors, comparing `miser`'s output value to the known answer of 1. The second number is the average of the standard deviations reported by `miser`, that is, its indicated error. In this example, we see that they are comparable, as expected, and that our Monte Carlo integration is being performed (each of five times) with an accuracy of about 2%.

Of course, in most real uses of `miser`, it would be better to perform the integration only once, with five times as many points, rather than five times.

```
/* Driver for routine miser */

#include <stdio.h>
#include <math.h>
#include "nr.h"
#include "nrutil.h"

long idum;      /* for ranno */

int ndim;       /* for func */
float xoff;

float func(float pt[])
{
    int j;
    float ans,sum;
```

```
    for (sum=0.0,j=1;j<=ndim;j++) sum += (100.0*SQR(pt[j]-xoff));
    ans=(sum < 80.0 ? exp(-sum) : 0.0);
    ans *= pow(5.64189,(double)ndim);
    return ans;
}

int main(void)
{
    unsigned long n;
    int j,nt,ntries;
    float ave,dith,sumav,sumsd,var,*regn;

    printf("IDUM=\n");
    scanf("%ld",&idum);
    if (idum > 0) idum = -idum;
    for (;;) {
        printf("ENTER N,NDIM,XOFF,DITH,NTRIES\n");
        if (scanf("%ld %d %f %f %d",&n,&ndim,&xoff,&dith,&ntries) == EOF) break;
        regn=vector(1,2*ndim);
        sumav=sumsd=0.0;
        for (nt=1;nt<=ntries;nt++) {
            for (j=1;j<=ndim;j++) {
                regn[j]=0.0;
                regn[ndim+j]=1.0;
            }
            miser(func,regn,ndim,n,dith,&ave,&var);
            sumav += SQR(ave-1.0);
            sumsd += sqrt(fabs(var));
        }
        sumav=sqrt(sumav/ntries);
        sumsd /= ntries;
        printf("Fractional error: actual,indicated= %12.6f% 12.6f\n",sumav,sumsd);
        free_vector(regn,1,2*ndim);
    }
    printf("Normal completion\n");
    return 0;
}
```

Chapter 8: Sorting

Chapter 8 of Numerical Recipes covers a variety of sorting tasks including sorting arrays into numerical order, preparing an index table for the order of an array, and preparing a rank table showing the rank order of each element in an array. `piksrt` *sorts a single array by the straight insertion method.* `piksr2` *sorts by the same method but makes the corresponding rearrangement of a second array as well.* `shell` *carries out a Shell sort.* `sort` *and* `sort2` *both do a quicksort, and they are related in the same way as* `piksrt` *and* `piksr2`, *that is,* `sort` *sorts a single array, and* `sort2` *sorts an array while correspondingly rearranging a second array.* `hpsort` *sorts an array by the heapsort algorithm, which is not quite as fast (on average) as quicksort, but requires no auxiliary storage.*

`indexx` *indexes an array. That is, it produces a second array that contains pointers to the elements of the original array in the order of their size.* `sort3` *uses* `indexx` *and illustrates its value by sorting one array while making corresponding rearrangements in two others.* `rank` *produces the rank table for an array of data. The rank table is a second array whose elements list the rank order of the corresponding elements of the original array.*

`select` *finds the Nth largest element in an array (e.g., the median). It uses the same partitioning idea as quicksort and is very fast.* `selip` *performs the same function without disturbing the array, and is correspondingly slower.* `hpsel` *finds the M largest elements of an array, without disturbing it.*

Finally, the routines `eclass` *and* `eclazz` *deal with equivalence classes.* `eclass` *gives the equivalence class of each element in an array based on a list of equivalent pairs that it is given as input.* `eclazz` *gives the same output but bases it on a procedure named* `equiv(j,k)` *that tells whether two array elements* j *and* k *are in the same equivalence class.*

⋆   ⋆   ⋆   ⋆

Routine `piksrt` sorts an array by straight insertion. Sample program `xpiksrt` provides it with a 100-element array from file `tarray.dat` which is listed in the Appendix to this chapter. The program prints both the original and the sorted array for comparison.

```
/* Driver for routine piksrt */

#include <stdio.h>
#include <stdlib.h>
#include "nr.h"
#include "nrutil.h"
```

```
#define MAXSTR 80
#define NP 100

int main(void)
{
    char txt[MAXSTR];
    int i,j;
    float *a;
    FILE *fp;

    a=vector(1,NP);
    if ((fp = fopen("tarray.dat","r")) == NULL)
        nrerror("Data file tarray.dat not found\n");
    fgets(txt,MAXSTR,fp);
    for (i=1;i<=NP;i++) fscanf(fp,"%f",&a[i]);
    fclose(fp);
    printf("original array:\n");
    for (i=0;i<=9;i++) {
        for (j=1;j<=10;j++) printf("%7.2f",a[10*i+j]);
        printf("\n");
    }
    piksrt(NP,a);
    printf("sorted array:\n");
    for (i=0;i<=9;i++) {
        for (j=1;j<=10;j++) printf("%7.2f",a[10*i+j]);
        printf("\n");
    }
    free_vector(a,1,NP);
    return 0;
}
```

piksr2 sorts an array, and simultaneously rearranges a second array (of the same size) correspondingly. In program xpiksr2, the first array a[i] is again taken from tarray.dat. The second is defined by b[i]=i. In other words, b is originally sorted and a is not. After a call to piksr2, the situation should be reversed. With a second call, this time with b as the first argument and a as the second, the two arrays should be returned to their original form.

```
/* Driver for routine piksr2 */

#include <stdio.h>
#include <stdlib.h>
#include "nr.h"
#include "nrutil.h"

#define NP 100
#define MAXSTR 80

int main(void)
{
    char txt[MAXSTR];
    int i,j;
    float *a,*b;
    FILE *fp;
```

```
a=vector(1,NP);
b=vector(1,NP);
if ((fp = fopen("tarray.dat","r")) == NULL)
    nrerror("Data file tarray.dat not found\n");
fgets(txt,MAXSTR,fp);
for (i=1;i<=NP;i++) fscanf(fp,"%f",&a[i]);
fclose(fp);
/* generate b-array */
for (i=1;i<=NP;i++) b[i]=i;
/* sort a and mix b */
piksr2(NP,a,b);
printf("\nAfter sorting a and mixing b, array a is:\n");
for (i=0;i<=9;i++) {
    for (j=1;j<=10;j++) printf("%7.2f",a[10*i+j]);
    printf("\n");
}
printf("\n... and array b is:\n");
for (i=0;i<=9;i++) {
    for (j=1;j<=10;j++) printf("%7.2f",b[10*i+j]);
    printf("\n");
}
printf("press return to continue ...\n");
(void) getchar();
/* sort b and mix a */
piksr2(NP,b,a);
printf("\nAfter sorting b and mixing a, array a is:\n");
for (i=0;i<=9;i++) {
    for (j=1;j<=10;j++) printf("%7.2f",a[10*i+j]);
    printf("\n");
}
printf("\n... and array b is:\n");
for (i=0;i<=9;i++) {
    for (j=1;j<=10;j++) printf("%7.2f",b[10*i+j]);
    printf("\n");
}
free_vector(b,1,NP);
free_vector(a,1,NP);
return 0;
}
```

Procedure `shell` does a Shell sort of a data array. The calling format is identical to that of `piksrt`, and so we use the same sample program, now called `xshell`.

```
/* Driver for routine shell */

#include <stdio.h>
#include <stdlib.h>
#include "nr.h"
#include "nrutil.h"

#define MAXSTR 80
#define NP 100

int main(void)
{
    char txt[MAXSTR];
    unsigned long i,j;
```

```
    float *a;
    FILE *fp;

    a=vector(1,NP);
    if ((fp = fopen("tarray.dat","r")) == NULL)
        nrerror("Data file tarray.dat not found\n");
    fgets(txt,MAXSTR,fp);
    for (i=1;i<=NP;i++) fscanf(fp,"%f",&a[i]);
    fclose(fp);
    printf("\nOriginal array:\n");
    for (i=0;i<=9;i++) {
        for (j=1;j<=10;j++) printf("%7.2f",a[10*i+j]);
        printf("\n");
    }
    shell(NP,a);
    printf("\nSorted array:\n");
    for (i=0;i<=9;i++) {
        for (j=1;j<=10;j++) printf("%7.2f",a[10*i+j]);
        printf("\n");
    }
    free_vector(a,1,NP);
    return 0;
}
```

By the same token, routines `sort` and `sort2` employ the same programs as routines `piksrt` and `piksr2`, respectively. (Here they are called `xsort` and `xsort2`.) Both routines use the quicksort algorithm. `sort`, however, works on a single array. `sort2` sorts one array while making corresponding rearrangements to a second.

```
/* Driver for routine sort */

#include <stdio.h>
#include <stdlib.h>
#include "nr.h"
#include "nrutil.h"

#define MAXSTR 80
#define NP 100

int main(void)
{
    char txt[MAXSTR];
    int i,j;
    float *a;
    FILE *fp;

    a=vector(1,NP);
    if ((fp = fopen("tarray.dat","r")) == NULL)
        nrerror("Data file tarray.dat not found\n");
    fgets(txt,MAXSTR,fp);
    for (i=1;i<=NP;i++) fscanf(fp,"%f",&a[i]);
    fclose(fp);
    printf("\noriginal array:\n");
    for (i=0;i<=9;i++) {
        for (j=1;j<=10;j++) printf("%7.2f",a[10*i+j]);
        printf("\n");
    }
```

```
        sort(NP,a);
        printf("\nsorted array:\n");
        for (i=0;i<=9;i++) {
            for (j=1;j<=10;j++) printf("%7.2f",a[10*i+j]);
            printf("\n");
        }
        free_vector(a,1,NP);
        return 0;
}

/* Driver for routine sort2 */

#include <stdio.h>
#include <stdlib.h>
#include "nr.h"
#include "nrutil.h"

#define MAXSTR 80
#define NP 100

int main(void)
{
    char txt[MAXSTR];
    int i,j;
    float *a,*b;
    FILE *fp;

    a=vector(1,NP);
    b=vector(1,NP);
    if ((fp = fopen("tarray.dat","r")) == NULL)
        nrerror("Data file tarray.dat not found\n");
    fgets(txt,MAXSTR,fp);
    for (i=1;i<=NP;i++) fscanf(fp,"%f",&a[i]);
    fclose(fp);
    /* generate b-array */
    for (i=1;i<=NP;i++) b[i]=i;
    /* sort a and mix b */
    sort2(NP,a,b);
    printf("\nAfter sorting a and mixing b, array a is:\n");
    for (i=0;i<=9;i++) {
        for (j=1;j<=10;j++) printf("%7.2f",a[10*i+j]);
        printf("\n");
    }
    printf("\n... and array b is:\n");
    for (i=0;i<=9;i++) {
        for (j=1;j<=10;j++) printf("%7.2f",b[10*i+j]);
        printf("\n");
    }
    printf("press return to continue...\n");
    (void) getchar();
    /* sort b and mix a */
    sort2(NP,b,a);
    printf("\nAfter sorting b and mixing a, array a is:\n");
    for (i=0;i<=9;i++) {
        for (j=1;j<=10;j++) printf("%7.2f",a[10*i+j]);
        printf("\n");
    }
```

```
    printf("\n... and array b is:\n");
    for (i=0;i<=9;i++) {
        for (j=1;j<=10;j++) printf("%7.2f",b[10*i+j]);
        printf("\n");
    }
    free_vector(b,1,NP);
    free_vector(a,1,NP);
    return 0;
}
```

hpsort sorts an array by the heapsort algorithm. Its calling sequence is exactly like that of piksrt and sort, so we again rely on the same sample program, now called xhpsort.

```
/* Driver for routine hpsort */

#include <stdio.h>
#include <stdlib.h>
#include "nr.h"
#include "nrutil.h"

#define MAXSTR 80
#define NP 100

int main(void)
{
    char txt[MAXSTR];
    int i,j;
    float *a;
    FILE *fp;

    a=vector(1,NP);
    if ((fp = fopen("tarray.dat","r")) == NULL)
        nrerror("Data file tarray.dat not found\n");
    fgets(txt,MAXSTR,fp);
    for (i=1;i<=NP;i++) fscanf(fp,"%f",&a[i]);
    fclose(fp);
    printf("\noriginal array:\n");
    for (i=0;i<=9;i++) {
        for (j=1;j<=10;j++) printf("%7.2f",a[10*i+j]);
        printf("\n");
    }
    hpsort(NP,a);
    printf("\nsorted array:\n");
    for (i=0;i<=9;i++) {
        for (j=1;j<=10;j++) printf("%7.2f",a[10*i+j]);
        printf("\n");
    }
    free_vector(a,1,NP);
    return 0;
}
```

The procedure indexx generates the index array for a given input array. The index array indx[j] gives, for each j, the index of the element of the input array which will assume position j if the array is sorted. That is, for an input array a, the sorted version of a will be a[indx[j]]. To demonstrate this, sample program xindexx produces an index for the array in `tarray.dat`. It then prints the array in the order `a[indx[j]]`, `j=1,..,100` for inspection.

```c
/* Driver for routine indexx */

#include <stdio.h>
#include <stdlib.h>
#include "nr.h"
#include "nrutil.h"

#define NP 100
#define MAXSTR 80

int main(void)
{
    char txt[MAXSTR];
    unsigned long i,j,*indx;
    float *a;
    FILE *fp;

    indx=lvector(1,NP);
    a=vector(1,NP);
    if ((fp = fopen("tarray.dat","r")) == NULL)
        nrerror("Data file tarray.dat not found\n");
    fgets(txt,MAXSTR,fp);
    for (i=1;i<=NP;i++) fscanf(fp,"%f",&a[i]);
    fclose(fp);
    indexx(NP,a,indx);
    printf("\noriginal array:\n");
    for (i=0;i<=9;i++) {
        for (j=1;j<=10;j++) printf("%7.2f",a[10*i+j]);
        printf("\n");
    }
    printf("\nsorted array:\n");
    for (i=0;i<=9;i++) {
        for (j=1;j<=10;j++) printf("%7.2f",a[indx[10*i+j]]);
        printf("\n");
    }
    free_vector(a,1,NP);
    free_lvector(indx,1,NP);
    return 0;
}
```

One use for `indexx` is the management of more than two arrays. `sort3`, for example, sorts one array while making corresponding reorderings of two other arrays. In sample program `xsort3`, the first array is taken as the first 64 elements of `tarray.dat` (see Appendix). The second and third arrays are taken to be the numbers 1 to 64 in forward order and reverse order, respectively. When the first array is ordered, the second and third are scrambled, but scrambled in exactly the same way. To prove this, a text message is assigned to a character array. Then the letters are scrambled according to the order of numbers found in the rearranged second array. They are subsequently unscrambled according to the order of numbers found in the rearranged third array. If `sort3` works properly, this ought to leave the message reading in the reverse order.

```
/* Driver for routine sort3 */

#include <stdio.h>
#include <stdlib.h>
#include <string.h>
#include <math.h>
#include "nr.h"
#include "nrutil.h"

#define NLEN 64

int main(void)
{
    int i,j;
    char dummy[NLEN],amsg[NLEN+1],bmsg[NLEN+1],cmsg[NLEN+1];
    float *a,*b,*c;
    FILE *fp;

    a=vector(1,NLEN);
    b=vector(1,NLEN);
    c=vector(1,NLEN);
    (void) strcpy(amsg,"I'd rather have a bottle in front of");
    (void) strcat(amsg," me than a frontal lobotomy.");
    printf("\noriginal message:\n%s\n",amsg);
    /* read array of random numbers */
    if ((fp = fopen("tarray.dat","r")) == NULL)
        nrerror("Data file tarray.dat not found\n");
    fgets(dummy,NLEN,fp);
    for (i=1;i<=NLEN;i++) fscanf(fp,"%f",&a[i]);
    fclose(fp);
    /* create array b and array c */
    for (i=1;i<=NLEN;i++) {
        b[i]=i;
        c[i]=NLEN+1-i;
    }
    /* sort array a while mixing b and c */
    sort3(NLEN,a,b,c);
    /* scramble message according to array b */
    bmsg[NLEN]=amsg[NLEN];  /* null terminating character */
    for (i=1;i<=NLEN;i++) {
        j=b[i];
        bmsg[i-1]=amsg[j-1];
    }
    printf("\nscrambled message:\n%s\n",bmsg);
    /* unscramble according to array c */
    cmsg[NLEN]=amsg[NLEN];
    for (i=1;i<=NLEN;i++) {
        j=c[i];
        cmsg[j-1]=bmsg[i-1];
    }
    printf("\nmirrored message:\n%s\n",cmsg);
    free_vector(c,1,NLEN);
    free_vector(b,1,NLEN);
    free_vector(a,1,NLEN);
    return 0;
}
```

rank is a procedure that is similar to indexx. Instead of producing an indexing array, though, it produces a rank table. For an array a[j] and rank table irank[j], entry j in irank will tell what index a[j] will have if a is sorted. irank actually takes its input information not from the array itself, but from the index array produced by indexx. Sample program xrank begins with the array from tarray.dat, and feeds it to indexx and rank. The table of ranks produced is listed. To check it, the array a is copied into an array b in the rank order suggested by irank. b should then be in proper order.

```c
/* Driver for routine rank */

#include <stdio.h>
#include <stdlib.h>
#include "nr.h"
#include "nrutil.h"

#define NP 100
#define MAXSTR 80

int main(void)
{
    char txt[MAXSTR];
    unsigned long i,j,k,l,*indx,*irank;
    float *a,b[11];
    FILE *fp;

    indx=lvector(1,NP);
    irank=lvector(1,NP);
    a=vector(1,NP);
    if ((fp = fopen("tarray.dat","r")) == NULL)
        nrerror("Data file tarray.dat not found\n");
    fgets(txt,MAXSTR,fp);
    for (i=1;i<=NP;i++) fscanf(fp,"%f",&a[i]);
    fclose(fp);
    indexx(NP,a,indx);
    rank(NP,indx,irank);
    printf("original array is:\n");
    for (i=0;i<=9;i++) {
        for (j=1;j<=10;j++) printf("%7.2f",a[10*i+j]);
        printf("\n");
    }
    printf("table of ranks is:\n");
    for (i=0;i<=9;i++) {
        for (j=1;j<=10;j++) printf("%7d",irank[10*i+j]);
        printf("\n");
    }
    printf("press return to continue...\n");
    (void) getchar();
    printf("array sorted according to rank table:\n");
    for (i=0;i<=9;i++) {
        for (j=1;j<=10;j++) {
            k=10*i+j;
            for (l=1;l<=NP;l++)
                if (irank[l] == k) b[j]=a[l];
        }
        for (j=1;j<=10;j++) printf("%7.2f",b[j]);
```

```
        printf("\n");
    }
    free_vector(a,1,NP);
    free_lvector(irank,1,NP);
    free_lvector(indx,1,NP);
    return 0;
}
```

Sample program xselect demonstrates the use of select to find the kth largest element in tarray.dat. As a cross-check, the answer is compared with that from the slower routine selip.

```
/* Driver for routine select */

#include <stdio.h>
#include <stdlib.h>
#include "nr.h"
#include "nrutil.h"

#define MAXSTR 80
#define NP 100

int main(void)
{
    char txt[MAXSTR];
    unsigned long i,k;
    float *a,*b,q,s;
    FILE *fp;

    a=vector(1,NP);
    b=vector(1,NP);
    if ((fp = fopen("tarray.dat","r")) == NULL)
        nrerror("Data file tarray.dat not found\n");
    fgets(txt,MAXSTR,fp);
    for (i=1;i<=NP;i++) fscanf(fp,"%f",&a[i]);
    fclose(fp);
    for (;;) {
        printf("Input k\n");
        if (scanf("%lu",&k) == EOF) break;
        for (i=1;i<=NP;i++) b[i]=a[i];
        s=selip(k,NP,a);
        q=select(k,NP,b);
        printf("Element in sort position %lu is %6.2f\n",k,q);
        printf("Cross-check from SELIP routine %6.2f\n",s);
    }
    free_vector(b,1,NP);
    free_vector(a,1,NP);
    printf("Normal completion\n");
    return 0;
}
```

The routine xselip tests selip by selecting the kth largest element of tarray.dat, for each $k = 1, \ldots, 100$. This is a very inefficient way of sorting the array. As a final check, xselip prints out a user-supplied request for the kth largest element of the array.

```
/* Driver for routine selip */

#include <stdio.h>
#include <stdlib.h>
#include "nr.h"
#include "nrutil.h"

#define MAXSTR 80
#define NP 100

int main(void)
{
    char txt[MAXSTR];
    unsigned long i,j,k;
    float *a,*b,q;
    FILE *fp;

    a=vector(1,NP);
    b=vector(1,NP);
    if ((fp = fopen("tarray.dat","r")) == NULL)
        nrerror("Data file tarray.dat not found\n");
    fgets(txt,MAXSTR,fp);
    for (i=1;i<=NP;i++) fscanf(fp,"%f",&a[i]);
    fclose(fp);
    printf("\noriginal array:\n");
    for (i=0;i<=9;i++) {
        for (j=1;j<=10;j++) printf("%7.2f",a[10*i+j]);
        printf("\n");
    }
    for (i=1;i<=NP;i++) b[i]=selip(i,100,a);
    printf("\nsorted array:\n");
    for (i=0;i<=9;i++) {
        for (j=1;j<=10;j++) printf("%7.2f",b[10*i+j]);
        printf("\n");
    }
    for (;;) {
        printf("Input k\n");
        if (scanf("%lu",&k) == EOF) break;
        q=selip(k,NP,a);
        printf("Element in sort position %lu is %6.2f\n",k,q);
    }
    free_vector(b,1,NP);
    free_vector(a,1,NP);
    printf("Normal completion\n");
    return 0;
}
```

Routine `hpsel` finds the k largest elements of an array, where (for efficiency) k should be much shorter than the length of the array. The sample program is much like that for `select`, except that all k largest elements (not just the kth) are printed out.

```
/* Driver for routine hpsel */

#include <stdio.h>
#include <stdlib.h>
#include "nr.h"
```

```
#include "nrutil.h"

#define MAXSTR 80
#define NP 100

int main(void)
{
    char txt[MAXSTR];
    unsigned long i,k;
    float *a,*heap,check;
    FILE *fp;

    a=vector(1,NP);
    heap=vector(1,NP);
    if ((fp = fopen("tarray.dat","r")) == NULL)
        nrerror("Data file tarray.dat not found\n");
    fgets(txt,MAXSTR,fp);
    for (i=1;i<=NP;i++) fscanf(fp,"%f",&a[i]);
    fclose(fp);
    for (;;) {
        printf("Input k\n");
        if (scanf("%lu",&k) == EOF) break;
        hpsel(k,NP,a,heap);
        check=select(NP+1-k,NP,a);
        printf("heap[1], check= %6.2f %6.2f\n",heap[1],check);
        printf("heap of numbers of size %lu\n",k);
        for (i=1;i<=k;i++) printf("%lu    %6.2f\n",i,heap[i]);
    }
    free_vector(heap,1,NP);
    free_vector(a,1,NP);
    printf("Normal completion\n");
    return 0;
}
```

Procedure `eclass` generates a list of equivalence classes for the elements of an input array, based on the arrays `lista[j]` and `listb[j]` which list equivalent pairs for each `j`. In sample program `xeclass`, these lists are

$$\text{lista:} \quad 1, 1, 5, 2, 6, 2, 7, 11, 3, 4, 12$$
$$\text{listb:} \quad 5, 9, 13, 6, 10, 14, 3, 7, 15, 8, 4$$

According to these lists, 1 is equivalent to 5, 1 is equivalent to 9, etc. If you work it out, you will find the following classes:

$$\text{class1:} \quad 1, 5, 9, 13$$
$$\text{class2:} \quad 2, 6, 10, 14$$
$$\text{class3:} \quad 3, 7, 11, 15$$
$$\text{class4:} \quad 4, 8, 12$$

The sample program prints out the classes and ought to agree with this list.

```
/* Driver for routine eclass */

#include <stdio.h>
#include "nr.h"
#include "nrutil.h"
```

```
#define M 11
#define N 15

int main(void)
{
    int i,j,k,lclas=0,nclass,*nf,*nflag,*nsav;
    static int lista[]={0,1,1,5,2,6,2,7,11,3,4,12},
            listb[]={0,5,9,13,6,10,14,3,7,15,8,4};

    nf=ivector(1,N);
    nflag=ivector(1,N);
    nsav=ivector(1,N);
    eclass(nf,N,lista,listb,M);
    for (i=1;i<=N;i++) nflag[i]=1;
    printf("\nNumbers from 1-%d divided according to\n",N);
    printf("their value modulo 4:\n\n");
    for (i=1;i<=N;i++) {
        nclass=nf[i];
        if (nflag[nclass]) {
            nflag[nclass]=0;
            lclas++;
            k=0;
            for (j=i;j<=N;j++)
                if (nf[j] == nf[i]) nsav[++k]=j;
            printf("Class %2d:      ",lclas);
            for (j=1;j<=k;j++) printf("%3d",nsav[j]);
            printf("\n");
        }
    }
    free_ivector(nsav,1,N);
    free_ivector(nflag,1,N);
    free_ivector(nf,1,N);
    return 0;
}
```

eclazz performs the same analysis but figures the equivalences from a boolean function equiv(i,j) that tells whether i and j are in the same equivalence class. In xeclazz, equiv is defined as true if (i % 4) and (j % 4) are the same. It is otherwise false.

```
/* Driver for routine eclazz */

#include <stdio.h>
#include "nr.h"
#include "nrutil.h"

#define N 15

int equiv(int i,int j)
{
    return (i % 4) == (j % 4);
}

int main(void)
{
    int i,j,k,lclas=0,nclass,*nf,*nflag,*nsav;
```

```
nf=ivector(1,N);
nflag=ivector(1,N);
nsav=ivector(1,N);
eclazz(nf,N,equiv);
for (i=1;i<=N;i++) nflag[i]=1;
printf("\nNumbers from 1-%d divided according to\n",N);
printf("their value modulo 4:\n");
for (i=1;i<=N;i++) {
    nclass=nf[i];
    if (nflag[nclass]) {
        nflag[nclass]=0;
        lclas++;
        k=0;
        for (j=i;j<=N;j++)
            if (nf[j] == nf[i]) nsav[++k]=j;
        printf("Class %2d:       ",lclas);
        for (j=1;j<=k;j++) printf("%3d",nsav[j]);
        printf("\n");
    }
}
free_ivector(nsav,1,N);
free_ivector(nflag,1,N);
free_ivector(nf,1,N);
return 0;
}
```

Appendix

File `tarray.dat`:

```
Test data for chapter 8:
 29.82 71.51  3.30 87.44 53.42 63.16 89.10 25.75 93.16 27.72
 71.58 48.34 53.11 18.34 27.13 60.31 83.34 22.81 66.84 52.91
 53.42 15.22  8.01 53.39 76.12 79.09 67.61 38.39 24.81 73.21
 13.42 52.10 34.86 99.83 38.46 81.59 61.75 79.62 93.39  3.21
 99.34 92.22 94.29  7.03  6.67 89.35 83.14  9.01 12.68 62.22
  2.95 85.02 95.82 73.96 49.29 77.72 36.65  3.48 48.98 71.83
  1.41  9.48 32.37 89.95 28.39 79.36 54.05 46.08 11.67 37.78
 77.17 74.33 10.13  4.62 49.95 68.40 19.40 34.06  4.11 98.40
 42.44 64.14 89.41 52.99 71.79  3.94 19.73 44.91 71.44 59.10
 27.54 15.67 67.95 55.61 26.05 25.01 82.09 89.67 57.08 38.27
```

Chapter 9: Root Finding and Sets of Equations

Chapter 9 of *Numerical Recipes* deals primarily with the problem of finding roots to equations, and treats the problem in greatest detail in one dimension. We begin with a general-purpose routine called `scrsho` that produces a crude graph of a given function on a specified interval. It is used for low-resolution plotting to investigate the properties of the function. With this in hand, we add bracketing routines `zbrac` and `zbrak`. The first of these takes a function and an interval, and expands the interval geometrically until it brackets a root. The second breaks the interval into N subintervals of equal size. It then reports any intervals that contain at least one root. Once bracketed, roots can be found by a number of other routines. `rtbis` finds such roots by bisection. `rtflsp` and `rtsec` use the method of false position and the secant method, respectively. `zriddr` implements Ridders' method. `zbrent` uses a combination of methods to give assured and relatively efficient convergence. `rtnewt` implements the Newton-Raphson root finding method, while `rtsafe` combines it with bisection to correct for its risky global convergence properties.

For finding the roots of polynomials, `laguer` is handy, and when combined with its driver `zroots` it can find all roots of a polynomial having complex coefficients. An alternative to `laguer` is provided by `zrhqr`, which finds the roots as eigenvalues of the companion matrix. When you have some tentative complex roots of a real polynomial, they can be polished by `qroot`, which employs Bairstow's method.

In multiple dimensions, root-finding requires some foresight. However, if you can identify the neighborhood of a root of a system of nonlinear equations, then `mnewt` will help you to zero in using Newton-Raphson. More powerful is to use the globally convergent multi-dimensional Newton method in `newt`, or Broyden's method `broydn`, which generalizes the secant method.

$$\star \quad \star \quad \star \quad \star$$

`scrsho` is a primitive graphing routine that will print graphs on virtually any terminal or printer. Sample program `xscrsho` demonstrates it by graphing the zero-order Bessel function J_0.

```
/* Driver for routine scrsho */

#include <stdio.h>
#include "nr.h"

static float fx(float x)
```

```
{
    return bessj0(x);
}

int main(void)
{
    scrsho(fx);
    return 0;
}
```

zbrac is a root-bracketing routine that works by expanding the range of an interval geometrically until it brackets a root. Sample program xzbrac applies it to the Bessel function J_0. It starts with the ten intervals $(1.0, 2.0)$, $(2.0, 3.0)$, etc., and expands each until it contains a root. Then it prints the interval limits, and the function J_0 evaluated at these limits. The two values of J_0 should have opposite signs.

```
/* Driver for routine zbrac */

#include <stdio.h>
#include "nr.h"

static float fx(float x)
{
    return bessj0(x);
}

int main(void)
{
    int succes,i;
    float x1,x2;

    printf("%21s %23s\n","bracketing values:","function values:");
    printf("%6s %10s %21s %12s\n","x1","x2","bessj0(x1)","bessj0(x2)");
    for (i=1;i<=10;i++) {
        x1=i;
        x2=x1+1.0;
        succes=zbrac(fx,&x1,&x2);
        if (succes) {
            printf("%7.2f %10.2f %6s %12.6f %12.6f \n",
                x1,x2," ",fx(x1),fx(x2));
        }
    }
    return 0;
}
```

zbrak is much like zbrac except that it takes an interval and subdivides it into N equal parts. It then identifies any of the subintervals that contain roots. Sample program xzbrak looks for roots of $J_0(x)$ between X1 = 1.0 and X2 = 50.0 by allowing zbrak to divide the interval into $N = 100$ parts. If there are no roots spaced closer than $\Delta x = 0.49$, then it will find brackets for all roots in this region. The limits of bracketing intervals, as well as function values at these limits, are printed, and again the function values at the end of each interval ought to be of opposite sign. There are 16 roots of J_0 between 1 and 50.

```
/* Driver for routine zbrak */

#include <stdio.h>
#include "nr.h"
#include "nrutil.h"

#define N 100
#define NBMAX 20
#define X1 1.0
#define X2 50.0

static float fx(float x)
{
    return bessj0(x);
}

int main(void)
{
    int i,nb=NBMAX;
    float *xb1,*xb2;

    xb1=vector(1,NBMAX);
    xb2=vector(1,NBMAX);
    zbrak(fx,X1,X2,N,xb1,xb2,&nb);
    printf("\nbrackets for roots of bessj0:\n");
    printf("%21s %10s %16s %10s\n","lower","upper","f(lower)","f(upper)");
    for (i=1;i<=nb;i++)
        printf("%s %2d  %10.4f %10.4f %3s %10.4f %10.4f\n",
            " root ",i,xb1[i],xb2[i]," ",
            fx(xb1[i]),fx(xb2[i]));
    free_vector(xb2,1,NBMAX);
    free_vector(xb1,1,NBMAX);
    return 0;
}
```

Routine `rtbis` begins with the brackets for a root and finds the root itself by bisection, The accuracy with which the root is found is determined by parameter `xacc`. Sample program `xrtbis` finds all the roots of Bessel function $J_0(x)$ between X1 = 1.0 and X2 = 50.0. In this case `xacc` is specified to be about 10^{-6} of the value of the root itself (actually, 10^{-6} of the center of the interval being bisected). The roots `root` are listed, as well as $J_0(\text{root})$ to verify their accuracy.

```
/* Driver for routine rtbis */

#include <stdio.h>
#include "nr.h"
#include "nrutil.h"

#define N 100
#define NBMAX 20
#define X1 1.0
#define X2 50.0

static float fx(float x)
{
    return bessj0(x);
```

```
}

int main(void)
{
    int i,nb=NBMAX;
    float xacc,root,*xb1,*xb2;

    xb1=vector(1,NBMAX);
    xb2=vector(1,NBMAX);
    zbrak(fx,X1,X2,N,xb1,xb2,&nb);
    printf("\nRoots of bessj0:\n");
    printf("%21s %15s\n","x","f(x)");
    for (i=1;i<=nb;i++) {
        xacc=(1.0e-6)*(xb1[i]+xb2[i])/2.0;
        root=rtbis(fx,xb1[i],xb2[i],xacc);
        printf("root %3d %14.6f %14.6f\n",i,root,fx(root));
    }
    free_vector(xb2,1,NBMAX);
    free_vector(xb1,1,NBMAX);
    return 0;
}
```

The next six sample programs are essentially identical to the one just discussed, except for the root-finder they employ. `xrtflsp` calls `rtflsp`, finding the root by "false position". `xrtsec` calls `rtsec` and uses the secant method. `xzriddr` uses Ridders' method, `zriddr`. `xzbrent` uses `zbrent` to give reliable and efficient convergence. The Newton-Raphson method implemented in `rtnewt` is demonstrated by `xrtnewt`, and `xrtsafe` calls `rtsafe`, which improves upon `rtnewt` by combining it with bisection to achieve better global convergence. The latter two programs include a procedure `funcd` that returns the value of the function and its derivative at a given x. In the case of test function $J_0(x)$ the derivative is $-J_1(x)$, and is conveniently in our collection of special functions.

```
/* Driver for routine rtflsp */

#include <stdio.h>
#include "nr.h"
#include "nrutil.h"

#define N 100
#define NBMAX 20
#define X1 1.0
#define X2 50.0

static float fx(float x)
{
    return bessj0(x);
}

int main(void)
{
    int i,nb=NBMAX;
    float xacc,root,*xb1,*xb2;

    xb1=vector(1,NBMAX);
    xb2=vector(1,NBMAX);
```

```
        zbrak(fx,X1,X2,N,xb1,xb2,&nb);
        printf("\nRoots of bessj0:\n");
        printf("%21s %15s\n","x","f(x)");
        for (i=1;i<=nb;i++) {
            xacc=(1.0e-6)*(xb1[i]+xb2[i])/2.0;
            root=rtflsp(fx,xb1[i],xb2[i],xacc);
            printf("root %3d %14.6f %14.6f\n",i,root,fx(root));
        }
        free_vector(xb2,1,NBMAX);
        free_vector(xb1,1,NBMAX);
        return 0;
}

/* Driver for routine rtsec */

#include <stdio.h>
#include "nr.h"
#include "nrutil.h"

#define N 100
#define NBMAX 20
#define X1 1.0
#define X2 50.0

static float fx(float x)
{
    return bessj0(x);
}

int main(void)
{
    int i,nb=NBMAX;
    float xacc,root,*xb1,*xb2;

    xb1=vector(1,NBMAX);
    xb2=vector(1,NBMAX);
    zbrak(fx,X1,X2,N,xb1,xb2,&nb);
    printf("\nRoots of bessj0:\n");
    printf("%21s %15s\n","x","f(x)");
    for (i=1;i<=nb;i++) {
        xacc=(1.0e-6)*(xb1[i]+xb2[i])/2.0;
        root=rtsec(fx,xb1[i],xb2[i],xacc);
        printf("root %3d %14.6f %14.6f\n",i,root,fx(root));
    }
    free_vector(xb2,1,NBMAX);
    free_vector(xb1,1,NBMAX);
    return 0;
}

/* Driver for routine zriddr */

#include <stdio.h>
#include "nr.h"
#include "nrutil.h"

#define N 100
#define NBMAX 20
```

```
#define X1 1.0
#define X2 50.0

static float fx(float x)
{
    return bessj0(x);
}

int main(void)
{
    int i,nb=NBMAX;
    float xacc,root,*xb1,*xb2;

    xb1=vector(1,NBMAX);
    xb2=vector(1,NBMAX);
    zbrak(fx,X1,X2,N,xb1,xb2,&nb);
    printf("\nRoots of bessj0:\n");
    printf("%21s %15s\n","x","f(x)");
    for (i=1;i<=nb;i++) {
        xacc=(1.0e-6)*(xb1[i]+xb2[i])/2.0;
        root=zriddr(fx,xb1[i],xb2[i],xacc);
        printf("root %3d %14.6f %14.6f\n",i,root,fx(root));
    }
    free_vector(xb2,1,NBMAX);
    free_vector(xb1,1,NBMAX);
    return 0;
}

/* Driver for routine zbrent */

#include <stdio.h>
#include "nr.h"
#include "nrutil.h"

#define N 100
#define NBMAX 20
#define X1 1.0
#define X2 50.0

static float fx(float x)
{
    return bessj0(x);
}

int main(void)
{
    int i,nb=NBMAX;
    float tol,root,*xb1,*xb2;

    xb1=vector(1,NBMAX);
    xb2=vector(1,NBMAX);
    zbrak(fx,X1,X2,N,xb1,xb2,&nb);
    printf("\nRoots of bessj0:\n");
    printf("%21s %15s\n","x","f(x)");
    for (i=1;i<=nb;i++) {
        tol=(1.0e-6)*(xb1[i]+xb2[i])/2.0;
        root=zbrent(fx,xb1[i],xb2[i],tol);
```

```
            printf("root %3d %14.6f %14.6f\n",i,root,fx(root));
        }
        free_vector(xb2,1,NBMAX);
        free_vector(xb1,1,NBMAX);
        return 0;
}

/* Driver for routine rtnewt */

#include <stdio.h>
#include "nr.h"
#include "nrutil.h"

#define N 100
#define NBMAX 20
#define X1 1.0
#define X2 50.0

static float fx(float x)
{
    return bessj0(x);
}

static void funcd(float x,float *fn, float *df)
{
    *fn=bessj0(x);
    *df = -bessj1(x);
}

int main(void)
{
    int i,nb=NBMAX;
    float xacc,root,*xb1,*xb2;

    xb1=vector(1,NBMAX);
    xb2=vector(1,NBMAX);
    zbrak(fx,X1,X2,N,xb1,xb2,&nb);
    printf("\nRoots of bessj0:\n");
    printf("%21s %15s\n","x","f(x)");
    for (i=1;i<=nb;i++) {
        xacc=(1.0e-6)*(xb1[i]+xb2[i])/2.0;
        root=rtnewt(funcd,xb1[i],xb2[i],xacc);
        printf("root %3d %14.6f %14.6f\n",i,root,fx(root));
    }
    free_vector(xb2,1,NBMAX);
    free_vector(xb1,1,NBMAX);
    return 0;
}

/* Driver for routine rtsafe */

#include <stdio.h>
#include "nr.h"
#include "nrutil.h"

#define N 100
#define NBMAX 20
```

```
#define X1 1.0
#define X2 50.0

static float fx(float x)
{
    return bessj0(x);
}

static void funcd(float x,float *fn, float *df)
{
    *fn=bessj0(x);
    *df = -bessj1(x);
}

int main(void)
{
    int i,nb=NBMAX;
    float xacc,root,*xb1,*xb2;

    xb1=vector(1,NBMAX);
    xb2=vector(1,NBMAX);
    zbrak(fx,X1,X2,N,xb1,xb2,&nb);
    printf("\nRoots of bessj0:\n");
    printf("%21s %15s\n","x","f(x)");
    for (i=1;i<=nb;i++) {
        xacc=(1.0e-6)*(xb1[i]+xb2[i])/2.0;
        root=rtsafe(funcd,xb1[i],xb2[i],xacc);
        printf("root %3d %14.6f %14.6f\n",i,root,fx(root));
    }
    free_vector(xb2,1,NBMAX);
    free_vector(xb1,1,NBMAX);
    return 0;
}
```

Routine `laguer` finds the roots of a polynomial with complex coefficients. The polynomial of degree M is specified by $M + 1$ coefficients which, in sample program `xlaguer`, are specified in the complex array a of dimension $M + 1$. The polynomial in this case is

$$F(x) = x^4 - (1 + 2i)x^2 + 2i$$

The four roots of this polynomial are $x = 1.0$, $x = -1.0$, $x = 1 + i$, and $x = -(1 + i)$. `laguer` proceeds on the basis of a trial root, and attempts to converge to true roots. The root to which it converges depends on the trial value. The program tries a series of complex trial values along the line in the imaginary plane from $-1.0 - i$ to $1.0 + i$. The actual roots to which it converges are compared to all previously found values, and if different, are printed.

```
/* Driver for routine laguer */

#include <stdio.h>
#include <math.h>
#include "nr.h"
#include "complex.h"

#define M 4      /* degree of polynomial */
#define MP1 (M+1)   /* no. of polynomial coefficients */
```

```
#define NTRY 21
#define NTRY1 NTRY+1
#define EPS 1.e-6

int main(void)
{
    fcomplex y[NTRY1],x;
    static fcomplex a[MP1] = {{0.0,2.0},
                {0.0,0.0},
                {-1.0,-2.0},
                {0.0,0.0},
                {1.0,0.0} };
    int i,iflag,its,j,n=0;

    printf("\nRoots of polynomial x^4-(1+2i)*x^2+2i\n");
    printf("\n%15s %15s %7s\n","Real","Complex","#iter");
    for (i=1;i<=NTRY;i++) {
        x=Complex((i-11.0)/10.0,(i-11.0)/10.0);
        laguer(a,M,&x,&its);
        if (n == 0) {
            n=1;
            y[1]=x;
            printf("%5d %12.6f %12.6f %5d\n",n,x.r,x.i,its);
        } else {
            iflag=0;
            for (j=1;j<=n;j++)
                if (Cabs(Csub(x,y[j])) <= EPS*Cabs(x)) iflag=1;
            if (iflag == 0) {
                y[++n]=x;
                printf("%5d %12.6f %12.6f %5d\n",n,x.r,x.i,its);
            }
        }
    }
    return 0;
}
```

zroots is a driver for laguer. Sample program xzroots exercises zroots, using the same polynomial as the previous routine. First it finds the four roots. Then it corrupts each one by multiplying by $1 + .01i$. Finally it uses zroots again to polish the corrupted roots by setting the boolean variable polish to TRUE.

```
/* Driver for routine zroots */

#include <stdio.h>
#include "nr.h"
#include "complex.h"

#define M 4
#define MP1 (M+1)
#define TRUE 1
#define FALSE 0

int main(void)
{
    int i,polish;
    fcomplex roots[MP1];
    static fcomplex a[MP1] = {{0.0,2.0},
```

```
                {0.0,0.0},
                {-1.0,-2.0},
                {0.0,0.0},
                {1.0,0.0} };

    printf("\nRoots of the polynomial x^4-(1+2i)*x^2+2i\n");
    polish=FALSE;
    zroots(a,M,roots,polish);
    printf("\nUnpolished roots:\n");
    printf("%14s %13s %13s\n","root #","real","imag.");
    for (i=1;i<=M;i++)
        printf("%11d %18.6f %12.6f\n",i,roots[i].r,roots[i].i);
    printf("\nCorrupted roots:\n");
    for (i=1;i<=M;i++)
        roots[i]=RCmul(1+0.01*i,roots[i]);
    printf("%14s %13s %13s\n","root #","real","imag.");
    for (i=1;i<=M;i++)
        printf("%11d %18.6f %12.6f\n",i,roots[i].r,roots[i].i);
    polish=TRUE;
    zroots(a,M,roots,polish);
    printf("\nPolished roots:\n");
    printf("%14s %13s %13s\n","root #","real","imag.");
    for (i=1;i<=M;i++)
        printf("%11d %18.6f %12.6f \n",i,roots[i].r,roots[i].i);
    return 0;
}
```

Routine zrhqr is an alternative to `laguer` for polynomials with real coefficients. Sample program xzrhqr tries it out on the polynomial $x^4 - 1$.

```
/* Driver for routine zrhqr */

#include <stdio.h>
#include "nr.h"
#include "nrutil.h"

#define M 4      /* degree of polynomial */
#define MP1 (M+1)   /* no. of polynomial coefficients */

int main(void)
{
    static float a[MP1] = {-1.0,0.0,0.0,0.0,1.0};
    float *rti,*rtr;
    int i;

    rti=vector(1,M);
    rtr=vector(1,M);
    printf("\nRoots of polynomial x^4-1\n");
    printf("\n%15s %15s\n","Real","Complex");
        zrhqr(a,M,rtr,rti);
        for (i=1;i<=M;i++) {
            printf("%5d %12.6f %12.6f\n",i,rtr[i],rti[i]);
        }
    free_vector(rtr,1,M);
    free_vector(rti,1,M);
    return 0;
}
```

`qroot` is used for finding quadratic factors of polynomials with real coefficients. In the case of sample program `xqroot`, the polynomial is

$$P(x) = x^6 - 6x^5 + 16x^4 - 24x^3 + 25x^2 - 18x + 10.$$

The program proceeds like that of `laguer`. Successive trial values for quadratic factors $x^2 + Bx + C$ (in the form of guesses for B and C) are made, and for each trial, `qroot` converges on correct values. If the B and C which are found are unlike any previous values, then they are printed. By this means, all three quadratic factors are located. The answer should be

$$P(x) = (x^2 + 1)(x^2 - 4x + 5)(x^2 - 2x + 2)$$

```
/* Driver for routine qroot */

#include <stdio.h>
#include <math.h>
#include "nr.h"
#include "nrutil.h"

#define N 6 /* degree of polynomial */
#define EPS 1.0e-6
#define NTRY 10
#define TINY 1.0e-5

int main(void)
{
    int i,j,nflag,nroot=0;
    static float p[N+1]={10.0,-18.0,25.0,-24.0,16.0,-6.0,1.0};
    float *b,*c;

    b=vector(1,NTRY);
    c=vector(1,NTRY);
    printf("\nP(x)=x^6-6x^5+16x^4-24x^3+25x^2-18x+10\n");
    printf("Quadratic factors x^2+bx+c\n\n");
    printf("%6s %10s %12s \n\n","factor","b","c");
    for (i=1;i<=NTRY;i++) {
        c[i]=0.5*i;
        b[i] = -0.5*i;
        qroot(p,N,&b[i],&c[i],EPS);
        if (nroot == 0) {
            printf("%4d %15.6f %12.6f\n",nroot,b[i],c[i]);
            nroot=1;
        } else {
            nflag=0;
            for (j=1;j<=nroot;j++)
                if ((fabs(b[i]-b[j]) < TINY)
                    && (fabs(c[i]-c[j]) < TINY))
                    nflag=1;
            if (nflag == 0) {
                printf("%4d %15.6f %12.6f\n",nroot,b[i],c[i]);
                ++nroot;
            }
        }
    }
    free_vector(c,1,NTRY);
```

```
    free_vector(b,1,NTRY);
    return 0;
}
```

mnewt looks for roots of multiple nonlinear equations. In order to run the sample program xmnewt we supply a procedure usrfun that returns the matrix alpha of partial derivatives of the functions with respect to each of the variables, and vector beta, containing the negatives of the function values. The sample program tries to find sets of variables that solve the four equations

$$-x_1^2 - x_2^2 - x_3^2 + x_4 = 0$$
$$x_1^2 + x_2^2 + x_3^2 + x_4^2 - 1 = 0$$
$$x_1 - x_2 = 0$$
$$x_2 - x_3 = 0$$

You will probably be able to find the two solutions to this set even without mnewt, noting that $x_1 = x_2$ and $x_2 = x_3$. If not, simply take the output from mnewt and plug it into these equations for verification. The output from mnewt should convince you of the need for good starting values (or that you should use newt instead!).

```
/* Driver for routine mnewt */

#include <stdio.h>
#include <math.h>
#include "nr.h"
#include "nrutil.h"

void usrfun(float *x,int n,float *fvec,float **fjac)
{
    int i;

    fjac[1][1] = -2.0*x[1];
    fjac[1][2] = -2.0*x[2];
    fjac[1][3] = -2.0*x[3];
    fjac[1][4]=1.0;
    for (i=1;i<=n;i++) fjac[2][i]=2.0*x[i];
    fjac[3][1]=1.0;
    fjac[3][2] = -1.0;
    fjac[3][3]=0.0;
    fjac[3][4]=0.0;
    fjac[4][1]=0.0;
    fjac[4][2]=1.0;
    fjac[4][3] = -1.0;
    fjac[4][4]=0.0;
    fvec[1] = -SQR(x[1])-SQR(x[2])-SQR(x[3])+x[4];
    fvec[2]=SQR(x[1])+SQR(x[2])+SQR(x[3])+SQR(x[4])-1.0;
    fvec[3]=x[1]-x[2];
    fvec[4]=x[2]-x[3];
}

#define NTRIAL 5
#define TOLX 1.0e-6
#define N 4
#define TOLF 1.0e-6
```

```c
int main(void)
{
    int i,j,k,kk;
    float xx,*x,*fvec,**fjac;

    fjac=matrix(1,N,1,N);
    fvec=vector(1,N);
    x=vector(1,N);
    for (kk=1;kk<=2;kk++) {
        for (k=1;k<=3;k++) {
            xx=0.2001*k*(2*kk-3);
            printf("Starting vector number %2d\n",k);
            for (i=1;i<=4;i++) {
                x[i]=xx+0.2*i;
                printf("%7s%1d%s %5.2f\n",
                    "x[",i,"] = ",x[i]);
            }
            printf("\n");
            for (j=1;j<=NTRIAL;j++) {
                mnewt(1,x,N,TOLX,TOLF);
                usrfun(x,N,fvec,fjac);
                printf("%5s %13s %13s\n","i","x[i]","f");
                for (i=1;i<=N;i++)
                    printf("%5d %14.6f %15.6f\n",
                        i,x[i],fvec[i]);
                printf("\npress RETURN to continue...\n");
                (void) getchar();
            }
        }
    }
    free_vector(x,1,N);
    free_vector(fvec,1,N);
    free_matrix(fjac,1,N,1,N);
    return 0;
}
```

Sample program `xnewt` tries out routine `newt` on the nonlinear system

$$f_1 = x_1^2 + x_2^2 - 2$$
$$f_2 = e^{(x_1-1)} + x_2^3 - 2$$

which has a root at $(x_1, x_2) = (1, 1)$. It attempts to find the root starting from the initial guess $(2, .5)$. (This is Example 6.5.1 in *Numerical Methods for Unconstrained Optimization and Nonlinear Equations* by Dennis and Schnabel.) `mnewt`, which applies Newton's method blindly without a globally convergent strategy, fails on this example. (Note that `newt` is also used in the solution of two point boundary value problems in Chapter 17.)

```c
/* Driver for routine newt */

#include <stdio.h>
#include <math.h>
#include "nr.h"
#include "nrutil.h"

void funcv(int n,float x[],float f[])
```

```
{
    f[1]=SQR(x[1])+SQR(x[2])-2.0;
    f[2]=exp(x[1]-1.0)+x[2]*SQR(x[2])-2.0;
}

#define N 2

int main(void)
{
    int i,check;
    float *x,*f;

    x=vector(1,N);
    f=vector(1,N);
    x[1]=2.0;
    x[2]=0.5;
    newt(x,N,&check,funcv);
    funcv(N,x,f);
    if (check) printf("Convergence problems.\n");
    printf("%7s %3s %12s\n","Index","x","f");
    for (i=1;i<=N;i++) printf("%5d %12.6f %12.6f\n",i,x[i],f[i]);
    free_vector(f,1,N);
    free_vector(x,1,N);
    return 0;
}
```

The sample program for broydn, which implements a multidimensional general-ization of the secant method, is identical to the one for newt:

```
/* Driver for routine broydn */

#include <stdio.h>
#include <math.h>
#include "nr.h"
#include "nrutil.h"

void funcv(int n,float x[],float f[])
{
    f[1]=SQR(x[1])+SQR(x[2])-2.0;
    f[2]=exp(x[1]-1.0)+x[2]*SQR(x[2])-2.0;
}

#define N 2

int main(void)
{
    int i,check;
    float *x,*f;

    x=vector(1,N);
    f=vector(1,N);
    x[1]=2.0;
    x[2]=0.5;
    broydn(x,N,&check,funcv);
    funcv(N,x,f);
    if (check) printf("Convergence problems.\n");
    printf("%7s %3s %12s\n","Index","x","f");
```

```
    for (i=1;i<=N;i++) printf("%5d %12.6f %12.6f\n",i,x[i],f[i]);
    free_vector(f,1,N);
    free_vector(x,1,N);
    return 0;
}
```

Chapter 10: Minimization and Maximization of Functions

Chapter 10 of Numerical Recipes deals with finding the maxima and minima of functions. The task has two parts, first the discovery of one or more bracketing intervals, and then the convergence to an extremum. mnbrak *begins with two specified abscissas of a function and searches in the "downhill" direction for brackets of a minimum.* golden *can then take a bracketing triplet and perform a golden section search to a specified precision, for the minimum itself. When you are not concerned with worst-case examples, but only very efficient average-case performance, Brent's method (routine* brent*) is recommended. In the event that means are at hand for calculating the function's derivative as well as its value, consider* dbrent.

Most multidimensional minimization strategies are based on the above one-dimensional algorithms. Our single example of an algorithm that is not so based is amoeba, *which utilizes the downhill simplex method. Among the ones that do use one-dimensional methods are* powell, frprmn, *and* dfpmin. *The first two make calls to* linmin, *a procedure that minimizes a function along a given direction in space.* linmin *in turn uses the one-dimensional algorithm* brent, *if derivatives are not known, or* dbrent *if they are.* powell *uses only function values and minimizes along an artfully chosen set of favorable directions.* frprmn *uses a Fletcher-Reeves-Polak-Ribiere minimization and requires the calculation of derivatives for the function.* dfpmin *uses a globally convergent variant of the Davidon-Fletcher-Powell variable metric method. This, too, requires calculation of derivatives, but the routine uses* lnsrch *for approximate one-dimensional minimizations.*

The chapter ends with two topics of somewhat different nature. The first is linear programming, which deals with the maximization of a linear combination of variables, subject to linear constraints. This problem is dealt with by the simplex method in routine simplx. *The second topic is the subject of large scale optimization, which is illustrated with the method of simulated annealing. One application of this is a combinatorial minimization problem, the "travelling salesman" problem in routine* anneal. *The second is the use of simulated annealing to find the global minimum in a continuous space, procedure* amebsa.

$\star$ $\star$ $\star$ $\star$

mnbrak searches a given function for a minimum. Given two values ax and bx of abscissa, it searches in the downward direction until it can find three new values

ax,bx,cx that bracket a minimum. fa,fb,fc are the values of the function at these
points. Sample program **xmnbrak** is a simple application of **mnbrak** applied to the
Bessel function J_0. It tries a series of starting values ax,bx each encompassing an
interval of length 1.0. **mnbrak** then finds several bracketing intervals of various minima
of J_0.

```
/* Driver for routine mnbrak */

#include <stdio.h>
#include "nr.h"

float func(float x)
{
    return bessj0(x);
}

int main(void)
{
    float ax,bx,cx,fa,fb,fc;
    int i;

    for (i=1;i<=10;i++) {
        ax=i*0.5;
        bx=(i+1.0)*0.5;
        mnbrak(&ax,&bx,&cx,&fa,&fb,&fc,func);
        printf("%14s %12s %12s\n","a","b","c");
        printf("%3s %14.6f %12.6f %12.6f\n","x",ax,bx,cx);
        printf("%3s %14.6f %12.6f %12.6f\n","f",fa,fb,fc);
    }
    return 0;
}
```

Routine golden continues the minimization process by taking a bracketing triplet
ax,bx,cx and performing a golden section search to isolate the contained minimum
to a stated precision TOL. Sample program **xgolden** again uses J_0 as the test function.
Using intervals (ax,bx) of length 1.0 it uses **mnbrak** to bracket all minima between
$x = 0.0$ and $x = 100.0$. Some minima are bracketed more than once. On each pass, the
bracketed solution is tracked down by golden. It is then compared to all previously
located minima, and if different it is added to the collection by incrementing **nmin**
(number of minima found) and adding the location **xmin** of the minima to the list
in array **amin**. As a check of golden, the routine prints out the value of J_0 at the
minimum, and also the value of J_1, which ought to be zero at extrema of J_0.

```
/* Driver for routine golden */

#include <stdio.h>
#include <math.h>
#include "nr.h"

#define TOL 1.0e-6
#define EQL 1.0e-3

float func(float x)
{
    return bessj0(x);
}
```

```
int main(void)
{
    int i,iflag,j,nmin=0;
    float ax,bx,cx,fa,fb,fc,xmin,gold,amin[21];

    printf("Minima of the function bessj0\n");
    printf("%10s %8s %17s %12s\n","min. #","x","bessj0(x)","bessj1(x)");
    for (i=1;i<=100;i++) {
        ax=i;
        bx=i+1.0;
        mnbrak(&ax,&bx,&cx,&fa,&fb,&fc,func);
        gold=golden(ax,bx,cx,func,TOL,&xmin);
        if (nmin == 0) {
            amin[1]=xmin;
            nmin=1;
            printf("%7d %15.6f %12.6f %12.6f\n",
                nmin,xmin,bessj0(xmin),bessj1(xmin));
        } else {
            iflag=0;
            for (j=1;j<=nmin;j++)
                if (fabs(xmin-amin[j]) <= EQL*xmin) iflag=1;
            if (iflag == 0) {
                amin[++nmin]=xmin;
                printf("%7d %15.6f %12.6f %12.6f\n",
                    nmin,xmin,bessj0(xmin),bessj1(xmin));
            }
        }
    }
    return 0;
}
```

There are two other routines presented that also take the bracketing triplet ax,bx,cx from mnbrak and find the contained minimum. They are brent and dbrent. The sample programs for these two, xbrent and xdbrent, are virtually identical to that used on golden. Note that dbrent is only used when the derivative can be calculated conveniently.

```
/* Driver for routine brent */

#include <stdio.h>
#include <math.h>
#include "nr.h"

#define TOL 1.0e-6
#define EQL 1.0e-4

float func(float x)
{
    return bessj0(x);
}

int main(void)
{
    int i,iflag,j,nmin=0;
    float ax,bx,cx,fa,fb,fc,xmin,bren,amin[21];
```

```
        printf("\nMinima of the function bessj0\n");
        printf("%10s %8s %17s %12s\n","min. #","x","bessj0(x)","bessj1(x)");
        for (i=1;i<=100;i++) {
            ax=i;
            bx=i+1.0;
            mnbrak(&ax,&bx,&cx,&fa,&fb,&fc,func);
            bren=brent(ax,bx,cx,func,TOL,&xmin);
            if (nmin == 0) {
                amin[1]=xmin;
                nmin=1;
                printf("%7d %15.6f %12.6f %12.6f\n",
                    nmin,xmin,bessj0(xmin),bessj1(xmin));
            } else {
                iflag=0;
                for (j=1;j<=nmin;j++)
                    if (fabs(xmin-amin[j]) <= (EQL*xmin)) iflag=1;
                if (iflag == 0) {
                    amin[++nmin]=xmin;
                    printf("%7d %15.6f %12.6f %12.6f\n",
                        nmin,xmin,bessj0(xmin),bessj1(xmin));
                }
            }
        }
        return 0;
}


/* Driver for routine dbrent */

#include <stdio.h>
#include <math.h>
#include "nr.h"

#define TOL 1.0e-6
#define EQL 1.0e-4

float dfunc(float x)
{
    return -bessj1(x);
}

float func(float x)
{
    return bessj0(x);
}

int main(void)
{
    int i,iflag,j,nmin=0;
    float ax,bx,cx,fa,fb,fc,xmin,dbr,amin[21];

    printf("\nMinima of the function bessj0\n");
    printf("%10s %8s %16s %12s %11s\n",
        "min. #","x","bessj0(x)","bessj1(x)","DBRENT");
    for (i=1;i<=100;i++) {
        ax=i;
        bx=i+1.0;
        mnbrak(&ax,&bx,&cx,&fa,&fb,&fc,func);
```

```
        dbr=dbrent(ax,bx,cx,func,dfunc,TOL,&xmin);
        if (nmin == 0) {
            amin[1]=xmin;
            nmin=1;
            printf("%7d %15.6f %12.6f %12.6f %12.6f\n",
                nmin,xmin,func(xmin),dfunc(xmin),dbr);
        } else {
            iflag=0;
            for (j=1;j<=nmin;j++)
                if (fabs(xmin-amin[j]) <= EQL*xmin) iflag=1;
            if (iflag == 0) {
                amin[++nmin]=xmin;
                printf("%7d %15.6f %12.6f %12.6f %12.6f\n",
                    nmin,xmin,func(xmin),dfunc(xmin),dbr);
            }
        }
    }
    return 0;
}
```

Numerical Recipes presents several methods for minimization in multiple dimensions. Among these, the downhill simplex method carried out by amoeba is the only one that does not treat the problem as a series of one-dimensional minimizations. As input, amoeba requires the coordinates of $N + 1$ vertices of a starting simplex in N-dimensional space, and the values y of the function at each of these vertices. Sample program xamoeba tries the method out on the exotic function

$$\texttt{famoeb} = 0.6 - J_0[(x - 0.5)^2 + (y - 0.6)^2 + (z - 0.7)^2]$$

which has a minimum at $(x, y, z) = (0.5, 0.6, 0.7)$. As vertices of the starting simplex, specified by the array p, we used $(0, 0, 0)$, $(1, 0, 0)$, $(0, 1, 0)$, and $(0, 0, 1)$. A vector x[i] is set successively to each vertex to allow the evaluation of function values y. This data is submitted to amoeba along with FTOL=1.0E-6 to specify the tolerance on the function value. The vertices and corresponding function values of the final simplex are printed out, and you can easily check whether the specified tolerance is met.

```
/* Driver for routine amoeba */

#include <stdio.h>
#include <math.h>
#include "nr.h"
#include "nrutil.h"

#define MP 4
#define NP 3
#define FTOL 1.0e-6

float func(float x[])
{
    return 0.6-bessj0(SQR(x[1]-0.5)+SQR(x[2]-0.6)+SQR(x[3]-0.7));
}

int main(void)
{
    int i,nfunc,j,ndim=NP;
```

```
        float *x,*y,**p;

        x=vector(1,NP);
        y=vector(1,MP);
        p=matrix(1,MP,1,NP);
        for (i=1;i<=MP;i++) {
            for (j=1;j<=NP;j++)
                x[j]=p[i][j]=(i == (j+1) ? 1.0 : 0.0);
            y[i]=func(x);
        }
        amoeba(p,y,ndim,FTOL,func,&nfunc);
        printf("\nNumber of function evaluations: %3d\n",nfunc);
        printf("Vertices of final 3-d simplex and\n");
        printf("function values at the vertices:\n\n");
        printf("%3s %10s %12s %12s %14s\n\n",
            "i","x[i]","y[i]","z[i]","function");
        for (i=1;i<=MP;i++) {
            printf("%3d ",i);
            for (j=1;j<=NP;j++) printf("%12.6f ",p[i][j]);
            printf("%12.6f\n",y[i]);
        }
        printf("\nTrue minimum is at (0.5,0.6,0.7)\n");
        free_matrix(p,1,MP,1,NP);
        free_vector(y,1,MP);
        free_vector(x,1,NP);
        return 0;
}
```

powell carries out one-dimensional minimizations along favorable directions in N-dimensional space. The function minimized in sample program xpowell is defined as

$$\text{func}(x, y, z) = \tfrac{1}{2} - J_0[(x - 1)^2 + (y - 2)^2 + (z - 3)^2].$$

The program provides powell with a starting point P of $(3/2, 3/2, 5/2)$ and a set of initial directions, here chosen to be the unit directions $(1, 0, 0)$, $(0, 1, 0)$, and $(0, 0, 1)$. powell performs its one-dimensional minimizations with linmin, which is discussed next.

```
/* Driver for routine powell */

#include <stdio.h>
#include <math.h>
#include "nr.h"
#include "nrutil.h"

#define NDIM 3
#define FTOL 1.0e-6

float func(float x[])
{
    return 0.5-bessj0(SQR(x[1]-1.0)+SQR(x[2]-2.0)+SQR(x[3]-3.0));
}

int main(void)
{
    int i,iter,j;
    float fret,**xi;
```

```
          static float p[]={0.0,1.5,1.5,2.5};

          xi=matrix(1,NDIM,1,NDIM);
          for (i=1;i<=NDIM;i++)
              for (j=1;j<=NDIM;j++)
                  xi[i][j]=(i == j ? 1.0 : 0.0);
          powell(p,xi,NDIM,FTOL,&iter,&fret,func);
          printf("Iterations: %3d\n\n",iter);
          printf("Minimum found at: \n");
          for (i=1;i<=NDIM;i++) printf("%12.6f",p[i]);
          printf("\n\nMinimum function value = %12.6f \n\n",fret);
          printf("True minimum of function is at:\n");
          printf("%12.6f %12.6f %12.6f\n",1.0,2.0,3.0);
          free_matrix(xi,1,NDIM,1,NDIM);
          return 0;
      }
```

linmin, as we have said, finds the minimum of a function func along a direction in N-dimensional space. To use it we specify a point P and a direction vector xi, both in N-space. linmin then does the bookkeeping required to treat the function as a function of position along this line, and minimizes the function with a conventional one-dimensional minimization routine. Sample program xlinmin feeds linmin the function

$$\text{func}(x, y, z) = (x - 1)^2 + (y - 1)^2 + (z - 1)^2$$

which has a minimum at $(x, y, z) = (1, 1, 1)$. It also chooses point P to be the origin $(0, 0, 0)$, and tries a series of directions

$$\left(\sqrt{2} \cos \left(\frac{\pi}{2} \frac{i}{10.0} \right), \quad \sqrt{2} \sin \left(\frac{\pi}{2} \frac{i}{10.0} \right), \quad 1.0 \right) \qquad i = 1, \ldots, 10$$

For each pass, the location of the minimum, and the value of the function at the minimum, are printed. Among the directions searched is the direction $(1, 1, 1)$. Along this direction, of course, the minimum function value should be zero and should occur at $(1, 1, 1)$.

```
/* Driver for routine linmin */

#include <stdio.h>
#include <math.h>
#include "nr.h"
#include "nrutil.h"

#define NDIM 3
#define PIO2 1.5707963

float func(float x[])
{
    int i;
    float f=0.0;

    for (i=1;i<=3;i++) f += (x[i]-1.0)*(x[i]-1.0);
    return f;
}

int main(void)
```

```
{
    int i,j;
    float fret,sr2,x,*p,*xi;

    p=vector(1,NDIM);
    xi=vector(1,NDIM);
    printf("\nMinimum of a 3-d quadratic centered\n");
    printf("at (1.0,1.0,1.0). Minimum is found\n");
    printf("along a series of radials.\n\n");
    printf("%9s %12s %12s %14s \n","x","y","z","minimum");
    for (i=0;i<=10;i++) {
        x=PIO2*i/10.0;
        sr2=sqrt(2.0);
        xi[1]=sr2*cos(x);
        xi[2]=sr2*sin(x);
        xi[3]=1.0;
        p[1]=p[2]=p[3]=0.0;
        linmin(p,xi,NDIM,&fret,func);
        for (j=1;j<=3;j++) printf("%12.6f ",p[j]);
        printf("%12.6f\n",fret);
    }
    free_vector(xi,1,NDIM);
    free_vector(p,1,NDIM);
    return 0;
}
```

f1dim accompanies linmin and is the routine that makes the N-dimensional function func effectively a one-dimensional function along a given line in N-space. There is little to check here, and our perfunctory demonstration of its use, in sample program xf1dim, simply plots f1dim as a one dimensional function, given the function

$$\texttt{func}(x,y,z) = (x-1)^2 + (y-1)^2 + (z-1)^2.$$

You get to choose the direction; then scrsho plots the function along this direction. Try the direction $(1,1,1)$ along which you should find a minimum value of func=0 at position $(1,1,1)$.

```
/* Driver for routine f1dim */

#include <stdio.h>
#include "nr.h"
#include "nrutil.h"

float func(float x[])
{
    int i;
    float f=0.0;

    for (i=1;i<=3;i++) f += (x[i]-1.0)*(x[i]-1.0);
    return f;
}

#define NDIM 3

int ncom;    /* defining declarations */
float *pcom,*xicom,(*nrfunc)(float []);
```

```
int main(void)
{
    ncom=NDIM;
    pcom=vector(1,ncom);
    xicom=vector(1,ncom);
    nrfunc=func;
    pcom[1]=pcom[2]=pcom[3]=0.0;
    printf("\nEnter vector direction along which to\n");
    printf("plot the function. Minimum is in the\n");
    printf("direction 1.0 1.0 1.0 - enter x y z:\n");
    scanf(" %f %f %f",&xicom[1],&xicom[2],&xicom[3]);
    scrsho(f1dim);
    free_vector(xicom,1,ncom);
    free_vector(pcom,1,ncom);
    return 0;
}
```

`frprmn` is another multidimensional minimizer that relies on the one-dimensional minimizations of `linmin`. It works, however, via the Fletcher-Reeves-Polak-Ribiere method and requires that routines be supplied for calculating both the function and its gradient. Sample program `xfrprmn`, for example, uses

$$\text{func}(x, y, z) = 1.0 - J_0(x - \tfrac{1}{2})J_0(y - \tfrac{1}{2})J_0(z - \tfrac{1}{2})$$

and

$$\frac{\partial\, \text{func}}{\partial x} = J_1(x - \tfrac{1}{2})J_0(y - \tfrac{1}{2})J_0(z - \tfrac{1}{2})$$

etc. A number of trial starting vectors are used, and each time, `frprmn` manages to find the minimum at $(1/2, 1/2, 1/2)$.

```
/* Driver for routine frprmn */

#include <stdio.h>
#include <math.h>
#include "nr.h"
#include "nrutil.h"

#define NDIM 3
#define FTOL 1.0e-6
#define PIO2 1.5707963

float func(float x[])
{
    return 1.0-bessj0(x[1]-0.5)*bessj0(x[2]-0.5)*bessj0(x[3]-0.5);
}

void dfunc(float x[],float df[])
{
    df[1]=bessj1(x[1]-0.5)*bessj0(x[2]-0.5)*bessj0(x[3]-0.5);
    df[2]=bessj0(x[1]-0.5)*bessj1(x[2]-0.5)*bessj0(x[3]-0.5);
    df[3]=bessj0(x[1]-0.5)*bessj0(x[2]-0.5)*bessj1(x[3]-0.5);
}

int main(void)
{
    int iter,k;
```

```
    float angl,fret,*p;

    p=vector(1,NDIM);
    printf("Program finds the minimum of a function\n");
    printf("with different trial starting vectors.\n");
    printf("True minimum is (0.5,0.5,0.5)\n");
    for (k=0;k<=4;k++) {
        angl=PIO2*k/4.0;
        p[1]=2.0*cos(angl);
        p[2]=2.0*sin(angl);
        p[3]=0.0;
        printf("\nStarting vector: (%6.4f,%6.4f,%6.4f)\n",
            p[1],p[2],p[3]);
        frprmn(p,NDIM,FTOL,&iter,&fret,func,dfunc);
        printf("Iterations: %3d\n",iter);
        printf("Solution vector: (%6.4f,%6.4f,%6.4f)\n",
            p[1],p[2],p[3]);
        printf("Func. value at solution %14f\n",fret);
    }
    free_vector(p,1,NDIM);
    return 0;
}
```

The routine dlinmin is an alternative to linmin for minimization routines that use derivative information as well as function values. The sample program for dlinmin is essentially the same as for linmin, given above.

```
/* Driver for routine dlinmin */

#include <stdio.h>
#include <math.h>
#include "nr.h"
#include "nrutil.h"

#define NDIM 3
#define PIO2 1.5707963

float func(float x[])
{
    int i;
    float f=0.0;

    for (i=1;i<=3;i++) f += (x[i]-1.0)*(x[i]-1.0);
    return f;
}

void dfunc(float x[],float df[])
{
    int i;

    for (i=1;i<=3;i++) df[i]=2.0*(x[i]-1.0);
}

int main(void)
{
    int i,j;
    float fret,sr2,x,*p,*xi;
```

```
    p=vector(1,NDIM);
    xi=vector(1,NDIM);
    printf("\nMinimum of a 3-d quadratic centered\n");
    printf("at (1.0,1.0,1.0). Minimum is found\n");
    printf("along a series of radials.\n\n");
    printf("%9s %12s %12s %14s \n","x","y","z","minimum");
    for (i=0;i<=10;i++) {
        x=PIO2*i/10.0;
        sr2=sqrt(2.0);
        xi[1]=sr2*cos(x);
        xi[2]=sr2*sin(x);
        xi[3]=1.0;
        p[1]=p[2]=p[3]=0.0;
        dlinmin(p,xi,NDIM,&fret,func,dfunc);
        for (j=1;j<=3;j++) printf("%12.6f ",p[j]);
        printf("%12.6f\n",fret);
    }
    free_vector(xi,1,NDIM);
    free_vector(p,1,NDIM);
    return 0;
}
```

Completeness requires that we provide a sample program for df1dim, which is presented in *Numerical Recipes* as a routine for converting the N-dimensional gradient procedure to one that provides the first derivative of the function along a specified line in N-dimensional space. It is exactly analogous to f1dim and the program xdf1dim is the same.

```
/* Driver for routine df1dim */

#include <stdio.h>
#include "nr.h"
#include "nrutil.h"

#define NDIM 3

int ncom;    /* defining declarations */
float *pcom,*xicom;
void (*nrdfun)(float [], float []);

void dfunc(float x[], float df[])
{
    int i;

    for (i=1;i<=3;i++) df[i]=(x[i]-1.0)*(x[i]-1.0);
}

int main(void)
{
    ncom=NDIM;
    pcom=vector(1,ncom);
    xicom=vector(1,ncom);
    nrdfun=dfunc;
    printf("\nEnter vector direction along which to\n");
    printf("plot the function. Minimum is in the\n");
    printf("direction 1.0 1.0 1.0 - enter x y z:\n\n");
```

```
    pcom[1]=pcom[2]=pcom[3]=0.0;
    scanf("%f %f %f",&xicom[1],&xicom[2],&xicom[3]);
    scrsho(df1dim);
    free_vector(xicom,1,ncom);
    free_vector(pcom,1,ncom);
    return 0;
}
```

dfpmin implements the Broyden-Fletcher-Goldfarb-Shanno variant of the David-on-Fletcher-Powell minimization by variable metric methods. It requires somewhat more intermediate storage than the preceding routine, but it works well with a globally convergent strategy that requires only approximate line minimizations, and can therefore be more efficient. Sample program xdfpmin works just as did the program for frprmn, including the fact that it requires a procedure for calculation of the derivative. The approximate line minimizations are carried out by lnsrch. In this case we try it on the somewhat harder problem (Stoer and Bulirsch, *Introduction to Numerical Analysis*, p. 308):

$$f = 10[(x_2^2(3 - x_1) - x_1^2(3 + x_1))]^2 + \frac{(2 + x_1)^2}{1 + (2 + x_1)^2}$$

which has a minimum at $(x_1, x_2) = (-2, \pm 0.89442719)$. We take the initial guess to be $(0.1, 4.2)$. (The original problem in Stoer and Bulirsch has 100 in the first term instead of 10. We have changed it because some machines may have trouble in single precision. Even this easier problem requires double precision if you try to solve it with powell.)

```
/* Driver for routine dfpmin */

#include <stdio.h>
#include "nr.h"
#include "nrutil.h"

static int nfunc,ndfunc;

float func(float x[])
{
    float x1p2sqr=SQR(2.0+x[1]);

    nfunc++;
    return 10.0*
        SQR(SQR(x[2])*(3.0-x[1])-SQR(x[1])*(3.0+x[1]))+
        x1p2sqr/(1.0+x1p2sqr);
}

void dfunc(float x[],float df[])
{
    float x1sqr=SQR(x[1]),x2sqr=SQR(x[2]),x1p2=x[1]+2.0;
    float x1p2sqr=SQR(x1p2);

    ndfunc++;
    df[1]=20.0*(x2sqr*(3.0-x[1])-x1sqr*(3.0+x[1]))*(-x2sqr-6.0*x[1]-3.0*x1sqr)+
        2.0*x1p2/(1.0+x1p2sqr)-2.0*x1p2*x1p2sqr/SQR((1.0+x1p2sqr));
    df[2]=40.0*(x2sqr*(3.0-x[1])-x1sqr*(3.0+x[1]))*x[2]*(3.0-x[1]));
}
```

```
#define NDIM 2
#define GTOL 1.0e-4

int main(void)
{
    int iter;
    float *p,fret;

    p=vector(1,NDIM);
    printf("True minimum is at (-2.0,+-0.89442719)\n");
    nfunc=ndfunc=0;
    p[1]=0.1;
    p[2]=4.2;
    printf("Starting vector: (%7.4f,%7.4f)\n",p[1],p[2]);
    dfpmin(p,NDIM,GTOL,&iter,&fret,func,dfunc);
    printf("Iterations: %3d\n",iter);
    printf("Func. evals: %3d\n",nfunc);
    printf("Deriv. evals: %3d\n",ndfunc);
    printf("Solution vector: (%9.6f,%9.6f)\n",p[1],p[2]);
    printf("Func. value at solution %14.6g\n",fret);
    free_vector(p,1,NDIM);
    return 0;
}
```

simplx is a procedure for dealing with problems in linear programming. In these problems the goal is to maximize a linear combination of N variables, subject to the constraint that none be negative, and that as a group they satisfy a number of other constraints. In order to clarify the subject, *Numerical Recipes* presents a sample problem in equations (10.8.6) and (10.8.7), translating the problem into tableau format in (10.8.18), and presenting a solution in equation (10.8.19). Sample program xsimplx carries out the calculations that lead to this solution.

```
/* Driver for routine simplx */

#include <stdio.h>
#include "nr.h"
#include "nrutil.h"

#define N 4
#define M 4
#define NP 5          /* NP >= N+1 */
#define MP 6          /* MP >= M+2 */
#define M1 2          /* M1+M2+M3 = M */
#define M2 1
#define M3 1
#define NM1M2 (N+M1+M2)

int main(void)
{
    int i,icase,j,*izrov,*iposv;
    static float c[MP][NP]=
        {0.0,1.0,1.0,3.0,-0.5,
        740.0,-1.0,0.0,-2.0,0.0,
        0.0,0.0,-2.0,0.0,7.0,
        0.5,0.0,-1.0,1.0,-2.0,
```

```
        9.0,-1.0,-1.0,-1.0,-1.0,
        0.0,0.0,0.0,0.0,0.0};
    float **a;
    static char *txt[NM1M2+1]=
        {" ","x1","x2","x3","x4","y1","y2","y3"};

    izrov=ivector(1,N);
    iposv=ivector(1,M);
    a=convert_matrix(&c[0][0],1,MP,1,NP);
    simplx(a,M,N,M1,M2,M3,&icase,izrov,iposv);
    if (icase == 1)
        printf("\nunbounded objective function\n");
    else if (icase == -1)
        printf("\nno solutions satisfy constraints given\n");
    else {
        printf("\n%11s"," ");
        for (i=1;i<=N;i++)
            if (izrov[i] <= NM1M2) printf("%10s",txt[izrov[i]]);
        printf("\n");
        for (i=1;i<=M+1;i++) {
            if (i == 1 || iposv[i-1] <= NM1M2) {
                if (i > 1)
                    printf("%s",txt[iposv[i-1]]);
                else
                    printf("  ");
                printf("%10.2f",a[i][1]);
                for (j=2;j<=N+1;j++)
                    if (izrov[j-1] <= NM1M2)
                        printf("%10.2f",a[i][j]);
                printf("\n");
            }
        }
    }
    free_convert_matrix(a,1,MP,1,NP);
    free_ivector(iposv,1,M);
    free_ivector(izrov,1,N);
    return 0;
}
```

anneal is a procedure for solving the travelling salesman problem — a problem that is included as a demonstration of the use of simulated annealing. Sample program xanneal has the function of setting up the initial route for the salesman and printing final results. For each of NCITY=10 cities, it chooses random coordinates x[i],y[i] using routine ran3, and puts an entry for each city in the array iptr[i]. The array indicates the order in which the cities will be visited. On the originally specified path, the cities are in the order i=1,...,10 so the sample program initially takes iptr[i]=i. (It is assumed that the salesman will return to the first city after visiting the last.) A call is then made to anneal, which attempts to find the shortest alternative route, which is recorded in the array. After finding a path that resists further improvement, the driver lists the modified itinerary.

```
/* Driver for routine anneal */

#include <stdio.h>
#include "nr.h"
#include "nrutil.h"
```

```
#define NCITY 10

int main(void)
{
    int i,ii,*iorder;
    long idum=(-111);
    float *x,*y;

    iorder=ivector(1,NCITY);
    x=vector(1,NCITY);
    y=vector(1,NCITY);
    for (i=1;i<=NCITY;i++) {
        x[i]=ran3(&idum);
        y[i]=ran3(&idum);
        iorder[i]=i;
    }
    anneal(x,y,iorder,NCITY);
    printf("*** System Frozen ***\n");
    printf("Final path:\n");
    printf("%8s %9s %12s\n","city","x","y");
    for (i=1;i<=NCITY;i++) {
        ii=iorder[i];
        printf("%4d %10.4f %10.4f\n",ii,x[ii],y[ii]);
    }
    free_vector(y,1,NCITY);
    free_vector(x,1,NCITY);
    free_ivector(iorder,1,NCITY);
    return 0;
}
```

If you have any doubts about the effectiveness of simulated annealing in dealing with otherwise intractable minimization problems with continuous control variables, the next example, using routine amebsa, ought to convince you of the merits of the technique. The 4-dimensional test function to be minimized, tfunk, is quite diabolical: It is bounded below by a parabolic function whose principal axes are in the ratio $1 : 3 : 10 : 30$, and bounded above by 3 (or more generally 1+AUG) times the lower bound. At most points, the function in fact takes on its upper bound; but within a radius 0.3 (that is, RAD) of any *integer* lattice point, it dives down to its lower bound in a narrow parabola. The narrow parabolas, in other words, form an infinite lattice of local minimum "traps," trying to capture the test point. The "surface" of this function would look like a piece of 4-dimensional Swiss cheese, scooped out into a paraboloid. Any minimization algorithm that goes strictly downhill will get captured by the holes, virtually always. Let's see how well simulated annealing does at finding the true minimum, located at the origin.

The program prompts for a starting temperature, t, and for the number of amoeba iterations, iter, that it should perform before reducing the temperature by 20%. The latter number, therefore, controls the annealing schedule. The test point is alway started at a position $\mathbf{p} = (10, 10, 10, 10)$, down at the bottom of a steep local minimum. Note how the program sets up the initial simplex and its function values, as required by the routine amebsa.

Try running the program first with t=1000, iter=20. For this low a value of t, the routine never gets out of its original local minimum, and it ends up with a function value of 101001. Next try the input values t=1 × 10⁶, iter=20. For most random seeds, the program now finds its way to the origin, with a final function value of 1, the true global minimum.

To see the effect of too rapid quenching, try t=1×10⁶, iter=2. For many random seeds, the program now gets part of the way towards the origin but then freezes into a local minimum. For other seeds, the program does find the true global minimum. You might enjoy experimenting with other input values to map out the regions of relatively secure and insecure global convergence.

```c
/* Driver for routine amebsa */

#include <stdio.h>
#include "nr.h"
#include "nrutil.h"

#define NP 4
#define MP 5
#define FTOL 1.0E-6
#define N 4
#define RAD 0.3
#define AUG 2.0

long idum=(-64);

float tfunk(float p[])
{
        int j;
        float q,r,sumd=0.0,sumr=0.0;
        static float wid[N+1]={0.0,1.0,3.0,10.0,30.0};

        for (j=1;j<=N;j++) {
            q=p[j]*wid[j];
            r=(float)(q >= 0 ? (int)(q+0.5) : -(int)(0.5-q));
            sumr += q*q;
            sumd += (q-r)*(q-r);
        }
        return 1+sumr*(1+(sumd > RAD*RAD ? AUG : AUG*sumd/(RAD*RAD)));
}

int main(void)
{
    int i,iiter,iter,j,jiter,ndim=NP,nit;
    float temptr,yb,ybb;
    float **p,*x,*y,*pb;
    static float xoff[NP+1]={0.0,10.0,10.0,10.0,10.0};

    p=matrix(1,MP,1,NP);
    x=vector(1,NP);
    y=vector(1,MP);
    pb=vector(1,NP);
    for (i=1;i<=MP;i++)
        for (j=1;j<=NP;j++) p[i][j]=0.0;
    for (;;) {
        for (j=2;j<=MP;j++) p[j][j-1]=1.0;
        for (i=1;i<=MP;i++) {
```

```
            for (j=1;j<=NP;j++) x[j]=(p[i][j] += xoff[j]);
            y[i]=tfunk(x);
        }
        yb=1.0e30;
        printf("Input t, iiter:\n");
        if (scanf("%f %d",&temptr,&iiter) == EOF) break;
        ybb=1.0e30;
        nit=0;
        for (jiter=1;jiter<=100;jiter++) {
            iter=iiter;
            temptr *= 0.8;
            amebsa(p,y,ndim,pb,&yb,FTOL,tfunk,&iter,temptr);
            nit += iiter-iter;
            if (yb < ybb) {
                ybb=yb;
                printf("%6d %10.3e ",nit,temptr);
                for (j=1;j<=NP;j++) printf("%10.5f ",pb[j]);
                printf("%15.7e\n",yb);
            }
            if (iter > 0) break;
        }
        printf("Vertices of final 3-D simplex and\n");
        printf("float values at the vertices:\n");
        printf("%3s %10s %12s %12s %14s\n\n",
            "i","x[i]","y[i]","z[i]","function");
        for (i=1;i<=MP;i++) {
            printf("%3d ",i);
            for (j=1;j<=NP;j++) printf("%12.6f ",p[i][j]);
            printf("%15.7e\n",y[i]);
        }
        printf("%3d ",99);
        for (j=1;j<=NP;j++) printf("%12.6f ",pb[j]);
        printf("%15.7e\n",yb);
    }
    free_vector(pb,1,NP);
    free_vector(y,1,MP);
    free_vector(x,1,NP);
    free_matrix(p,1,MP,1,NP);
    printf("Normal completion\n");
    return 0;
}
```

Chapter 11: Eigensystems

*In Chapter 11 of Numerical Recipes, we deal with the problem of finding
eigenvectors and eigenvalues of matrices, first dealing with symmetric matrices, and then with more general cases. For real symmetric matrices of
small-to-moderate size, the routine* jacobi *is recommended as a simple and
foolproof scheme of finding eigenvalues and eigenvectors. Routine* eigsrt *may
be used to reorder the output of* jacobi *into descending order of eigenvalue.
A more efficient (but operationally more complicated) procedure is to reduce
the symmetric matrix to tridiagonal form before doing the eigenvalue analysis.* tred2 *uses the Householder scheme to perform this reduction and is used
in conjunction with* tqli. tqli *determines the eigenvalues and eigenvectors
of a real, symmetric, tridiagonal matrix.*

*For nonsymmetric matrices, we offer only routines for finding eigenvalues,
and not eigenvectors. To ameliorate problems with roundoff error,* balanc
*makes the corresponding rows and columns of the matrix have comparable
norms while leaving eigenvalues unchanged. Then the matrix is reduced to
Hessenberg form by Gaussian elimination using* elmhes. *Finally* hqr *applies
the QR algorithm to find the eigenvalues of the Hessenberg matrix.*

$$\star \quad \star \quad \star \quad \star$$

jacobi is a reliable scheme for finding both the eigenvalues and eigenvectors of a
symmetric matrix. It is not the most efficient scheme available, but it is simple and
trustworthy, and it is recommended for problems of small-to-moderate order. Sample
program xjacobi defines three matrices a,b,c for use by jacobi. They are of order
3, 5, and 10 respectively and are, each in turn, sent to jacobi. For each matrix, the
eigenvalues and eigenvectors are reported. Then, an eigenvector test takes place in
which the original matrix is applied to the purported eigenvector, and the ratio of
the result to the vector itself is found. The ratio should, of course, be the eigenvalue.

```
/* Driver for routine jacobi */

#include <stdio.h>
#include "nr.h"
#include "nrutil.h"

#define NP 10
#define NMAT 3

int main(void)
{
    int i,j,k,kk,l,ll,nrot;
    static float a[3][3]=
```

```
            {1.0,2.0,3.0,
            2.0,2.0,3.0,
            3.0,3.0,3.0};
    static float b[5][5]=
            {-2.0,-1.0,0.0,1.0,2.0,
            -1.0,-1.0,0.0,1.0,2.0,
            0.0,0.0,0.0,1.0,2.0,
            1.0,1.0,1.0,1.0,2.0,
            2.0,2.0,2.0,2.0,2.0};
    static float c[NP][NP]=
            {5.0,4.3,3.0,2.0,1.0,0.0,-1.0,-2.0,-3.0,-4.0,
            4.3,5.1,4.0,3.0,2.0,1.0,0.0,-1.0,-2.0,-3.0,
            3.0,4.0,5.0,4.0,3.0,2.0,1.0,0.0,-1.0,-2.0,
            2.0,3.0,4.0,5.0,4.0,3.0,2.0,1.0,0.0,-1.0,
            1.0,2.0,3.0,4.0,5.0,4.0,3.0,2.0,1.0,0.0,
            0.0,1.0,2.0,3.0,4.0,5.0,4.0,3.0,2.0,1.0,
            -1.0,0.0,1.0,2.0,3.0,4.0,5.0,4.0,3.0,2.0,
            -2.0,-1.0,0.0,1.0,2.0,3.0,4.0,5.0,4.0,3.0,
            -3.0,-2.0,-1.0,0.0,1.0,2.0,3.0,4.0,5.0,4.0,
            -4.0,-3.0,-2.0,-1.0,0.0,1.0,2.0,3.0,4.0,5.0};
float *d,*r,**v,**e;
static int num[4]={0,3,5,10};

d=vector(1,NP);
r=vector(1,NP);
v=matrix(1,NP,1,NP);
for (i=1;i<=NMAT;i++) {
    if (i == 1) e=convert_matrix(&a[0][0],1,num[i],1,num[i]);
    else if (i == 2) e=convert_matrix(&b[0][0],1,num[i],1,num[i]);
    else if (i == 3) e=convert_matrix(&c[0][0],1,num[i],1,num[i]);
    jacobi(e,num[i],d,v,&nrot);
    printf("matrix number %2d\n",i);
    printf("number of JACOBI rotations: %3d\n",nrot);
    printf("eigenvalues: \n");
    for (j=1;j<=num[i];j++) {
        printf("%12.6f",d[j]);
        if ((j % 5) == 0) printf("\n");
    }
    printf("\neigenvectors:\n");
    for (j=1;j<=num[i];j++) {
        printf("%9s %3d \n","number",j);
        for (k=1;k<=num[i];k++) {
            printf("%12.6f",v[k][j]);
            if ((k % 5) == 0) printf("\n");
        }
        printf("\n");
    }
    /* eigenvector test */
    printf("eigenvector test\n");
    for (j=1;j<=num[i];j++) {
        for (l=1;l<=num[i];l++) {
            r[l]=0.0;
            for (k=1;k<=num[i];k++) {
                if (k > l) {
                    kk=l;
                    ll=k;
                } else {
```

```
                        kk=k;
                        11=1;
                    }
                    r[1] += (e[11][kk]*v[k][j]);
                }
            }
            printf("vector number %3d\n",j);
            printf("%11s %14s %10s\n",
                "vector","mtrx*vec.","ratio");
            for (1=1;1<=num[i];1++)
                printf("%12.6f %12.6f %12.6f\n",
                    v[1][j],r[1],r[1]/v[1][j]);
        }
        printf("press RETURN to continue...\n");
        (void) getchar();
        free_convert_matrix(e,1,num[i],1,num[i]);
    }
    free_matrix(v,1,NP,1,NP);
    free_vector(r,1,NP);
    free_vector(d,1,NP);
    return 0;
}
```

eigsrt reorders the output of jacobi so that the eigenvectors are in the order of decreasing eigenvalue. Sample program xeigsrt uses matrix c from the previous program to illustrate. This 10×10 matrix is passed to jacobi and the ten eigenvectors are found. They are printed, along with their eigenvalues, in the order that jacobi returns them. Then the matrices d and v from jacobi, which contain the eigenvalues and eigenvectors, are passed to eigsrt, and ought to return in descending order of eigenvalue. The result is printed for inspection.

```
/* Driver for routine eigsrt */

#include <stdio.h>
#include "nr.h"
#include "nrutil.h"

#define NP 10

int main(void)
{
    int i,j,nrot;
    static float c[NP][NP]=
        {5.0,4.3,3.0,2.0,1.0,0.0,-1.0,-2.0,-3.0,-4.0,
         4.3,5.1,4.0,3.0,2.0,1.0,0.0,-1.0,-2.0,-3.0,
         3.0,4.0,5.0,4.0,3.0,2.0,1.0,0.0,-1.0,-2.0,
         2.0,3.0,4.0,5.0,4.0,3.0,2.0,1.0,0.0,-1.0,
         1.0,2.0,3.0,4.0,5.0,4.0,3.0,2.0,1.0,0.0,
         0.0,1.0,2.0,3.0,4.0,5.0,4.0,3.0,2.0,1.0,
         -1.0,0.0,1.0,2.0,3.0,4.0,5.0,4.0,3.0,2.0,
         -2.0,-1.0,0.0,1.0,2.0,3.0,4.0,5.0,4.0,3.0,
         -3.0,-2.0,-1.0,0.0,1.0,2.0,3.0,4.0,5.0,4.0,
         -4.0,-3.0,-2.0,-1.0,0.0,1.0,2.0,3.0,4.0,5.0};
    float *d,**v,**e;

    d=vector(1,NP);
```

```
    v=matrix(1,NP,1,NP);
    e=convert_matrix(&c[0][0],1,NP,1,NP);
    printf("****** Finding Eigenvectors ******\n");
    jacobi(e,NP,d,v,&nrot);
    printf("unsorted eigenvectors:\n");
    for (i=1;i<=NP;i++) {
        printf("eigenvalue %3d = %12.6f\n",i,d[i]);
        printf("eigenvector:\n");
        for (j=1;j<=NP;j++) {
            printf("%12.6f",v[j][i]);
            if ((j % 5) == 0) printf("\n");
        }
        printf("\n");
    }
    printf("\n****** Sorting Eigenvectors ******\n\n");
    eigsrt(d,v,NP);
    printf("sorted eigenvectors:\n");
    for (i=1;i<=NP;i++) {
        printf("eigenvalue %3d = %12.6f\n",i,d[i]);
        printf("eigenvector:\n");
        for (j=1;j<=NP;j++) {
            printf("%12.6f",v[j][i]);
            if ((j % 5) == 0) printf("\n");
        }
        printf("\n");
    }
    free_convert_matrix(e,1,NP,1,NP);
    free_matrix(v,1,NP,1,NP);
    free_vector(d,1,NP);
    return 0;
}
```

tred2 reduces a real symmetric matrix to tridiagonal form. Sample program xtred2 again uses matrix c from the earlier programs, and copies it into matrix a. Matrix a is sent to tred2, while c is saved for a check of the transformation matrix that tred2 returns in a. The program prints the diagonal and off-diagonal elements of the reduced matrix. It then forms the matrix f defined by $F = A^T C A$ to prove that f is tridiagonal and that the listed diagonal and off-diagonal elements are correct.

```
/* Driver for routine tred2 */

#include <stdio.h>
#include "nr.h"
#include "nrutil.h"

#define NP 10

int main(void)
{
    int i,j,k,l,m;
    float *d,*e,**a,**f;
    static float c[NP][NP]=
        { 5.0, 4.3, 3.0, 2.0, 1.0, 0.0,-1.0,-2.0,-3.0,-4.0,
          4.3, 5.1, 4.0, 3.0, 2.0, 1.0, 0.0,-1.0,-2.0,-3.0,
          3.0, 4.0, 5.0, 4.0, 3.0, 2.0, 1.0, 0.0,-1.0,-2.0,
          2.0, 3.0, 4.0, 5.0, 4.0, 3.0, 2.0, 1.0, 0.0,-1.0,
```

```
          1.0, 2.0, 3.0, 4.0, 5.0, 4.0, 3.0, 2.0, 1.0, 0.0,
          0.0, 1.0, 2.0, 3.0, 4.0, 5.0, 4.0, 3.0, 2.0, 1.0,
         -1.0, 0.0, 1.0, 2.0, 3.0, 4.0, 5.0, 4.0, 3.0, 2.0,
         -2.0,-1.0, 0.0, 1.0, 2.0, 3.0, 4.0, 5.0, 4.0, 3.0,
         -3.0,-2.0,-1.0, 0.0, 1.0, 2.0, 3.0, 4.0, 5.0, 4.0,
         -4.0,-3.0,-2.0,-1.0, 0.0, 1.0, 2.0, 3.0, 4.0, 5.0};

    d=vector(1,NP);
    e=vector(1,NP);
    a=matrix(1,NP,1,NP);
    f=matrix(1,NP,1,NP);
    for (i=1;i<=NP;i++)
        for (j=1;j<=NP;j++) a[i][j]=c[i-1][j-1];
    tred2(a,NP,d,e);
    printf("diagonal elements\n");
    for (i=1;i<=NP;i++) {
        printf("%12.6f",d[i]);
        if ((i % 5) == 0) printf("\n");
    }
    printf("off-diagonal elements\n");
    for (i=2;i<=NP;i++) {
        printf("%12.6f",e[i]);
        if ((i % 5) == 0) printf("\n");
    }
    /* Check transformation matrix */
    for (j=1;j<=NP;j++) {
        for (k=1;k<=NP;k++) {
            f[j][k]=0.0;
            for (l=1;l<=NP;l++) {
                for (m=1;m<=NP;m++)
                    f[j][k] += a[l][j]*c[l-1][m-1]*a[m][k];
            }
        }
    }
    /* How does it look? */
    printf("tridiagonal matrix\n");
    for (i=1;i<=NP;i++) {
        for (j=1;j<=NP;j++) printf("%7.2f",f[i][j]);
        printf("\n");
    }
    free_matrix(f,1,NP,1,NP);
    free_matrix(a,1,NP,1,NP);
    free_vector(e,1,NP);
    free_vector(d,1,NP);
    return 0;
}
```

tqli finds the eigenvectors and eigenvalues for a real, symmetric, tridiagonal matrix. Sample program xtqli operates with matrix c again, and uses tred2 to reduce it to tridiagonal form as before. More specifically, c is copied into matrix a, which is sent to tred2. From tred2 come two vectors d,e which are the diagonal and subdiagonal elements of the tridiagonal matrix. d and e are made arguments of tqli, as is a, the returned transformation matrix from tred2. On output from tqli, d is replaced with eigenvalues, and a with corresponding eigenvectors. These are checked as in the program for jacobi. That is, the original matrix c is applied to each eigenvector, and the result is divided (element by element) by the eigenvector. Look for a result equal to the eigenvalue. (Note: in some cases, the vector element is zero or nearly so. These cases are flagged with the words "div. by zero".)

```c
/* Driver for routine tqli */

#include <stdio.h>
#include <math.h>
#include "nr.h"
#include "nrutil.h"

#define NP 10
#define TINY 1.0e-6

int main(void)
{
    int i,j,k;
    float *d,*e,*f,**a;
    static float c[NP][NP]=
        { 5.0, 4.3, 3.0, 2.0, 1.0, 0.0,-1.0,-2.0,-3.0,-4.0,
          4.3, 5.1, 4.0, 3.0, 2.0, 1.0, 0.0,-1.0,-2.0,-3.0,
          3.0, 4.0, 5.0, 4.0, 3.0, 2.0, 1.0, 0.0,-1.0,-2.0,
          2.0, 3.0, 4.0, 5.0, 4.0, 3.0, 2.0, 1.0, 0.0,-1.0,
          1.0, 2.0, 3.0, 4.0, 5.0, 4.0, 3.0, 2.0, 1.0, 0.0,
          0.0, 1.0, 2.0, 3.0, 4.0, 5.0, 4.0, 3.0, 2.0, 1.0,
         -1.0, 0.0, 1.0, 2.0, 3.0, 4.0, 5.0, 4.0, 3.0, 2.0,
         -2.0,-1.0, 0.0, 1.0, 2.0, 3.0, 4.0, 5.0, 4.0, 3.0,
         -3.0,-2.0,-1.0, 0.0, 1.0, 2.0, 3.0, 4.0, 5.0, 4.0,
         -4.0,-3.0,-2.0,-1.0, 0.0, 1.0, 2.0, 3.0, 4.0, 5.0};

    d=vector(1,NP);
    e=vector(1,NP);
    f=vector(1,NP);
    a=matrix(1,NP,1,NP);
    for (i=1;i<=NP;i++)
        for (j=1;j<=NP;j++) a[i][j]=c[i-1][j-1];
    tred2(a,NP,d,e);
    tqli(d,e,NP,a);
    printf("\nEigenvectors for a real symmetric matrix\n");
    for (i=1;i<=NP;i++) {
        for (j=1;j<=NP;j++) {
            f[j]=0.0;
            for (k=1;k<=NP;k++)
                f[j] += (c[j-1][k-1]*a[k][i]);
        }
        printf("%s %3d %s %10.6f\n","eigenvalue",i," =",d[i]);
        printf("%11s %14s %9s\n","vector","mtrx*vect.","ratio");
        for (j=1;j<=NP;j++) {
            if (fabs(a[j][i]) < TINY)
                printf("%12.6f %12.6f %12s\n",
                    a[j][i],f[j],"div. by 0");
            else
                printf("%12.6f %12.6f %12.6f\n",
                    a[j][i],f[j],f[j]/a[j][i]);
        }
        printf("Press ENTER to continue...\n");
        (void) getchar();
```

```
    }
    free_matrix(a,1,NP,1,NP);
    free_vector(f,1,NP);
    free_vector(e,1,NP);
    free_vector(d,1,NP);
    return 0;
}
```

balanc reduces error in eigenvalue problems involving non-symmetric matrices. It does this by adjusting corresponding rows and columns to have comparable norms, without changing eigenvalues. Sample program xbalanc prepares the following array a for balanc

$$\begin{pmatrix} 1 & 100 & 1 & 100 & 1 \\ 1 & 1 & 1 & 1 & 1 \\ 1 & 100 & 1 & 100 & 1 \\ 1 & 1 & 1 & 1 & 1 \\ 1 & 100 & 1 & 100 & 1 \end{pmatrix}$$

The norms of the five rows and five columns are printed out. It is clear from the array that three of the rows and two of the columns have much larger norms than the others. After balancing with balanc, the norms are recalculated, and this time the row, column pairs should be much more nearly equal.

```
/* Driver for routine balanc */

#include <stdio.h>
#include <math.h>
#include "nr.h"
#include "nrutil.h"

#define NP 5

int main(void)
{
    int i,j;
    float *c,*r,**a;

    c=vector(1,NP);
    r=vector(1,NP);
    a=matrix(1,NP,1,NP);
    for (i=1;i<=NP;i++)
        for (j=1;j<=NP;j++)
            a[i][j] = (((i & 1) && !(j & 1)) ? 100.0 : 1.0);
    /* Write norms */
    for (i=1;i<=NP;i++) {
        r[i]=c[i]=0.0;
        for (j=1;j<=NP;j++) {
            r[i] += fabs(a[i][j]);
            c[i] += fabs(a[j][i]);
        }
    }
    printf("rows:\n");
    for (i=1;i<=NP;i++) printf("%12.2f",r[i]);
    printf("\ncolumns:\n");
    for (i=1;i<=NP;i++) printf("%12.2f",c[i]);
```

```
printf("\n\n***** Balancing matrix *****\n\n");
balanc(a,NP);
/* Write norms */
for (i=1;i<=NP;i++) {
    r[i]=c[i]=0.0;
    for (j=1;j<=NP;j++) {
        r[i] += fabs(a[i][j]);
        c[i] += fabs(a[j][i]);
    }
}
printf("rows:\n");
for (i=1;i<=NP;i++) printf("%12.2f",r[i]);
printf("\ncolumns:\n");
for (i=1;i<=NP;i++) printf("%12.2f",c[i]);
printf("\n");
free_matrix(a,1,NP,1,NP);
free_vector(r,1,NP);
free_vector(c,1,NP);
return 0;
}
```

elmhes reduces a general matrix to Hessenberg form using Gaussian elimination. It is particularly valuable for real, non-symmetric matrices. Sample program xelmhes employs balanc and elmhes to get a non-symmetric and grossly unbalanced matrix into Hessenberg form. The matrix a is

$$\begin{pmatrix} 1 & 2 & 300 & 4 & 5 \\ 2 & 3 & 400 & 5 & 6 \\ 3 & 4 & 5 & 6 & 7 \\ 4 & 5 & 600 & 7 & 8 \\ 5 & 6 & 700 & 8 & 9 \end{pmatrix}$$

After printing the original matrix, the program feeds it to balanc and prints the balanced version. This is submitted to elmhes and the result is printed. Notice that the elements of a with $i>j+1$ are all set to zero by the program, because elmhes returns random values in this part of the matrix. Therefore, the output of the test program is guaranteed to *look* Hessenberg. To check that it is the *correct* Hessenberg matrix, one has to go further and compute its eigenvalues, as will be done in the example program after this one. We include here the expected results for comparison.

Balanced Matrix:

1.00	2.00	37.50	4.00	5.00
2.00	3.00	50.00	5.00	6.00
24.00	32.00	5.00	48.00	56.00
4.00	5.00	75.00	7.00	8.00
5.00	6.00	87.50	8.00	9.00

Reduced to Hessenberg Form:

1.00e+00	3.94e+01	9.62e+00	3.33e+00	4.00e+00
2.40e+01	2.73e+01	1.16e+02	4.80e+01	4.80e+01
0.00e+00	8.55e+01	-4.78e+00	-1.33e+00	-2.00e+00
0.00e+00	0.00e+00	5.19e+00	1.45e+00	2.17e+00
0.00e+00	0.00e+00	0.00e+00	6.95e-08	4.77e-07

```
/* Driver for routine elmhes */

#include <stdio.h>
#include "nr.h"
#include "nrutil.h"

#define NP 5

int main(void)
{
    int i,j;
    static float b[NP][NP]=
        {1.0,2.0,300.0,4.0,5.0,
        2.0,3.0,400.0,5.0,6.0,
        3.0,4.0,5.0,6.0,7.0,
        4.0,5.0,600.0,7.0,8.0,
        5.0,6.0,700.0,8.0,9.0};
    float **a;

    a=convert_matrix(&b[0][0],1,NP,1,NP);
    printf("***** original matrix *****\n");
    for (i=1;i<=NP;i++) {
        for (j=1;j<=NP;j++) printf("%12.2f",a[i][j]);
        printf("\n");
    }
    printf("***** balance matrix *****\n");
    balanc(a,NP);
    for (i=1;i<=NP;i++) {
        for (j=1;j<=NP;j++) printf("%12.2f",a[i][j]);
        printf("\n");
    }
    printf("***** reduce to hessenberg form *****\n");
    elmhes(a,NP);
    for (j=1;j<=NP-2;j++)
        for (i=j+2;i<=NP;i++)
            a[i][j]=0.0;
    for (i=1;i<=NP;i++) {
        for (j=1;j<=NP;j++) printf("%12.2e",a[i][j]);
        printf("\n");
    }
    free_convert_matrix(a,1,NP,1,NP);
    return 0;
}
```

hqr, finally, is a routine for finding the eigenvalues of a Hessenberg matrix using the QR algorithm. The 5×5 matrix a specified in the array a is treated just as you would expect to treat any general real non-symmetric matrix. It is fed to balanc for balancing, to elmhes for reduction to Hessenberg form, and to hqr for eigenvalue determination. The eigenvalues may be complex-valued, and both real and imaginary parts are given. The original matrix has enough strategically placed zeros in it that you should have no trouble finding the eigenvalues by hand. Alternatively, you may check them against the list below:

Matrix:

1.00	2.00	.00	.00	.00

```
     -2.00        3.00         .00         .00         .00
      3.00        4.00       50.00         .00         .00
     -4.00        5.00      -60.00        7.00         .00
     -5.00        6.00      -70.00        8.00       -9.00
```

Eigenvalues:

```
        real            imag.
     50.000000        0.000000
      2.000000       -1.732052
      2.000000        1.732052
      6.999999        0.000000
     -9.000000        0.000000
```

```c
/* Driver for routine hqr */

#include <stdio.h>
#include "nr.h"
#include "nrutil.h"

#define NP 5

int main(void)
{
    int i,j;
    static float c[NP][NP]=
        {1.0,2.0,0.0,0.0,0.0,
         -2.0,3.0,0.0,0.0,0.0,
         3.0,4.0,50.0,0.0,0.0,
         -4.0,5.0,-60.0,7.0,0.0,
         -5.0,6.0,-70.0,8.0,-9.0};
    float *wr,*wi,**a;

    wr=vector(1,NP);
    wi=vector(1,NP);
    a=convert_matrix(&c[0][0],1,NP,1,NP);
    printf("matrix:\n");
    for (i=1;i<=NP;i++) {
        for (j=1;j<=NP;j++) printf("%12.2f",a[i][j]);
        printf("\n");
    }
    balanc(a,NP);
    elmhes(a,NP);
    hqr(a,NP,wr,wi);
    printf("eigenvalues:\n");
    printf("%11s %16s \n","real","imag.");
    for (i=1;i<=NP;i++) printf("%15f %14f\n",wr[i],wi[i]);
    free_convert_matrix(a,1,NP,1,NP);
    free_vector(wi,1,NP);
    free_vector(wr,1,NP);
    return 0;
}
```

Chapter 12: Fast Fourier Transform

Chapter 12 of Numerical Recipes deals with the fast Fourier transform (FFT). Routine four1 *performs the FFT on a complex data array.* twofft *does the same transform on two real-valued data arrays (at the same time) and returns two complex-valued transforms. Finally,* realft *finds the Fourier transform of a single real-valued array. Two related transforms are the sine transform and the cosine transform, which comes in two incarnations. The corresponding routines are* sinft, cosft1 *and* cosft2. *For FFTs in two or more dimensions the routine* fourn *is supplied. Real data in two or three dimensions is more efficiently handled by* rlft3, *while huge data sets on external media can be FFT'd by* fourfs.

$$\star \quad \star \quad \star \quad \star$$

Routine four1 performs the fast Fourier transform on a complex-valued array of data points. Example program xfour1 has five tests for this transform. First, it checks the following four symmetries (where $h(t)$ is the data and $H(n)$ is the transform):

1. If $h(t)$ is real-valued and even, then $H(n) = H(N - n)$ and H is real.

2. If $h(t)$ is imaginary-valued and even, then $H(n) = H(N-n)$ and H is imaginary.

3. If $h(t)$ is real-valued and odd, then $H(n) = -H(N - n)$ and H is imaginary.

4. If $h(t)$ is imaginary-valued and odd, then $H(n) = -H(N-n)$ and H is real.

The fifth test is that if a data array is Fourier transformed twice in succession, the resulting array should be identical to the original.

```
/* Driver for routine four1 */

#include <stdio.h>
#include <math.h>
#include "nr.h"
#include "nrutil.h"

void prntft(float data[],unsigned long nn)
{
    unsigned long n;

    printf("%4s %13s %13s %12s %13s\n",
        "n","real(n)","imag.(n)","real(N-n)","imag.(N-n)");
    printf("   0 %14.6f %12.6f %12.6f %12.6f\n",
        data[1],data[2],data[1],data[2]);
    for (n=3;n<=nn+1;n+=2) {
        printf("%4lu %14.6f %12.6f %12.6f %12.6f\n",
```

```
                ((n-1)/2),data[n],data[n+1],
                data[2*nn+2-n],data[2*nn+3-n]);
    }
    printf(" press return to continue ...\n");
    (void) getchar();
    return;
}

#define NN 32
#define NN2 (2*NN)

int main(void)
{
    long i;
    int isign;
    float *data,*dcmp;

    data=vector(1,NN2);
    dcmp=vector(1,NN2);
    printf("h(t)=real-valued even-function\n");
    printf("h(n)=h(N-n) and real?\n");
    for (i=1;i<NN2;i+=2) {
        data[i]=1.0/(SQR((float) (i-NN-1)/NN)+1.0);
        data[i+1]=0.0;
    }
    isign=1;
    four1(data,NN,isign);
    prntft(data,NN);
    printf("h(t)=imaginary-valued even-function\n");
    printf("h(n)=h(N-n) and imaginary?\n");
    for (i=1;i<NN2;i+=2) {
        data[i+1]=1.0/(SQR((float) (i-NN-1)/NN)+1.0);
        data[i]=0.0;
    }
    isign=1;
    four1(data,NN,isign);
    prntft(data,NN);
    printf("h(t)=real-valued odd-function\n");
    printf("h(n) = -h(N-n) and imaginary?\n");
    for (i=1;i<NN2;i+=2) {
        data[i]=((float) (i-NN-1)/NN)/(SQR((float) (i-NN-1)/NN)+1.0);
        data[i+1]=0.0;
    }
    data[1]=0.0;
    isign=1;
    four1(data,NN,isign);
    prntft(data,NN);
    printf("h(t)=imaginary-valued odd-function\n");
    printf("h(n) = -h(N-n) and real?\n");
    for (i=1;i<NN2;i+=2) {
        data[i+1]=((float) (i-NN-1)/NN)/(SQR((float) (i-NN-1)/NN)+1.0);
        data[i]=0.0;
    }
    data[2]=0.0;
    isign=1;
    four1(data,NN,isign);
    prntft(data,NN);
```

```
      /* transform, inverse-transform test */
      for (i=1;i<NN2;i+=2) {
          data[i]=1.0/(SQR(0.5*(i-NN-1.0)/NN)+1.0);
          dcmp[i]=data[i];
          data[i+1]=(0.25*(i-NN-1.0)/NN)*exp(-SQR(0.5*(i-NN-1)/NN));
          dcmp[i+1]=data[i+1];
      }
      isign=1;
      four1(data,NN,isign);
      isign = -1;
      four1(data,NN,isign);
      printf("%23s %33s \n","original data:","double fourier transform:");
      printf("\n %3s %15s %12s %12s %12s \n",
          "k","real h(k)","imag h(k)","real h(k)","imag h(k)");
      for (i=1;i<NN;i+=2)
          printf("%4lu %14.6f %12.6f %12.6f %12.6f\n",
              (i+1)/2,dcmp[i],dcmp[i+1],data[i]/NN,data[i+1]/NN);
      free_vector(dcmp,1,NN2);
      free_vector(data,1,NN2);
      return 0;
}
```

twofft is a routine that performs an efficient FFT of two real arrays at once by packing them into a complex array and transforming with four1. Sample program xtwofft generates two periodic data sets, out of phase with one another, and performs a transform and an inverse transform on each. It will be difficult to judge whether the transform itself gives the right answer, but if the inverse transform gets you back to the easily recognized original, you may be fairly confident that the routine works.

```
/* Driver for routine twofft */

#include <stdio.h>
#include <math.h>
#include "nr.h"
#include "nrutil.h"

#define N 32
#define N2 (2*N)
#define PER 8
#define PI 3.1415926

void prntft(float data[],unsigned long nn)
{
    unsigned long n;

    printf("%4s %13s %13s %12s %13s\n",
        "n","real(n)","imag.(n)","real(N-n)","imag.(N-n)");
    printf("  0 %14.6f %12.6f %12.6f %12.6f\n",
        data[1],data[2],data[1],data[2]);
    for (n=3;n<=nn+1;n+=2) {
        printf("%4lu %14.6f %12.6f %12.6f %12.6f\n",
            ((n-1)/2),data[n],data[n+1],
            data[2*nn+2-n],data[2*nn+3-n]);
    }
    printf(" press return to continue ...\n");
    (void) getchar();
```

```
    return;
}

int main(void)
{
    unsigned long i;
    int isign;
    float *data1,*data2,*fft1,*fft2;

    data1=vector(1,N);
    data2=vector(1,N);
    fft1=vector(1,N2);
    fft2=vector(1,N2);
    for (i=1;i<=N;i++) {
        data1[i]=floor(0.5+cos(i*2.0*PI/PER));
        data2[i]=floor(0.5+sin(i*2.0*PI/PER));
    }
    twofft(data1,data2,fft1,fft2,N);
    printf("Fourier transform of first function:\n");
    prntft(fft1,N);
    printf("Fourier transform of second function:\n");
    prntft(fft2,N);
    /* Invert transform */
    isign = -1;
    four1(fft1,N,isign);
    printf("inverted transform  =  first function:\n");
    prntft(fft1,N);
    four1(fft2,N,isign);
    printf("inverted transform  =  second function:\n");
    prntft(fft2,N);
    free_vector(fft2,1,N2);
    free_vector(fft1,1,N2);
    free_vector(data2,1,N);
    free_vector(data1,1,N);
    return 0;
}
```

realft performs the Fourier transform of a single real-valued data array. Sample routine xrealft takes this function to be sinusoidal, and allows you to choose the period. After transforming, it simply plots the magnitude of each element of the transform. If the period you choose is a power of two, the transform will be nonzero in a single bin; otherwise there will be leakage to adjacent channels. xrealft follows every transform by an inverse transform to make sure the original function is recovered.

```
/* Driver for routine realft */

#include <stdio.h>
#include <math.h>
#include "nr.h"
#include "nrutil.h"

#define EPS 1.0e-3
#define NP 32
#define WIDTH 50.0
#define PI 3.1415926
```

```
int main(void)
{
    int i,j,n=NP/2,nlim;
    float big,per,scal,small,*data,*size;

    data=vector(1,NP);
    size=vector(1,NP/2+1);
    for (;;) {
        printf("Period of sinusoid in channels (2-%2d)\n",NP);
        scanf("%f",&per);
        if (per <= 0.0) break;
        for (i=1;i<=NP;i++)
            data[i]=cos(2.0*PI*(i-1)/per);
        realft(data,NP,1);
        big = -1.0e10;
        for (i=2;i<=n;i++) {
            size[i]=sqrt(SQR(data[2*i-1])+SQR(data[2*i]));
            if (size[i] > big) big=size[i];
        }
        size[1]=fabs(data[1]);
        if (size[1] > big) big=size[1];
        size[n+1]=fabs(data[2]);
        if (size[n+1] > big) big=size[n+1];
        scal=WIDTH/big;
        for (i=1;i<=n;i++) {
            nlim=(int) (0.5+scal*size[i]+EPS);
            printf("%4d ",i);
            for (j=1;j<=nlim+1;j++) printf("*");
            printf("\n");
        }
        printf("press RETURN to continue ...\n");
        (void) getchar();
        realft(data,NP,-1);
        big = -1.0e10;
        small=1.0e10;
        for (i=1;i<=NP;i++) {
            if (data[i] < small) small=data[i];
            if (data[i] > big) big=data[i];
        }
        scal=WIDTH/(big-small);
        for (i=1;i<=NP;i++) {
            nlim=(int) (0.5+scal*(data[i]-small)+EPS);
            printf("%4d ",i);
            for (j=1;j<=nlim+1;j++) printf("*");
            printf("\n");
        }
    }
    free_vector(size,1,NP/2+1);
    free_vector(data,1,NP);
    return 0;
}
```

sinft performs a sine-transform of a real-valued array. The necessity for such a transform arises in solution methods for partial differential equations with certain kinds of boundary conditions (see Chapter 19). The sample program xsinft works exactly as the previous program. Notice that in this program no distinction needs to be made between the transform and its inverse. They are identical.

```
/* Driver for routine sinft */

#include <stdio.h>
#include <math.h>
#include "nr.h"
#include "nrutil.h"

#define EPS 1.0e-3
#define NP 16
#define WIDTH 30.0
#define PI 3.1415926

int main(void)
{
    float big,per,scal,small,*data;
    int i,j,nlim;

    data=vector(1,NP);
    for (;;) {
        printf("\nPeriod of sinusoid in channels (3-%2d)\n",NP);
        scanf("%f",&per);
        if (per <= 0.0) break;
        for (i=1;i<=NP;i++)
            data[i]=sin(2.0*PI*(i-1)/per);
        sinft(data,NP);
        big = -1.0e10;
        small=1.0e10;
        for (i=1;i<=NP;i++) {
            if  (data[i] < small) small=data[i];
            if  (data[i] > big) big=data[i];
        }
        scal=WIDTH/(big-small);
        for (i=1;i<=NP;i++) {
            nlim=(int) (scal*(data[i]-small)+EPS+0.5);
            printf("%4d ",i);
            for (j=1;j<=nlim+1;j++) printf("*");
            printf("\n");
        }
        printf("press RETURN to continue ...\n");
        (void) getchar();
        sinft(data,NP);
        big = -1.0e10;
        small=1.0e10;
        for (i=1;i<=NP;i++) {
            if (data[i] < small) small=data[i];
            if (data[i] > big) big=data[i];
        }
        scal=WIDTH/(big-small);
        for (i=1;i<=NP;i++) {
            nlim=(int) (scal*(data[i]-small)+EPS+0.5);
            printf("%4d ",i);
            for (j=1;j<=nlim+1;j++) printf("*");
            printf("\n");
        }
```

```
    }
    free_vector(data,1,NP);
    return 0;
}
```

cosft1 is a companion procedure to sinft that does the cosine transform. It also plays a role in partial differential equation solutions. Program xcosft1 is again the same as xrealft.

```
/* Driver for routine cosft1 */

#include <stdio.h>
#include <math.h>
#include "nr.h"
#include "nrutil.h"

#define EPS 1.0e-3
#define NP 17
#define WIDTH 30.0
#define PI 3.1415926

int main(void)
{
    int i,j,nlim;
    float big,per,scal,small,*data;

    data=vector(1,NP);
    for (;;) {
        printf("Period of cosine in channels (2-%2d)\n",NP);
        scanf("%f",&per);
        if (per <= 0.0) break;
        for (i=1;i<=NP;i++)
            data[i]=cos(2.0*PI*(i-1)/per);
        cosft1(data,NP-1);
        big = -1.0e10;
        small=1.0e10;
        for (i=1;i<=NP;i++) {
            if (data[i] < small) small=data[i];
            if (data[i] > big) big=data[i];
        }
        scal=WIDTH/(big-small);
        for (i=1;i<=NP;i++) {
            nlim=(int) (0.5+scal*(data[i]-small)+EPS);
            printf("%4d %6.2f ",i,data[i]);
            for (j=1;j<=nlim+1;j++) printf("*");
            printf("\n");
        }
        printf("press RETURN to continue ...\n");
        (void) getchar();
        cosft1(data,NP-1);
        big = -1.0e10;
        small=1.0e10;
        for (i=1;i<=NP;i++) {
            if (data[i] < small) small=data[i];
            if (data[i] > big) big=data[i];
        }
        scal=WIDTH/(big-small);
```

```
        for (i=1;i<=NP;i++) {
            nlim=(int) (0.5+scal*(data[i]-small)+EPS);
            printf("%4d ",i);
            for (j=1;j<=nlim+1;j++) printf("*");
            printf("\n");
        }
    }
    free_vector(data,1,NP);
    return 0;
}
```

cosft2 carries out a cosine transform on a "staggered" grid. Its demonstration program, xcosft2, is again the same as xrealft, except that the data array is filled on the staggered grid points.

```
/* Driver for routine cosft2 */

#include <stdio.h>
#include <math.h>
#include "nr.h"
#include "nrutil.h"

#define EPS 1.0e-3
#define NP 16
#define WIDTH 30.0
#define PI 3.1415926

int main(void)
{
    int i,j,nlim;
    float big,per,scal,small,*data;

    data=vector(1,NP);
    for (;;) {
        printf("Period of cosine in channels (2-%2d)\n",NP);
        scanf("%f",&per);
        if (per <= 0.0) break;
        for (i=1;i<=NP;i++)
            data[i]=cos(2.0*PI*(i-0.5)/per);
        cosft2(data,NP,1);
        big = -1.0e10;
        small=1.0e10;
        for (i=1;i<=NP;i++) {
            if (data[i] < small) small=data[i];
            if (data[i] > big) big=data[i];
        }
        scal=WIDTH/(big-small);
        for (i=1;i<=NP;i++) {
            nlim=(int) (0.5+scal*(data[i]-small)+EPS);
            printf("%4d %6.2f ",i,data[i]);
            for (j=1;j<=nlim+1;j++) printf("*");
            printf("\n");
        }
        printf("press RETURN to continue ...\n");
        (void) getchar();
        cosft2(data,NP,-1);
        big = -1.0e10;
```

```
            small=1.0e10;
            for (i=1;i<=NP;i++) {
                if (data[i] < small) small=data[i];
                if (data[i] > big) big=data[i];
            }
            scal=WIDTH/(big-small);
            for (i=1;i<=NP;i++) {
                nlim=(int) (0.5+scal*(data[i]-small)+EPS);
                printf("%4d ",i);
                for (j=1;j<=nlim+1;j++) printf("*");
                printf("\n");
            }
        }
        free_vector(data,1,NP);
        return 0;
    }
```

fourn is a routine for performing N-dimensional Fourier transforms. We have used it in sample program xfourn to transform a 3-dimensional complex data array of dimensions $4 \times 8 \times 16$. The function analyzed is simply an array of random values. The test conducted here is to perform a 3-dimensional transform and inverse transform in succession, and to compare the result with the original array. Ratios are provided for convenience. In the program, the 3-dimensional complex array is stored in a 1-dimensional real array data2 as described in *Numerical Recipes*. The array data1 is a saved copy of data2 for the comparison at the end.

```
/* Driver for routine fourn */

#include <stdio.h>
#include "nr.h"
#include "nrutil.h"

#define NDIM 3
#define NDAT2 1024

int main(void)
{
    int isign;
    long idum=(-23);
    unsigned long i,j,k,l,ndum=2,*nn;
    float *data1,*data2;

    nn=lvector(1,NDIM);
    data1=vector(1,NDAT2);
    data2=vector(1,NDAT2);
    for (i=1;i<=NDIM;i++) nn[i]=(ndum <<= 1);
    for (i=1;i<=nn[3];i++)
        for (j=1;j<=nn[2];j++)
            for (k=1;k<=nn[1];k++) {
                l=k+(j-1)*nn[1]+(i-1)*nn[2]*nn[1];
                l=(l<<1)-1;
                /* real part of component */
                data2[l]=data1[l]=2*ran1(&idum)-1;
                /* imaginary part of component */
                l++;
                data2[l]=data1[l]=2*ran1(&idum)-1;
```

```
          }
      isign=1;
      fourn(data2,nn,NDIM,isign);
      /* here would be any processing to be done in Fourier space */
      isign = -1;
      fourn(data2,nn,NDIM,isign);
      printf("Double 3-dimensional transform\n\n");
      printf("%22s %24s %20s\n",
          "Double transf.","Original data","Ratio");
      printf("%10s %13s %12s %13s %11s %13s\n\n",
          "real","imag.","real","imag.","real","imag.");
      for (i=1;i<=4;i++) {
          k=2*(j=2*i);
          l=k+(j-1)*nn[1]+(i-1)*nn[2]*nn[1];
          l=(l<<1)-1;
          printf("%12.2f %12.2f %10.2f %12.2f %14.2f %12.2f\n",
              data2[l],data2[l+1],data1[l],data1[l+1],
              data2[l]/data1[l],data2[l+1]/data1[l+1]);
      }
      printf("\nThe product of transform lengths is: %4lu\n",nn[1]*nn[2]*nn[3]);
      free_vector(data2,1,NDAT2);
      free_vector(data1,1,NDAT2);
      free_lvector(nn,1,NDIM);
      return 0;
}
```

When your multidimensional data consists of real, rather than complex, values, it is faster (and more efficient in memory use) to use an FFT routine that is optimized for real data. rlft3 is such a routine. The main book already gives, at least in outline, three sample programs using this routine. Here we content ourselves with an example that simply fills a three dimensional real array with random values, performs the Fourier transform followed by its inverse, and then verifies that (up to a known constant multiplicative constant) the original matrix is recovered.

The function compare is a simple utility for comparing all the values of an array for equality with another array to within a certain tolerance (allowing for roundoff errors). If the tolerance 0.0008 looks large to you, it is because we are comparing numbers whose magnitude is as large as $\sim 2 \times 10^3$, so the fractional tolerance is actually parts in 10^{-7}

```
/* Driver for routine rlft3 */

#include <stdio.h>
#include <math.h>
#include "nr.h"
#include "nrutil.h"
#define IPRNT 20

static unsigned long compare(char *string, float ***arr1, float ***arr2,
    unsigned long len1, unsigned long len2, unsigned long len3, float eps)
{
    unsigned long err=0,i1,i2,i3;
    float a1,a2;

    printf("%s\n",string);
    for (i1=1;i1<=len1;i1++)
```

```
            for (i2=1;i2<=len2;i2++)
                for (i3=1;i3<=len3;i3++) {
                    a1=arr1[i1][i2][i3];
                    a2=arr2[i1][i2][i3];
                    if ((a2 == 0.0 && fabs(a1-a2) > eps) || (fabs((a1-a2)/a2) > eps)) {
                        if (++err <= IPRNT)
                            printf("%d %d %d %12.6f %12.6f\n",
                                    i1,i2,i3,a1,a2);
                    }
                }
    return err;
}

#define NX 32
#define NY 8
#define NZ 16

#define EPS 0.0008

int main(void)
{
    long idum=(-3);
    unsigned long err,i,j,k,l,nn1=NX,nn2=NY,nn3=NZ;
    float fnorm,***data1,***data2,**speq1;

    fnorm=(float)nn1*(float)nn2*(float)nn3/2.0;
    data1=f3tensor(1,nn1,1,nn2,1,nn3);
    data2=f3tensor(1,nn1,1,nn2,1,nn3);
    speq1=matrix(1,nn1,1,nn2<<1);
    for (i=1;i<=nn1;i++)
        for (j=1;j<=nn2;j++)
            for (k=1;k<=nn3;k++)
                data2[i][j][k]=fnorm*(data1[i][j][k]=2*ran1(&idum)-1);
    rlft3(data1,speq1,nn1,nn2,nn3,1);
    /* here would be any processing in Fourier space */
    rlft3(data1,speq1,nn1,nn2,nn3,-1);
    err=compare("data",data1,data2,nn1,nn2,nn3,EPS);
    if (err) {
        printf("Comparison error at tolerance %12.6f\n",EPS);
        printf("Total number of errors is %d\n",err);
    }
    else {
        printf("Data compares OK to tolerance %12.6f\n",EPS);
    }
    free_matrix(speq1,1,nn1,1,nn2<<1);
    free_f3tensor(data2,1,nn1,1,nn2,1,nn3);
    free_f3tensor(data1,1,nn1,1,nn2,1,nn3);
    return (err > 0);
}
```

If your need to take the Fourier transform of a data set that is simply too long to hold in your computer's memory, but which can be read from an external medium such as a tape or disk, then the routine fourfs can be used. The following example program creates a "pretend" such data set corresponding to a complex, 3-dimensional array (it actually fits perfectly well into the arrays data1 and data2), writes its first half to one sequential file and its second half to another. fourfs is then called to take the the Fourier transform of the sequential files. To demonstrate that the correct answer is obtained, we use the in-memory routine `fourn` to take the same transform, and compare the answers.

Note that, for multidimensional arrays, the output of `fourfs` is the transpose of that from `fourn`. This is taken into account when the comparison is made.

The example program then inverts the Fourier transform. Since we are starting with the transpose of the inverse, the components of the array `idim` must be reversed for `fourfs`, but restored for `fourn`. A final comparison of values is made.

```c
/* Driver for routine fourfs */

#include <stdio.h>
#include <math.h>
#include "nr.h"
#include "nrutil.h"

#define SWAP(a,b) {iswap=(a);(a)=(b);(b)=iswap;}
#define FSWAP(a,b) {flswap=(a);(a)=(b);(b)=flswap;}

#define NX 8
#define NY 32
#define NZ 4
#define NDAT (2*NX*NY*NZ)

#define TEMPFILE1 "frfstmp1"
#define TEMPFILE2 "frfstmp2"
#define TEMPFILE3 "frfstmp3"
#define TEMPFILE4 "frfstmp4"

#if defined(MSDOS) || defined(_MSDOS) || defined(_MSDOS_) || defined(__MSDOS__)
#define BINREADWRITE "wb+"
#else
#define BINREADWRITE "w+"
#endif

char *fnames[4]={TEMPFILE1,TEMPFILE2,TEMPFILE3,TEMPFILE4};

int main(void)
{
    int cc,nnv=3,nwrite;
    long idum=(-23);
    unsigned long dim[4],i,j,k,l,ll,iswap;
    float diff,smax,sum,sum1=0.0,sum2=0.0,tot,*data1,*data2,*data1p,*data2p;
    FILE *flswap,*file[5];

    data1=vector(1,NDAT);
    data2=vector(1,NDAT);
    dim[1]=NX;
    dim[2]=NY;
    dim[3]=NZ;
    tot=(float)NX*(float)NY*(float)NZ;
    for (j=1;j<=4;j++)
        if ((file[j]=fopen(fnames[j-1],BINREADWRITE)) == NULL)
            nrerror("Couldn't open temporary file");
    for (i=1;i<=dim[3];i++)
```

```
      for (j=1;j<=dim[2];j++)
          for (k=1;k<=dim[1];k++) {
              l=k+(j-1)*dim[1]+(i-1)*dim[2]*dim[1];
              l=(l<<1)-1;
              data2[l]=data1[l]=2*ran1(&idum)-1;
              l++;
              data2[l]=data1[l]=2*ran1(&idum)-1;
          }
nwrite=NDAT >> 1;
cc=fwrite(&data1[1],sizeof(float),nwrite,file[1]);
if (cc != nwrite) nrerror("write error in xfourfs");
cc=fwrite(&data1[nwrite+1],sizeof(float),nwrite,file[2]);
if (cc != nwrite) nrerror("write error in xfourfs");
rewind(file[1]);
rewind(file[2]);
printf("*************** now doing fourfs *********\n");
fourfs(file,dim,nnv,1);
for (j=1;j<=4;j++) rewind(file[j]);
cc=fread(&data1[1],sizeof(float),nwrite,file[3]);
if (cc != nwrite) nrerror("read error in xfourfs");
cc=fread(&data1[nwrite+1],sizeof(float),nwrite,file[4]);
if (cc != nwrite) nrerror("read error in xfourfs");
printf("*************** now doing fourn *********\n");
fourn(data2,dim,nnv,1);
sum=smax=0.0;
for (i=1;i<=dim[3];i++)
    for (j=1;j<=dim[2];j++)
        for (k=1;k<=dim[1];k++) {
            l=k+(j-1)*dim[1]+(i-1)*dim[2]*dim[1];
            l=(l<<1)-1;
            ll=i+(j-1)*dim[3]+(k-1)*dim[3]*dim[2];
            ll=(ll<<1)-1;
            diff=sqrt(SQR(data2[ll]-data1[l])+SQR(data2[ll+1]-data1[l+1]));
            sum2 += SQR(data1[l])+SQR(data1[l+1]);
            sum += diff;
            if (diff > smax) smax=diff;
        }
sum2=sqrt(sum2/tot);
sum=sum/tot;
printf("(r.m.s.) value, (max,ave) discrepancy= %12.7f %12.7f %12.7f\n",
    sum2,smax,sum);
/* now check the inverse transforms */
SWAP(dim[1],dim[3]);
/* This step swap step is conceptually a reversal, but for
three dimensions a swap accomplishes that. */
FSWAP(file[1],file[3])
FSWAP(file[4],file[2])
for (j=1;j<=4;j++) rewind(file[j]);
printf("*************** now doing fourfs *********\n");
fourfs(file,dim,nnv,-1);
for (j=1;j<=4;j++) rewind(file[j]);
cc=fread(&data1[1],sizeof(float),nwrite,file[3]);
if (cc != nwrite) nrerror("read error in xfourfs");
cc=fread(&data1[nwrite+1],sizeof(float),nwrite,file[4]);
if (cc != nwrite) nrerror("read error in xfourfs");
SWAP(dim[1],dim[3]);
printf("*************** now doing fourn *********\n");
```

```
    fourn(data2,dim,nnv,-1);
    sum=smax=0.0;
    data1p=data1;
    data2p=data2;
    for (j=1;j<=NDAT;j+=2) {
        sum1 += SQR(data2p[1])+SQR(data2p[2]);
        diff=sqrt(SQR(data2p[1]-data1p[1])+SQR(data2p[2]-data1p[2]));
        sum += diff;
        if (diff > smax) smax=diff;
        data1p += 2;
        data2p += 2;
    }
    sum=sum/tot;
    sum1=sqrt(sum1/tot);
    printf("(r.m.s.) value, (max,ave) discrepancy= %12.7f %12.7f %12.7f\n",
        sum1,smax,sum);
    printf("ratio of r.m.s. values, expected ratio= %12.6f %12.6f\n",
        sum1/sum2,sqrt(tot));
    for (j=1;j<=4;j++)
        if (fclose(file[j]) == EOF) nrerror("Couldn't close temporary file");
    free_vector(data2,1,NDAT);
    free_vector(data1,1,NDAT);
    for (j=1;j<=4;j++)
        if (remove(fnames[j-1])) nrerror("Couldn't delete temporary file");
    return 0;
}
```

Chapter 13: Fourier and Spectral Applications

Chapter 13 of *Numerical Recipes* covers mainly Fourier transform spectral methods, particularly the transform of discretely sampled data. Two common uses of the Fourier transform are the convolution of data with a response function, and the computation of the correlation of two data sets. These operations are carried out by `convlv` and `correl` respectively. Other applications of Fourier methods in the main book include data filtering, power spectrum estimation (`spctrm`, or `evlmem` with `memcof`), and linear prediction (`predic` with `fixrts` and `memcof`). For power spectrum estimation of unevenly sampled data, you can use `period`, or its "fast" cousin for large data sets, `fasper`. An application with a somewhat different flavor is the high accuracy computation of Fourier integrals with `dftint`. Finally, we provide routines for wavelet transforms: `wt1` for a one-dimensional wavelet transform, `wtn` for a multidimensional wavelet transform, and (in `daub4`, `pwtset`, and `pwt`) a small selection of wavelet filters, out of which the wavelets themselves are constructed.

$\star$ $\star$ $\star$ $\star$

Procedure `convlv` performs the convolution of a data set with a response function using an FFT. Sample program `xconvlv` uses two functions that take on only the values 0.0 and 1.0. The data array `data[i]` has sixteen values, and is zero everywhere except between `i=6` and `i=10` where it is 1.0. The response function `respns[i]` has nine values and is zero except between `i=3` and `i=6` where it is 1.0. The expected value of the convolution is determined simply by flipping the response function end-to-end, moving it to the left by the desired shift, and counting how many non-zero channels of `respns` fall on non-zero channels of `data`. In this way, you should be able to verify the result from the program. The sample program, incidentally, does the calculation by this direct method for the purpose of comparison.

```
/* Driver for routine convlv */

#include <stdio.h>
#include "nr.h"
#include "nrutil.h"

#define N 16        /* data array size */
#define M 9         /* response function dimension - must be odd */
#define N2 (2*N)

int main(void)
{
    unsigned long i,j;
```

```
    int isign;
    float cmp,*data,*respns,*resp,*ans;

    data=vector(1,N);
    respns=vector(1,N);
    resp=vector(1,N);
    ans=vector(1,N2);
    for (i=1;i<=N;i++)
        if ((i >= N/2-N/8) && (i <= N/2+N/8))
            data[i]=1.0;
        else
            data[i]=0.0;
    for (i=1;i<=M;i++) {
        if ((i > 2) && (i < 7))
            respns[i]=1.0;
        else
            respns[i]=0.0;
        resp[i]=respns[i];
    }
    isign=1;
    convlv(data,N,resp,M,isign,ans);
    /* compare with a direct convolution */
    printf("%3s %14s %13s\n","i","CONVLV","Expected");
    for (i=1;i<=N;i++) {
        cmp=0.0;
        for (j=1;j<=M/2;j++) {
            cmp += data[((i-j-1+N) % N)+1]*respns[j+1];
            cmp += data[((i+j-1) % N)+1]*respns[M-j+1];
        }
        cmp += data[i]*respns[1];
        printf("%3ld %15.6f %12.6f\n",i,ans[i],cmp);
    }
    free_vector(ans,1,N2);
    free_vector(resp,1,N);
    free_vector(respns,1,N);
    free_vector(data,1,N);
    return 0;
}
```

correl calculates the correlation function of two data sets. Sample program xcorrel defines data1[i] as an array of 64 values that are all zero except from i=25 to i=39, where they are one. data2[i] is defined in the same way. Therefore, the correlation being performed is an autocorrelation. The sample routine compares the result of the calculation as performed by correl with that found by a direct calculation. In this case the calculation may be done manually simply by successively shifting data2 with respect to data1 and counting the number of nonzero channels of the two that overlap.

```
/* Driver for routine correl */

#include <stdio.h>
#include "nr.h"
#include "nrutil.h"

#define N 64
#define N2 (2*N)
```

```
int main(void)
{
    unsigned long i,j;
    float cmp,*data1,*data2,*ans;

    data1=vector(1,N);
    data2=vector(1,N);
    ans=vector(1,N2);
    for (i=1;i<=N;i++) {
        if ((i > N/2-N/8) && (i < N/2+N/8))
            data1[i]=1.0;
        else
            data1[i]=0.0;
        data2[i]=data1[i];
    }
    correl(data1,data2,N,ans);
    /* Calculate directly */
    printf("%3s %14s %18s\n","n","CORREL","direct calc.");
    for (i=0;i<=16;i++) {
        cmp=0.0;
        for (j=1;j<=N;j++)
            cmp += data1[((i+j-1) % N)+1]*data2[j];
        printf("%3ld %15.6f %15.6f\n",i,ans[i+1],cmp);
    }
    free_vector(ans,1,N2);
    free_vector(data2,1,N);
    free_vector(data1,1,N);
    return 0;
}
```

spctrm does a spectral estimate of a data set by reading it in as segments, windowing, Fourier transforming, and accumulating the power spectrum. Data segments may or may not be overlapped at the decision of the user. In sample program xspctrm the spectral data is read in from a file called spctrl.dat containing 1200 numbers and included on the diskette. It is analyzed first with overlap and then without. The results are tabulated side by side for comparison.

```
/* Driver for routine spctrm */

#include <stdio.h>
#include <stdlib.h>
#include "nr.h"
#include "nrutil.h"

#define M 16
#define TRUE 1
#define FALSE 0

int main(void)
{
    int j,k,ovrlap;
    float *p,*q;
    FILE *fp;

    p=vector(1,M);
```

```
    q=vector(1,M);
    if ((fp = fopen("spctrl.dat","r")) == NULL)
        nrerror("Data file spctrl.dat not found\n");
    k=8;
    ovrlap=TRUE;
    spctrm(fp,p,M,k,ovrlap);
    rewind(fp);
    k=16;
    ovrlap=FALSE;
    spctrm(fp,q,M,k,ovrlap);
    fclose(fp);
    printf("\nSpectrum of data in file spctrl.dat\n");
    printf("%13s %s %5s %s\n"," ","overlapped "," ","non-overlapped");
    for (j=1;j<=M;j++)
        printf("%3d %5s %13f %5s %13f\n",j," ",p[j]," ",q[j]);
    free_vector(q,1,M);
    free_vector(p,1,M);
    return 0;
}
```

memcof is used in two different capacities in *Numerical Recipes*. In combination
with evlmem it is used to perform spectral analysis by the maximum entropy method.
memcof finds the coefficients for a model spectrum, the magnitude squared of the inverse of a polynomial series. Sample program xmemcof determines the coefficients for
1000 numbers from the file spctrl.dat and simply prints the results for comparison
to the following table:

```
Coefficients for spectral estimation of spctrl.dat
 a[ 1] =     1.261539
 a[ 2] =    -0.007695
 a[ 3] =    -0.646778
 a[ 4] =    -0.280603
 a[ 5] =     0.163693
 a[ 6] =     0.347674
 a[ 7] =     0.111247
 a[ 8] =    -0.337141
 a[ 9] =    -0.358043
 a[10] =     0.378774
    a0 =     0.003511
```

```
/* Driver for routine memcof */

#include <stdio.h>
#include <stdlib.h>
#include "nr.h"
#include "nrutil.h"

#define N 1000
#define M 10

int main(void)
{
    int i;
    float pm,*cof,*data;
    FILE *fp;
```

```
        cof=vector(1,M);
        data=vector(1,N);
        if ((fp = fopen("spctrl.dat","r")) == NULL)
            nrerror("Data file spctrl.dat not found\n");
        for (i=1;i<=N;i++) fscanf(fp,"%f",&data[i]);
        fclose(fp);
        memcof(data,N,M,&pm,cof);
        printf("Coefficients for spectral estimation of spctrl.dat\n\n");
        for (i=1;i<=M;i++) printf("a[%2d] = %12.6f \n",i,cof[i]);
        printf("\na0 =%12.6f\n",pm);
        free_vector(data,1,N);
        free_vector(cof,1,M);
        return 0;
}
```

Although `evlmem` appears later in Chapter 13, we show its sample program here, because it is so closely tied to the previous program. `evlmem` uses coefficients from `memcof` to generate a spectral estimate. The example `xevlmem` uses the data from `spctrl.dat` and prints the spectral estimate. You may compare the result to:

```
Power spectrum estimate of data in spctrl.dat
     f*delta        power
     0.000000     0.026023
     0.031250     0.029266
     0.062500     0.193087
     0.093750     0.139241
     0.125000    29.915518
     0.156250     0.003878
     0.187500     0.000633
     0.218750     0.000334
     0.250000     0.000437
     0.281250     0.001331
     0.312500     0.000780
     0.343750     0.000451
     0.375000     0.000784
     0.406250     0.001381
     0.437500     0.000649
     0.468750     0.000775
     0.500000     0.001716
```

```
/* Driver for routine evlmem */

#include <stdio.h>
#include <stdlib.h>
#include "nr.h"
#include "nrutil.h"

#define N 1000
#define M 10
#define NFDT 16

int main(void)
{
    int i;
    float fdt,pm,*cof,*data;
    FILE *fp;
```

```
    cof=vector(1,M);
    data=vector(1,N);
    if ((fp = fopen("spctrl.dat","r")) == NULL)
        nrerror("Data file spctrl.dat not found\n");
    for (i=1;i<=N;i++) fscanf(fp,"%f",&data[i]);
    fclose(fp);
    memcof(data,N,M,&pm,cof);
    printf("Power spectrum estimate of data in spctrl.dat\n");
    printf("    f*delta       power\n");
    for (fdt=0.0;fdt<=0.5;fdt+=0.5/NFDT)
        printf("%12.6f %12.6f\n",fdt,evlmem(fdt,cof,M,pm));
    free_vector(data,1,N);
    free_vector(cof,1,M);
    return 0;
}
```

Notice that once `memcof` has determined coefficients, we may evaluate the estimate at any intervals we wish. Notice also that we have built a spectral peak into the noisy data in `spctrl.dat`.

The second use of `memcof` is to perform linear prediction in combination with `predic` and `fixrts`. `memcof` produces the linear prediction coefficients from the data set. `fixrts` massages the coefficients so that all roots of the characteristic polynomial fall inside the unit circle of the complex domain, thus insuring stability of the prediction algorithm. Finally, `predic` predicts future data points based on the modified coefficients. Sample program `xfixrts` demonstrates the operation of `fixrts`. The coefficients provided in the array `d[i]` are those appropriate to the polynomial $(z - 1)^6 = 1$. This equation has six roots on a circle of radius one, centered at $(1.0, 0.0)$ in the complex plane. Some of these lie within the unit circle and some outside. The ones outside are moved by `fixrts` according to $z_i \rightarrow 1/z_i{}^*$. You can easily figure these out by hand and check the results. Also, the sample routine calculates $(z - 1)^6$ for each of the adjusted roots, and thereby shows which have been changed and which have not.

```
/* Driver for routine fixrts */

#include <stdio.h>
#include "nr.h"
#include "complex.h"

#define NPOLES 6
#define NP1 (NPOLES+1)
#define ONE Complex(1.0,0.0)
#define TRUE 1

int main(void)
{
    int i,polish=TRUE;
    static float d[NP1]=
        {0.0,6.0,-15.0,20.0,-15.0,6.0,0.0};
    fcomplex zcoef[NP1],zeros[NP1],z1,z2;

    /* finding roots of (z-1.0)^6=1.0 */
    /* first write roots */
    zcoef[NPOLES]=ONE;
```

```
    for (i=NPOLES-1;i>=0;i--)
        zcoef[i] = Complex(-d[NPOLES-i],0.0);
    zroots(zcoef,NPOLES,zeros,polish);
    printf("Roots of (z-1.0)^6 = 1.0\n");
    printf("%24s %27s \n","Root","(z-1.0)^6");
    for (i=1;i<=NPOLES;i++) {
        z1=Csub(zeros[i],ONE);
        z2=Cmul(z1,z1);
        z1=Cmul(z1,z2);
        z1=Cmul(z1,z1);
        printf("%6d %12.6f %12.6f %12.6f %12.6f\n",
            i,zeros[i].r,zeros[i].i,z1.r,z1.i);
    }
    /* now fix them to lie within unit circle */
    fixrts(d,NPOLES);
    /* check results */
    zcoef[NPOLES]=ONE;
    for (i=NPOLES-1;i>=0;i--)
        zcoef[i] = Complex(-d[NPOLES-i],0.0);
    zroots(zcoef,NPOLES,zeros,polish);
    printf("\nRoots reflected in unit circle\n");
    printf("%24s %27s \n","Root","(z-1.0)^6");
    for (i=1;i<=NPOLES;i++) {
        z1=Csub(zeros[i],ONE);
        z2=Cmul(z1,z1);
        z1=Cmul(z1,z2);
        z1=Cmul(z1,z1);
        printf("%6d %12.6f %12.6f %12.6f %12.6f\n",
            i,zeros[i].r,zeros[i].i,z1.r,z1.i);
    }
    return 0;
}
```

`predic` carries out the job of performing the prediction. The function chosen for investigation in sample program `xpredic` is

$$F(n) = \exp(-n/\texttt{npts}) \sin(2\pi n/50) + \exp(-2n/\texttt{npts}) \sin(2.2\pi n/50)$$

the sum of two sine waves of similar period and exponentially decaying amplitudes. On the basis of 500 data points, and working with coefficients representing ten poles, the routine predicts 20 future points. The quality of this prediction may be judged by comparing these 20 points with the evaluations of $F(n)$ that are provided.

```
/* Driver for routine predic */

#include <stdio.h>
#include <math.h>
#include "nr.h"
#include "nrutil.h"

#define NPTS 500
#define NPOLES 10
#define NFUT 20
#define PI 3.1415926

float f(int n,int npts)
{
```

```
        return exp(-(1.0*n)/npts)*sin(2.0*PI*n/50.0)
            +exp(-(2.0*n)/npts)*sin(2.2*PI*n/50.0);
}

int main(void)
{
    int i;
    float dum,*d,*future,*data;

    d=vector(1,NPOLES);
    future=vector(1,NFUT);
    data=vector(1,NPTS);
    for (i=1;i<=NPTS;i++)
        data[i]=f(i,NPTS);
    memcof(data,NPTS,NPOLES,&dum,d);
    fixrts(d,NPOLES);
    predic(data,NPTS,d,NPOLES,future,NFUT);
    printf("%6s %11s %12s\n","I","Actual","PREDIC");
    for (i=1;i<=NFUT;i++)
        printf("%6d %12.6f %12.6f\n",i,f(i+NPTS,NPTS),future[i]);
    free_vector(data,1,NPTS);
    free_vector(future,1,NFUT);
    free_vector(d,1,NPOLES);
    return 0;
}
```

Spectral analysis of unevenly sampled data, or of evenly sampled data with missing values, can be performed by the routine `period`, which constructs the Lomb normalized periodogram. In the following example, we construct a data array of length 90 that consists of a regular sampling of length 100 with 10 (random) deletions. The sampled function is a pure cosine, plus Gaussian white noise of somewhat larger amplitude. A call to `period` yields the significance level with which the signal is found to be non-Gaussian (very significant in this case!), and the output spectrum. We print out a few values in the neighborhood of the largest peak found. You can verify that the peak is indeed nicely centered on the frequency of the input cosine.

```
/* Driver for routine period */

#include <stdio.h>
#include <math.h>
#include "nr.h"
#include "nrutil.h"

#define NP 90
#define NPR 11
#define TWOPI 6.2831853

int main(void)
{
    long idum=(-4);
    int j=0,jmax,n,nout;
    float prob,*px,*py,*x,*y;

    x=vector(1,NP);
    y=vector(1,NP);
    px=vector(1,2*NP);
```

```
    py=vector(1,2*NP);
    for (n=1;n<=NP+10;n++) {
        if (n != 3 && n != 4 && n != 6 && n != 21 &&
            n != 38 && n != 51 && n != 67 && n != 68 &&
            n != 83 && n != 93) {
            x[++j]=n;
            y[j]=0.75*cos(0.6*x[j])+gasdev(&idum);
        }
    }
    period(x,y,j,4.0,1.0,px,py,2*NP,&nout,&jmax,&prob);
    printf("period results for test signal (cos(0.6x) + noise):\n");
    printf("nout,jmax,prob=%d %d %12.6g\n",nout,jmax,prob);
    for (n=LMAX(1,jmax-NPR/2);n<=LMIN(nout,jmax+NPR/2);n++)
        printf("%d %12.6f %12.6f\n",n,TWOPI*px[n],py[n]);
    free_vector(py,1,2*NP);
    free_vector(px,1,2*NP);
    free_vector(y,1,NP);
    free_vector(x,1,NP);
    return 0;
}
```

The routine `fasper` has the same capabilities as `period`, but performs its underlying calculations using FFT routines. For large data arrays, `fasper` is very much faster than `period`. Here, we give a demonstration program that is closely equivalent to the one given for the previous routine. About the only difference is that `fasper` requires a larger workspace, here 4096, onto which it "extirpolates" the given data, as described in the main book.

```
/* Driver for routine fasper */

#include <stdio.h>
#include <math.h>
#include "nr.h"
#include "nrutil.h"

#define NP 90
#define MP 4096
#define NPR 11
#define TWOPI 6.2831853

int main(void)
{
    long idum=(-4);
    unsigned long j=0,jmax,n,nout;
    float prob,*px,*py,*x,*y;

    x=vector(1,NP);
    y=vector(1,NP);
    px=vector(1,MP);
    py=vector(1,MP);
    for (n=1;n<=NP+10;n++) {
        if (n != 3 && n != 4 && n != 6 && n != 21 &&
            n != 38 && n != 51 && n != 67 && n != 68 &&
            n != 83 && n != 93) {
            x[++j]=n;
            y[j]=0.75*cos(0.6*x[j])+gasdev(&idum);
```

```
        }
    }
    fasper(x,y,j,4.0,1.0,px,py,MP,&nout,&jmax,&prob);
    printf("fasper results for test signal (cos(0.6x) + noise):\n");
    printf("nout,jmax,prob=%ld %ld %12.6g\n",nout,jmax,prob);
    for (n=LMAX(1,jmax-NPR/2);n<=LMIN(nout,jmax+NPR/2);n++)
        printf("%ld %12.6f %12.6f\n",n,TWOPI*px[n],py[n]);
    free_vector(py,1,MP);
    free_vector(px,1,MP);
    free_vector(y,1,NP);
    free_vector(x,1,NP);
    return 0;
}
```

We test `dftint` on the integrals

$$\int_a^b \cos(cx+d)\cos\omega x\,dx = \left[\frac{\sin[(\omega-c)x-d]}{2(\omega-c)} + \frac{\sin[(\omega+c)x+d]}{2(\omega+c)}\right]_a^b$$

$$\int_a^b \cos(cx+d)\sin\omega x\,dx = \left[-\frac{\cos[(\omega-c)x-d]}{2(\omega-c)} - \frac{\cos[(\omega+c)x+d]}{2(\omega+c)}\right]_a^b$$

where $\omega \neq \pm c$. Sample program `xdftint` prompts for values of c, d, a, b and ω. It then compares the results of `dftint` with the analytic values of the integrals.

```
/* Driver for routine dftint */

#include <stdio.h>
#include <math.h>
#include "nr.h"

static float c,d;

float coscxd(float x)
{
    return cos(c*x+d);
}

#define ci(x) (sin((w-c)*x-d)/(2.0*(w-c))+sin((w+c)*x+d)/(2.0*(w+c)))
#define si(x) (-cos((w-c)*x-d)/(2.0*(w-c))-cos((w+c)*x+d)/(2.0*(w+c)))

void getans(float w,float a,float b,float *cans,float *sans)
{
    *cans=ci(b)-ci(a);
    *sans=si(b)-si(a);
}

int main(void)
{
    float a,b,cans,cosint,sans,sinint,w;

    printf(" Omega  Integral cosine*test func  Err");
    printf("     Integral sine*test func   Err\n");
    for (;;) {
        printf("input c,d:\n");
        if (scanf("%f %f",&c,&d) == EOF) break;
        for (;;) {
```

```
                printf("input a,b:\n");
                if (scanf("%f %f",&a,&b) == EOF) break;
                if (a == b) break;
                    for (;;) {
                        printf("input w:\n");
                        if (scanf("%f",&w) == EOF) break;
                        if (w < 0.0) break;
                        dftint(coscxd,a,b,w,&cosint,&sinint);
                        getans(w,a,b,&cans,&sans);
                        printf("%15.6e %15.6e %15.6e %15.6e %15.6e\n",
                            w,cans,cosint-cans,sans,sinint-sans);
                    }
            }
        }
        printf("normal completion\n");
        return 0;
    }
```

The wavelet transform routine `wt1` can be used either with the lowest-order
wavelet filter, as provided by `daub4`, or else with the higher-order filters provided
by the combination of `pwtset` and `pwt`. The following sample program allows for
both possibilities by prompting for the input order as `k`. For $k = -4$ the program
uses `daub4`, while for `k` positive it uses `pwtset(k)`. Also prompted for is a quantity
`frac`: The program computes the wavelet transform of a function consisting of a
parabolic "bump" on an otherwise zero background. It then truncates the wavelet
transform to zeros, except for a fraction `frac` of the components. After reconstructing
the function from the truncated data with an inverse wavelet transform, it finds the
maximum deviation between the original function and its approximate reconstruction. One finds that with a comparatively small number of wavelet components, one
is able to reconstruct the original bump to very good approximation.

```
/* Driver for routine wt1 */
#include <stdio.h>
#include <math.h>
#include "nr.h"
#include "nrutil.h"

#define NMAX 512
#define NCEN 333
#define NWID 33

int main(void)
{
    unsigned long i,nused;
    int itest,k;
    float *u,*v,*w,frac,thresh,tmp;

    u=vector(1,NMAX);
    v=vector(1,NMAX);
    w=vector(1,NMAX);
    for (;;) {
        printf("Enter k (4, -4, 12, or 20) and frac (0.0 to 1.0):\n");
        if (scanf("%d %f",&k,&frac) == EOF) break;
        frac=FMIN(1.0,FMAX(0.0,frac));
        itest=(k == -4 ? 1 : 0);
```

```
        if (k < 0) k = -k;
        if (k != 4 && k != 12 && k != 20) continue;
        for (i=1;i<=NMAX;i++)
            w[i]=v[i]=(i > NCEN-NWID && i < NCEN+NWID ?
                ((float)(i-NCEN+NWID)*(float)(NCEN+NWID-i))/(NWID*NWID) : 0.0);
        if (!itest) pwtset(k);
        wt1(v,NMAX,1,itest ? daub4 : pwt);
        for (i=1;i<=NMAX;i++) u[i]=fabs(v[i]);
        thresh=select((int)((1.0-frac)*NMAX),NMAX,u);
        nused=0;
        for (i=1;i<=NMAX;i++) {
            if (fabs(v[i]) <= thresh)
                v[i]=0.0;
            else
                nused++;
        }
        wt1(v,NMAX,-1,itest ? daub4 : pwt);
        for (thresh=0.0,i=1;i<=NMAX;i++)
            if ((tmp=fabs(v[i]-w[i])) > thresh) thresh=tmp;
        printf("k,NMAX,nused= %d %d %d\n",k,NMAX,nused);
        printf("discrepancy= %12.6f\n",thresh);
    }
    free_vector(w,1,NMAX);
    free_vector(v,1,NMAX);
    free_vector(u,1,NMAX);
    return 0;
}
```

As an example of using **wtn**, the *n*-dimensional wavelet transform routine, we set up the two-dimensional matrix described in the caption to Figure 13.10.5 of *Numerical Recipes*, then take its wavelet transform (using the DAUB12 wavelet), then undo the transform and show that the original matrix is recovered. In real applications, of course, what matters is what we have left out: processes that would be applied to the matrix while it is in wavelet space. The simplest such process would be truncation, encoding, and transmission of the matrix in a wavelet-compressed form, for reconstruction at another location. Another kind of process would be truncation followed by some operations of linear algebra optimized for the now-sparse form.

```
/* Driver for routine wtn */
#include <stdio.h>
#include <math.h>
#include "nr.h"
#include "nrutil.h"

#define EPS 1.0e-06
#define NX 8
#define NY 16

int main(void)
{
    unsigned long i,j,l,nerror=0,ntot=NX*NY;
    float *a,*aorg;
    static unsigned long ndim[]={0,NX,NY};

    aorg=vector(1,ntot);
    a=vector(1,ntot);
```

```
pwtset(12);
for (i=1;i<=NX;i++)
    for (j=1;j<=NY;j++) {
        l=i+(j-1)*NX;
        aorg[l]=a[l]=(i == j ? -1.0 : 1.0/sqrt(fabs((float)(i-j))));
    }
wtn(a,ndim,2,1,pwt);
/* here, one might set the smallest components to zero, encode and transmit
the remaining components as a compressed form of the "image" */
wtn(a,ndim,2,-1,pwt);
for (l=1;l<=ntot;l++) {
    if (fabs(aorg[l]-a[l]) >= EPS) {
        printf("Compare Error at element %ld\n",l);
        nerror++;
    }
}
if (nerror) printf("Number of comparision errors: %ld\n",nerror);
else printf("transform-inverse transform check OK\n");
free_vector(a,1,ntot);
free_vector(aorg,1,ntot);
return nerror;
}
```

Chapter 14: Statistical Description of Data

Chapter 14 of *Numerical Recipes* covers the subject of descriptive statistics, the representation of data in terms of its statistical properties, and the use of such properties to compare data sets. We start with a procedure to characterize data sets: `moment` returns the average, average deviation, standard deviation, variance, skewness, and kurtosis of a data array.

Most of the remaining procedures compare data sets. `ttest` compares the means of two data sets having the same variance; `tutest` does the same for two sets having different variance; and `tptest` does it for paired samples, correcting for covariance. `ftest` is a test of whether two data arrays have significantly different variance. The question of whether two distributions are different is treated by four procedures (pertaining to whether the data is binned or continuous, and whether data is compared to a model distribution or to other data). Specifically,

1. `chsone` compares binned data to a model distribution.

2. `chstwo` compares two binned data sets.

3. `ksone` compares the cumulative distribution function of an unbinned data set to a given function.

4. `kstwo` compares the cumulative distribution functions of two unbinned data sets.

The next set of procedures tests for associations between nominal variables. `cntab1` and `cntab2` both check for associations in a two-dimensional contingency table, the first calculating on the basis of χ^2, and the second by evaluating entropies. Linear correlation is represented by Pearson's r, or the linear correlation coefficient, which is calculated with routine `pearsn`. Alternatively, the data can be investigated with a nonparametric or rank correlation, using `spear` to find Spearman's rank correlation r_s. Kendall's τ uses rank ordering of ordinal data to test for monotonic correlations. `kendl1` does this for two data arrays of the same size, while `kendl2` applies it to contingency tables.

We provide two procedures for testing two-dimensional data sets. `ks2d1s` performs a Kolmogorov-Smirnov test of data to a given model (generalization of `ksone`), while `ks2d2s` does the same for testing one data set against another (generalization of `kstwo`).

One final routine `savgol` makes no attempt to describe or compare data statistically. It calculates Savitzky-Golay smoothing coefficients that can be used to smooth out statistical fluctuations, usually for the purpose of visual *presentation.*

★   ★   ★   ★

Procedure moment calculates successive moments of a given distribution of data. The example program xmoment creates an unusual distribution, one that has a sinusoidal distribution of values (over a half-period of the sine, so the distribution is a symmetrical peak). We have worked out the moments of such a distribution theoretically and recorded them in the program for comparison. The data is discrete and will only approximate these values.

```
/* Driver for routine moment */

#include <stdio.h>
#include <math.h>
#include "nr.h"
#include "nrutil.h"

#define PI 3.14159265
#define NPTS 5000
#define NBIN 100
#define NPPNB (NPTS+NBIN)

int main(void)
{
    int i=0,k,nlim;
    float adev,ave,curt,sdev,skew,vrnce,x,*data;

    data=vector(1,NPPNB);
    for (x=PI/NBIN;x<=PI;x+=PI/NBIN) {
        nlim=(int) (0.5+sin(x)*PI/2.0*NPTS/NBIN);
        for (k=1;k<=nlim;k++) data[++i]=x;
    }
    printf("moments of a sinusoidal distribution\n\n");
    moment(data,i,&ave,&adev,&sdev,&vrnce,&skew,&curt);
    printf("%39s %11s\n\n","calculated","expected");
    printf("%s %17s %12.4f %12.4f\n","Mean :"," ",ave,PI/2.0);
    printf("%s %4s %12.4f %12.4f\n",
        "Average Deviation :"," ",adev,(PI/2.0)-1.0);
    printf("%s %3s %12.4f %12.4f\n",
        "Standard Deviation :"," ",sdev,0.683667);
    printf("%s %13s %12.4f %12.4f\n",
        "Variance :"," ",vrnce,0.467401);
    printf("%s %13s %12.4f %12.4f\n",
        "Skewness :"," ",skew,0.0);
    printf("%s %13s %12.4f %12.4f\n",
        "Kurtosis :"," ",curt,-0.806249);
    free_vector(data,1,NPPNB);
    return 0;
}
```

Student's *t*-test is a test of two data sets for significantly different means. It is applied by xttest to two Gaussian data sets data1 and data2 that are generated by gasdev. data2 is originally given an artificial shift of its mean to the right of that of data1, by NSHFT/2 units of EPS. Then data1 is successively shifted NSHFT times to the right by EPS and compared to data2 by ttest. At about step NSHFT/2, the two distributions should superpose and indicate populations with the same mean. Notice that the two populations have the same variance (i.e. 1.0), as required by ttest.

```
/* Driver for routine ttest */

#include <stdio.h>
#include "nr.h"
#include "nrutil.h"

#define NPTS 1024
#define MPTS 512
#define EPS 0.02
#define NSHFT 10

int main(void)
{
    long idum=(-5);
    int i,j;
    float prob,t,*data1,*data2;

    data1=vector(1,NPTS);
    data2=vector(1,MPTS);
    /* Generate gaussian distributed data */
    printf("%6s %8s %16s\n","shift","t","probability");
    for (i=1;i<=NPTS;i++) data1[i]=gasdev(&idum);
    for (i=1;i<=MPTS;i++) data2[i]=(NSHFT/2.0)*EPS+gasdev(&idum);
    for (i=1;i<=NSHFT+1;i++) {
        ttest(data1,NPTS,data2,MPTS,&t,&prob);
        printf("%6.2f %10.2f %10.2f\n",(i-1)*EPS,t,prob);
        for (j=1;j<=NPTS;j++) data1[j] += EPS;
    }
    free_vector(data2,1,MPTS);
    free_vector(data1,1,NPTS);
    return 0;
}
```

avevar is an auxiliary routine for ttest. It finds the average and variance of a data set. Sample program xavevar generates a series of Gaussian distributions for $i=1,\ldots,11$, and gives each a shift of $(i-1)$EPS and a variance of i^2. This progression allows you easily to check the operation of avevar "by eye".

```
/* Driver for routine avevar */

#include <stdio.h>
#include "nr.h"
#include "nrutil.h"

#define NPTS 1000
#define EPS 0.1

int main(void)
{
    int i,j;
    long idum=(-5);
    float ave,shift,vrnce,*data;

    data=vector(1,NPTS);
```

```
    /* generate gaussian distributed data */
    printf("\n%9s %11s %12s\n","shift","average","variance");
    for (i=1;i<=11;i++) {
        shift=(i-1)*EPS;
        for (j=1;j<=NPTS;j++)
            data[j]=shift+i*gasdev(&idum);
        avevar(data,NPTS,&ave,&vrnce);
        printf("%8.2f %11.2f %12.2f\n",shift,ave,vrnce);
    }
    free_vector(data,1,NPTS);
    return 0;
}
```

tutest also does Student's *t*-test, but applies to the comparison of means of two distributions with different variance. The example **xtutest** employs the comparison method used in **ttest** but gives the two distributions **data1** and **data2** variances of 1.0 and 4.0 respectively.

```
/* Driver for routine tutest */

#include <stdio.h>
#include <math.h>
#include "nr.h"
#include "nrutil.h"

#define NPTS 5000
#define MPTS 1000
#define EPS 0.02
#define VAR1 1.0
#define VAR2 4.0
#define NSHFT 10

int main(void)
{
    long idum=(-51773);
    int i,j;
    float fctr1,fctr2,prob,t,*data1,*data2;

    data1=vector(1,NPTS);
    data2=vector(1,MPTS);
    /* Generate two gaussian distributions of different variance */
    fctr1=sqrt(VAR1);
    for (i=1;i<=NPTS;i++) data1[i]=fctr1*gasdev(&idum);
    fctr2=sqrt(VAR2);
    for (i=1;i<=MPTS;i++)
        data2[i]=NSHFT/2.0*EPS+fctr2*gasdev(&idum);
    printf("\nDistribution #1 : variance = %6.2f\n",VAR1);
    printf("Distribution #2 : variance = %6.2f\n\n",VAR2);
    printf("%7s %8s %16s\n","shift","t","probability");
    for (i=1;i<=NSHFT+1;i++) {
        tutest(data1,NPTS,data2,MPTS,&t,&prob);
        printf("%6.2f %10.2f %11.2f\n",(i-1)*EPS,t,prob);
        for (j=1;j<=NPTS;j++) data1[j] += EPS;
    }
    free_vector(data2,1,MPTS);
    free_vector(data1,1,NPTS);
    return 0;
```

}

    tptest goes a step further, and compares two distributions not only having different variances, but also perhaps having point by point correlations. The example xtptest creates two situations, one with correlated and one with uncorrelated distributions. It does this by way of three data sets. data1 is a simple Gaussian distribution of zero mean and unit variance. data2 is data1 plus some additional Gaussian fluctuations of smaller amplitude. data3 is similar to data2 but generated with independent calls to gasdev so that its fluctuations ought not to have any correlation with those of data1. data1 is then given an offset with respect to the others and they are successively shifted as in previous routines. At each step of the shift tptest was applied. Our results are given below:

| | Correlated: | | Uncorrelated: | |
Shift	t	Probability	t	Probability
0.01	3.1199	0.0019	0.6287	0.5298
0.02	2.3399	0.0197	0.4716	0.6374
0.03	1.5600	0.1194	0.3144	0.7534
0.04	0.7800	0.4358	0.1572	0.8751
0.05	0.0000	1.0000	0.0000	1.0000
0.06	−0.7800	0.4358	−0.1572	0.8751
0.07	−1.5600	0.1194	−0.3144	0.7534
0.08	−2.3399	0.0197	−0.4716	0.6374
0.09	−3.1199	0.0019	−0.6287	0.5298
0.10	−3.8999	0.0001	−0.7859	0.4323
0.11	−4.6799	0.0000	−0.9431	0.3460

```
/* Driver for routine tptest */

#include <stdio.h>
#include "nr.h"
#include "nrutil.h"

#define NPTS 500
#define EPS 0.01
#define NSHFT 11
#define ANOISE 0.3

int main(void)
{
    long idum=(-5);
    int i,j;
    float ave1,ave2,ave3;
    float offset,prob1,prob2,shift,t1,t2;
    float var1,var2,var3,*data1,*data2,*data3;

    data1=vector(1,NPTS);
    data2=vector(1,NPTS);
    data3=vector(1,NPTS);
    printf("%29s %31s\n","Correlated:","Uncorrelated:");
    printf("%7s %11s %17s %11s %17s\n",
        "Shift","t","Probability","t","Probability");
    offset=(NSHFT/2)*EPS;
    for (j=1;j<=NPTS;j++) {
        data1[j]=gasdev(&idum);
        data2[j]=data1[j]+ANOISE*gasdev(&idum);
```

```
            data3[j]=gasdev(&idum);
            data3[j] += ANOISE*gasdev(&idum);
        }
        avevar(data1,NPTS,&ave1,&var1);
        avevar(data2,NPTS,&ave2,&var2);
        avevar(data3,NPTS,&ave3,&var3);
        for (j=1;j<=NPTS;j++) {
            data1[j] -= ave1-offset;
            data2[j] -= ave2;
            data3[j] -= ave3;
        }
        for (i=1;i<=NSHFT;i++) {
            shift=i*EPS;
            for (j=1;j<=NPTS;j++) {
                data2[j] += EPS;
                data3[j] += EPS;
            }
            tptest(data1,data2,NPTS,&t1,&prob1);
            tptest(data1,data3,NPTS,&t2,&prob2);
            printf("%6.2f %14.4f %12.4f %16.4f %12.4f\n",
                shift,t1,prob1,t2,prob2);
        }
        free_vector(data3,1,NPTS);
        free_vector(data2,1,NPTS);
        free_vector(data1,1,NPTS);
        return 0;
    }
```

The F-test (routine `ftest`) is a test for differing variances between two distributions. For demonstration purposes, sample program `xftest` generates two distributions `data1` and `data2` having Gaussian distributions of unit variance. The values of a third array `data3` are then set by multiplying `data2` by a series of values `factr` which takes its variance from 1.0 to 1.1 in ten equal steps. The effect of this on the F-test can be evaluated from the probabilities `prob`.

```
/* Driver for routine ftest */

#include <stdio.h>
#include <math.h>
#include "nr.h"
#include "nrutil.h"

#define NPTS 1000
#define MPTS 500
#define EPS 0.01
#define NVAL 11

int main(void)
{
    long idum=(-13);
    int i,j;
    float f,factor,prob,vrnce,*data1,*data2,*data3;

    data1=vector(1,NPTS);
    data2=vector(1,MPTS);
    data3=vector(1,MPTS);
```

```
/* Generate two gaussian distributions with different variances */
printf("\n%16s %5.2f\n","Variance 1 = ",1.0);
printf("%13s %11s %16s\n","Variance 2","Ratio","Probability");
for (j=1;j<=NPTS;j++) data1[j]=gasdev(&idum);
for (j=1;j<=MPTS;j++) data2[j]=gasdev(&idum);
for (i=1;i<=NVAL;i++) {
    vrnce=1.0+(i-1)*EPS;
    factor=sqrt(vrnce);
    for (j=1;j<=MPTS;j++) data3[j]=factor*data2[j];
    ftest(data1,NPTS,data3,MPTS,&f,&prob);
    printf("%11.4f %13.4f %13.4f\n",vrnce,f,prob);
}
free_vector(data3,1,MPTS);
free_vector(data2,1,MPTS);
free_vector(data1,1,NPTS);
return 0;
}
```

chsone and chstwo compare two distributions on the basis of a χ^2 test to see if they are different. chsone, specifically, compares a data distribution to an expected distribution. Sample program xchsone generates an exponential distribution bins[i] of data using routine expdev. It then creates an array ebins[i] which is the expected result (a smooth exponential decay in the absence of statistical fluctuations). ebins and bins are compared by chsone to give χ^2 and a probability that they represent the same distribution.

```
/* Driver for routine chsone */

#include <stdio.h>
#include <math.h>
#include "nr.h"
#include "nrutil.h"

#define NBINS 10
#define NPTS 2000

int main(void)
{
    long idum=(-15);
    int i,ibin,j;
    float chsq,df,prob,x,*bins,*ebins;

    bins=vector(1,NBINS);
    ebins=vector(1,NBINS);
    for (j=1;j<=NBINS;j++) bins[j]=0.0;
    for (i=1;i<=NPTS;i++) {
        x=expdev(&idum);
        ibin=(int) (x*NBINS/3.0)+1;
        if (ibin <= NBINS) ++bins[ibin];
    }
    for (i=1;i<=NBINS;i++)
        ebins[i]=3.0*NPTS/NBINS*exp(-3.0*(i-0.5)/NBINS);
    chsone(bins,ebins,NBINS,0,&df,&chsq,&prob);
    printf("%15s %15s\n","expected","observed");
    for (i=1;i<=NBINS;i++)
        printf("%14.2f %15.2f\n",ebins[i],bins[i]);
```

```
        printf("\n%19s %10.4f\n","chi-squared:",chsq);
        printf("%19s %10.4f\n","probability:",prob);
        free_vector(ebins,1,NBINS);
        free_vector(bins,1,NBINS);
        return 0;
}
```

chstwo compares two binned distributions bins1 and bins2, again using a χ^2 test. Sample program xchstwo prepares these distributions both in the same way. Each is composed of 2000 random numbers, drawn from an exponential deviate, and placed into 10 bins. The two data sets are then analyzed by chstwo to calculate χ^2 and probability prob.

```
/* Driver for routine chstwo */

#include <stdio.h>
#include <math.h>
#include "nr.h"
#include "nrutil.h"

#define NBINS 10
#define NPTS 2000

int main(void)
{
    long idum=(-17);
    int i,ibin,j;
    float chsq,df,prob,x,*bins1,*bins2;

    bins1=vector(1,NBINS);
    bins2=vector(1,NBINS);
    for (j=1;j<=NBINS;j++) {
        bins1[j]=0.0;
        bins2[j]=0.0;
    }
    for (i=1;i<=NPTS;i++) {
        x=expdev(&idum);
        ibin=(int) (x*NBINS/3.0+1);
        if (ibin <= NBINS) ++bins1[ibin];
        x=expdev(&idum);
        ibin=(int) (x*NBINS/3.0+1);
        if (ibin <= NBINS) ++bins2[ibin];
    }
    chstwo(bins1,bins2,NBINS,0,&df,&chsq,&prob);
    printf("\n%15s %15s\n","dataset 1","dataset 2");
    for (i=1;i<=NBINS;i++)
        printf("%13.2f %15.2f\n",bins1[i],bins2[i]);
    printf("\n%18s %12.4f\n","chi-squared:",chsq);
    printf("%18s %12.4f\n","probability:",prob);
    free_vector(bins2,1,NBINS);
    free_vector(bins1,1,NBINS);
    return 0;
}
```

The Kolmogorov-Smirnov test used in ksone and kstwo applies to unbinned distributions with a single independent variable. ksone uses the K-S criterion to compare a single data set to an expected distribution, and `kstwo` uses it to compare two data sets. Sample program `xksone` creates data sets with Gaussian distributions and with stepwise increasing variance, and compares their cumulative distribution function to the expected result for a Gaussian distribution of unit variance. This result is the error function and is generated by routine `erff`. Increasing variance in the test distribution should reduce the likelihood that it was drawn from the same distribution represented by the comparison function.

```
/* Driver for routine ksone */

#include <stdio.h>
#include <math.h>
#include "nr.h"
#include "nrutil.h"

#define NPTS 1000
#define EPS 0.1

float func(float x)
{
    return 1.0 - erfcc(x/sqrt(2.0));
}

int main(void)
{
    long idum=(-5);
    int i,j;
    float d,factr,prob,varnce,*data;

    data=vector(1,NPTS);
    printf("%19s %16s %15s\n\n",
        "variance ratio","k-s statistic","probability");
    for (i=1;i<=11;i++) {
        varnce=1.0+(i-1)*EPS;
        factr=sqrt(varnce);
        for (j=1;j<=NPTS;j++)
            data[j]=factr*fabs(gasdev(&idum));
        ksone(data,NPTS,func,&d,&prob);
        printf("%16.6f %16.6f %16.6f \n",varnce,d,prob);
    }
    free_vector(data,1,NPTS);
    return 0;
}
```

`kstwo` compares the cumulative distribution functions of two unbinned data sets, `data1` and `data2`. In sample program `xkstwo`, they are both Gaussian distributions, but `data2` is given a stepwise increase of variance. In other respects, `xkstwo` is like `xksone`.

```
/* Driver for routine kstwo */

#include <stdio.h>
#include <math.h>
#include "nr.h"
#include "nrutil.h"
```

```
#define N1 2000
#define N2 1000
#define EPS 0.1

int main(void)
{
    long idum=(-1357);
    int i,j;
    float d,factr,prob,varnce,*data1,*data2;

    data1=vector(1,N1);
    data2=vector(1,N2);
    for (i=1;i<=N1;i++) data1[i]=gasdev(&idum);
    printf("%18s %15s %14s\n",
        "variance ratio","k-s statistic","probability");
    for (i=1;i<=11;i++) {
        varnce=1.0+(i-1)*EPS;
        factr=sqrt(varnce);
        for (j=1;j<=N2;j++)
                data2[j]=factr*gasdev(&idum);
        kstwo(data1,N1,data2,N2,&d,&prob);
        printf("%15.6f %15.6f %15.6f\n",varnce,d,prob);
    }
    free_vector(data2,1,N2);
    free_vector(data1,1,N1);
    return 0;
}
```

probks is an auxiliary routine for `ksone` and `kstwo` which calculates the function $Q_{ks}(\lambda)$ used to evaluate the probability that the two distributions being compared are the same. There is no independent means of producing this function, so in sample program xprobks we have chosen simply to graph it. Our output is reproduced below.

```
/* Driver for routine probks */

#include <stdio.h>
#include "nr.h"

#define NPTS 20
#define EPS 0.1
#define ISCAL 40

int main(void)
{
    int i,j,jmax;
    char txt[ISCAL+1];
    float alam,aval;

    printf("probability function for kolmogorov-smirnov statistic\n\n");
    printf("%7s %10s %13s\n","lambda","value:","graph:");
    for (i=1;i<=NPTS;i++) {
        alam=i*EPS;
        aval=probks(alam);
        jmax=(int) (0.5+(ISCAL-1)*aval);
        for (j=0;j<ISCAL;j++) {
            if (j < jmax)
                txt[j]='*';
```

```
          else
              txt[j]=' ';
      }
      txt[ISCAL]='\0';
      printf("%8.6f %10.6f        %s\n",alam,aval,txt);
  }
  return 0;
}
```

```
Probability func. for Kolmogorov-Smirnov statistic
  lambda     value:    graph:
 0.100000   1.000000   ****************************************
 0.200000   1.000000   ****************************************
 0.300000   0.999991   ****************************************
 0.400000   0.997192   ****************************************
 0.500000   0.963945   ***************************************
 0.600000   0.864283   **********************************
 0.700000   0.711235   ****************************
 0.800000   0.544142   *********************
 0.900000   0.392731   ****************
 1.000000   0.270000   ***********
 1.100000   0.177718   *******
 1.200000   0.112250   ****
 1.300000   0.068092   ***
 1.400000   0.039682   **
 1.500000   0.022218   *
 1.600000   0.011952
 1.700000   0.006177
 1.800000   0.003068
 1.900000   0.001464
 2.000000   0.000671
```

Procedure cntab1 analyzes a two-dimensional contingency table and returns several parameters describing any association between its nominal variables. Sample program xcntab1 supplies a table from a file table1.dat that is listed in the Appendix to this chapter. The table shows the rate of certain accidents, tabulated on a monthly basis. These data are listed, as well as their statistical properties, by xcntab1. We found the results to be:

```
            chi-squared                5026.29
            degrees of freedom           88.00
            probability                   0.0000
            cramer-v                      0.0772
            contingency coeff.            0.2134
```

```
/* Driver for routine cntab1 */

#include <stdio.h>
#include <stdlib.h>
#include "nr.h"
#include "nrutil.h"

#define NDAT 9
#define NMON 12
#define MAXSTR 80
```

```
int main(void)
{
    int i,j,**nmbr;
    float ccc,chisq,cramrv,df,prob;
    char dummy[MAXSTR],fate[NDAT+1][16],mon[NMON+1][6],txt[16];
    FILE *fp;

    nmbr=imatrix(1,NDAT,1,NMON);
    if ((fp = fopen("table1.dat","r")) == NULL)
        nrerror("Data file table1.dat not found\n");
    fgets(dummy,MAXSTR,fp);
    fgets(dummy,MAXSTR,fp);
    fscanf(fp,"%16c",txt);
    txt[15]='\0';
    for (i=1;i<=12;i++) fscanf(fp," %s",mon[i]);
    fgets(dummy,MAXSTR,fp);
    fgets(dummy,MAXSTR,fp);
    for (i=1;i<=NDAT;i++) {
        fscanf(fp,"%16[^0123456789]",fate[i]);
        fate[i][15]='\0';
        for (j=1;j<=12;j++)
            fscanf(fp,"%d ",&nmbr[i][j]);
    }
    fclose(fp);
    printf("\n%s",txt);
    for (i=1;i<=12;i++) printf("%5s",mon[i]);
    printf("\n\n");
    for (i=1;i<=NDAT;i++) {
        printf("%s",fate[i]);
        for (j=1;j<=12;j++) printf("%5d",nmbr[i][j]);
        printf("\n");
    }
    cntab1(nmbr,NDAT,NMON,&chisq,&df,&prob,&cramrv,&ccc);
    printf("\n%15s chi-squared      %20.2f\n"," ",chisq);
    printf("%15s degrees of freedom%20.2f\n"," ",df);
    printf("%15s probability       %20.4f\n"," ",prob);
    printf("%15s cramer-v          %20.4f\n"," ",cramrv);
    printf("%15s contingency coeff.%20.4f\n"," ",ccc);
    free_imatrix(nmbr,1,NDAT,1,NMON);
    return 0;
}
```

The test looks for any association between accidents and the months in which they occur. table1.dat clearly shows some. Drownings, for example, happen mostly in the summer. cntab2 carries out a similar analysis on table1.dat but measures associations on the basis of entropy. Sample program xcntab2 prints out the following entropies for the table:

entropy of table	4.0368
entropy of x-distribution	1.5781
entropy of y-distribution	2.4820
entropy of y given x	2.4588
entropy of x given y	1.5548
dependency of y on x	0.0094
dependency of x on y	0.0147
symmetrical dependency	0.0114

```
/* Driver for routine cntab2 */

#include <stdio.h>
#include <stdlib.h>
#include "nr.h"
#include "nrutil.h"

#define NI 9
#define NMON 12
#define MAXSTR 80

int main(void)
{
    float h,hx,hxgy,hy,hygx,uxgy,uxy,uygx;
    int i,j,**nmbr;
    char dummy[MAXSTR],fate[NI+1][16],mon[NMON+1][6],txt[16];
    FILE *fp;

    nmbr=imatrix(1,NI,1,NMON);
    if ((fp = fopen("table1.dat","r")) == NULL)
        nrerror("Data file table1.dat not found\n");
    fgets(dummy,MAXSTR,fp);
    fgets(dummy,MAXSTR,fp);
    fscanf(fp,"%16c",txt);
    txt[15]='\0';
    for (i=1;i<=12;i++) fscanf(fp," %s",mon[i]);
    fgets(dummy,MAXSTR,fp);
    fgets(dummy,MAXSTR,fp);
    for (i=1;i<=NI;i++) {
        fscanf(fp,"%16[^0123456789]",fate[i]);
        fate[i][15]='\0';
        for (j=1;j<=12;j++)
            fscanf(fp,"%d ",&nmbr[i][j]);
    }
    fclose(fp);
    printf("\n%s",txt);
    for (i=1;i<=12;i++) printf("%5s",mon[i]);
    printf("\n\n");
    for (i=1;i<=NI;i++) {
        printf("%s",fate[i]);
        for (j=1;j<=12;j++) printf("%5d",nmbr[i][j]);
        printf("\n");
    }
    cntab2(nmbr,NI,NMON,&h,&hx,&hy,&hygx,&hxgy,&uygx,&uxgy,&uxy);
    printf("\n       entropy of table          %10.4f\n",h);
    printf("        entropy of x-distribution %10.4f\n",hx);
    printf("        entropy of y-distribution %10.4f\n",hy);
    printf("        entropy of y given x      %10.4f\n",hygx);
    printf("        entropy of x given y      %10.4f\n",hxgy);
    printf("        dependency of y on x      %10.4f\n",uygx);
    printf("        dependency of x on y      %10.4f\n",uxgy);
    printf("        symmetrical dependency    %10.4f\n",uxy);
    free_imatrix(nmbr,1,NI,1,NMON);
    return 0;
}
```

The dependencies of x on y and y on x indicate the degree to which the type of accident can be predicted by knowing the month, or vice-versa.

pearsn makes an examination of two ordinal or continuous variables to find linear correlations. It returns a linear correlation coefficient r, a probability of correlation prob, and Fisher's z. Sample program xpearsn sets up data pairs in arrays dose and spore which show hypothetical data for the spore count from plants exposed to various levels of γ-rays. The results of applying pearsn to this data set are compared with the correct results by the program.

```
/* Driver for routine pearsn */

#include <stdio.h>
#include "nr.h"

#define N 10

int main(void)
{
    int i;
    float prob,r,z;
    static float dose[N+1]=
        {0.0,56.1,64.1,70.0,66.6,82.0,91.3,90.0,99.7,115.3,110.0};
    static float spore[N+1]=
        {0.0,0.11,0.40,0.37,0.48,0.75,0.66,0.71,1.20,1.01,0.95};

    printf("\nEffect of Gamma Rays on Man-in-the-Moon Marigolds\n");
    printf("%16s %23s\n","Count Rate (cpm)","Pollen Index");
    for (i=1;i<=N;i++)
        printf("%10.2f %25.2f \n",dose[i],spore[i]);
    pearsn(dose,spore,N,&r,&prob,&z);
    printf("\n%30s %16s\n","PEARSN","Expected");
    printf("%s %8s %9f %15f\n","Corr. Coeff."," ",r,0.9069586);
    printf("%s %9s %9f %15f\n","Probability"," ",prob,0.2926505e-3);
    printf("%s %10s %9f %15f\n","Fisher's z"," ",z,1.510110);
    return 0;
}
```

Rank order correlation may be done with spear to compare two distributions data1 and data2 for correlation. Correlations are reported both in terms of d, the sum-squared difference in ranks, and rs, Spearman's rank correlation parameter. Sample program xspear applies the calculation to the data in table table2.dat (see Appendix) which shows the solar flux incident on various cities during different months of the year. It then checks for correlations between columns of the table, considering each column as a separate data set. In this fashion it looks for correlations between the July solar flux and that of other months. The probability of such correlations are shown by probd and probrs. Our results are:

```
Check correlation of sampled U.S. solar radiation (july with other months)
month        d      st. dev.      probd     spearman-r    probrs
 jul       0.00    -4.358899    0.000013    1.000000     0.000000
 aug     122.00    -3.958458    0.000075    0.908132     0.000000
 sep     218.00    -3.643896    0.000269    0.835967     0.000004
 oct     384.00    -3.098495    0.001945    0.710843     0.000443
 nov     390.50    -3.077642    0.002086    0.706060     0.000503
 dec     622.00    -2.318075    0.020445    0.531803     0.015806
 jan     644.50    -2.244251    0.024816    0.514866     0.020181
```

feb	483.50	-2.772503	0.005563	0.636056	0.002573
mar	497.00	-2.728208	0.006368	0.625894	0.003158
apr	405.50	-3.027925	0.002462	0.694654	0.000677
may	264.00	-3.492371	0.000479	0.801205	0.000022
jun	121.50	-3.960099	0.000075	0.908509	0.000000

```c
/* Driver for routine spear */

#include <stdio.h>
#include <stdlib.h>
#include "nr.h"
#include "nrutil.h"

#define NDAT 20
#define NMON 12
#define MAXSTR 80

int main(void)
{
    int i,j;
    float d,probd,probrs,rs,zd,*ave,*data1,*data2,*zlat,**rays;
    char dummy[MAXSTR],txt[MAXSTR],city[NDAT+1][17],mon[NMON+1][5];
    FILE *fp;

    ave=vector(1,NDAT);
    data1=vector(1,NDAT);
    data2=vector(1,NDAT);
    zlat=vector(1,NDAT);
    rays=matrix(1,NDAT,1,NMON);
    if ((fp = fopen("table2.dat","r")) == NULL)
        nrerror("Data file table2.dat not found\n");
    fgets(dummy,MAXSTR,fp);
    fgets(txt,MAXSTR,fp);
    fscanf(fp,"%*15c");
    for (i=1;i<=NMON;i++) fscanf(fp," %s",mon[i]);
    fgets(dummy,MAXSTR,fp);
    fgets(dummy,MAXSTR,fp);
    for (i=1;i<=NDAT;i++) {
        fscanf(fp,"%[^0123456789]",city[i]);
        city[i][16]='\0';
        for (j=1;j<=NMON;j++) fscanf(fp,"%f",&rays[i][j]);
        fscanf(fp,"%f %f ",&ave[i],&zlat[i]);
    }
    fclose(fp);
    printf("%s\n",txt);
    printf("%16s"," ");
    for (i=1;i<=12;i++) printf("%4s",mon[i]);
    printf("\n");
    for (i=1;i<=NDAT;i++) {
        printf("%s",city[i]);
        for (j=1;j<=12;j++)
            printf("%4d",(int) (0.5+rays[i][j]));
        printf("\n");
    }
    /* Check temperature correlations between different months */
    printf("\nAre sunny summer places also sunny winter places?\n");
    printf("Check correlation of sampled U.S. solar radiation\n");
```

```
        printf("(july with other months)\n\n");
        printf("%s %9s %14s %11s %15s %10s\n","month","d",
            "st. dev.","probd","spearman-r","probrs");
        for (i=1;i<=NDAT;i++) data1[i]=rays[i][1];
        for (j=1;j<=12;j++) {
            for (i=1;i<=NDAT;i++) data2[i]=rays[i][j];
            spear(data1,data2,NDAT,&d,&zd,&probd,&rs,&probrs);
            printf("%4s %12.2f %12.6f %12.6f %13.6f %12.6f\n",
                mon[j],d,zd,probd,rs,probrs);
        }
        free_matrix(rays,1,NDAT,1,NMON);
        free_vector(zlat,1,NDAT);
        free_vector(data2,1,NDAT);
        free_vector(data1,1,NDAT);
        free_vector(ave,1,NDAT);
        return 0;
}
```

crank is an auxiliary routine for **spear** and is used in conjunction with **sort2**. The latter sorts an array, and **crank** then assigns ranks to each data entry, including the midranking of ties. Sample program xcrank uses the solar flux data of **table2.dat** (see Appendix) to illustrate. Each column of the solar flux table is replaced by the rank order of its entries. You can check the rank order chart against the chart of original values to verify the ordering.

```
/* Driver for routine crank */

#include <stdio.h>
#include "nr.h"
#include "nrutil.h"

#define NDAT 20
#define NMON 12
#define MAXSTR 80

int main(void)
{
    int i,j;
    float *data,*order,*s,**rays;
    char dummy[MAXSTR],txt[MAXSTR],city[NDAT+1][17],mon[NMON+1][5];
    FILE *fp;

    data=vector(1,NDAT);
    order=vector(1,NDAT);
    s=vector(1,NMON);
    rays=matrix(1,NDAT,1,NMON);
    if ((fp = fopen("table2.dat","r")) == NULL)
        nrerror("Data file table2.dat not found\n");
    fgets(dummy,MAXSTR,fp);
    fgets(txt,MAXSTR,fp);
    fscanf(fp,"%*15c");
    for (i=1;i<=NMON;i++) fscanf(fp," %s",mon[i]);
    fgets(dummy,MAXSTR,fp);
    fgets(dummy,MAXSTR,fp);
    for (i=1;i<=NDAT;i++) {
        fscanf(fp,"%[^0123456789]",city[i]);
```

```
            city[i][16]='\0';
            for (j=1;j<=NMON;j++) fscanf(fp,"%f",&rays[i][j]);
            fgets(dummy,MAXSTR,fp);
    }
    fclose(fp);
    printf("%s\n%16s",txt," ");
    for (i=1;i<=12;i++) printf(" %s",mon[i]);
    printf("\n");
    for (i=1;i<=NDAT;i++) {
        printf("%s",city[i]);
        for (j=1;j<=12;j++)
            printf("%4d",(int) (0.5+rays[i][j]));
        printf("\n");
    }
    printf(" press return to continue ...\n");
    getchar();
    /* Replace solar flux in each column by rank order */
    for (j=1;j<=12;j++) {
        for (i=1;i<=NDAT;i++) {
            data[i]=rays[i][j];
            order[i]=i;
        }
        sort2(NDAT,data,order);
        crank(NDAT,data,&s[j]);
        for (i=1;i<=NDAT;i++)
            rays[(int) (0.5+order[i])][j]=data[i];
    }
    printf("%16s"," ");
    for (i=1;i<=12;i++) printf(" %s",mon[i]);
    printf("\n");
    for (i=1;i<=NDAT;i++) {
        printf("%s",city[i]);
        for (j=1;j<=12;j++)
            printf("%4d",(int) (0.5+rays[i][j]));
        printf("\n");
    }
    free_matrix(rays,1,NDAT,1,NMON);
    free_vector(s,1,NMON);
    free_vector(order,1,NDAT);
    free_vector(data,1,NDAT);
    return 0;
}
```

kend11 and kend12 test for monotonic correlations of ordinal data. They differ in that kend11 compares two data sets of the same rank, while kend12 operates on a contingency table. Sample program xkend11, for example, uses kend11 to look for pair correlations in our five random number routines. That is to say, it tests for randomness by seeing if two consecutive numbers from the generator have a monotonic correlation. It uses the random number generators ran0, ..., ran4, one at a time, to generate 200 pairs of random numbers each. Then kend11 tests for correlation of the pairs, and a chart is made showing Kendall's τ, the standard deviation from the null hypotheses, and the probability. For a better test of the generators, you may wish to increase the number of pairs NDAT. It would also be a good idea to see how your result depends on the value of the seed idum.

```
/* Driver for routine kendl1 */

#include <stdio.h>
#include "nr.h"
#include "nrutil.h"

#define NDAT 200

int main(void)
{
    int i,j;
    long idum;
    float prob,tau,z,*data1,*data2;
    static char *txt[5]={"RAN0","RAN1","RAN2","RAN3","RAN4"};

    data1=vector(1,NDAT);
    data2=vector(1,NDAT);
    /* Look for correlations in RAN0, RAN1, RAN2, RAN3 and RAN4 */
    printf("\nPair correlations of RAN0 ... RAN4\n\n");
    printf("%9s %17s %16s %18s\n",
        "Program","Kendall tau","Std. Dev.","Probability");
    for (i=1;i<=5;i++) {
        idum=(-1357);
        for (j=1;j<=NDAT;j++) {
            if (i == 1) {
                data1[j]=ran0(&idum);
                data2[j]=ran0(&idum);
            } else if (i == 2) {
                data1[j]=ran1(&idum);
                data2[j]=ran1(&idum);
            } else if (i == 3) {
                data1[j]=ran2(&idum);
                data2[j]=ran2(&idum);
            } else if (i == 4) {
                data1[j]=ran3(&idum);
                data2[j]=ran3(&idum);
            } else if (i == 5) {
                data1[j]=ran4(&idum);
                data2[j]=ran4(&idum);
            }
        }
        kendl1(data1,data2,NDAT,&tau,&z,&prob);
        printf("%8s %17.6f %17.6f %17.6f\n",txt[i-1],tau,z,prob);
    }
    free_vector(data2,1,NDAT);
    free_vector(data1,1,NDAT);
    return 0;
}
```

Here is the output from the above test:

```
Pair correlations of RAN0 ... RAN4
Program       Kendall tau       Std. Dev.       Probability
  RAN0         -0.010452       -0.219801         0.826026
  RAN1         -0.023719       -0.498780         0.617934
  RAN2          0.130251        2.739064         0.006161
  RAN3         -0.062814       -1.320922         0.186527
```

```
RAN4        0.023015        0.483986        0.628396
```

Sample program `xkendl2`, for routine `kendl2`, prepares a contingency table based on the routines `irbit1` and `irbit2`. You may recall that these routines generate random binary sequences. The program checks the sequences by breaking them into groups of three bits. Each group is treated as a three-bit binary number. Two consecutive groups then act as indices into an 8×8 contingency table that records how many times each possible sequence of six bits (two groups) occurs. For each random bit generator, NDAT=1000 samples are taken. Then the contingency table `tab[k][l]` is analyzed by `kendl2` to find Kendall's τ, the standard deviation, and the probability. Notice that Kendall's τ can only be applied when both variables are ordinal (here, the numbers 0 to 7), and that the test is specifically for monotonic correlations. In this case we are actually testing whether the larger 3-bit binary numbers tend to be followed by others of their own kind. Within the program, we have expressed this roughly as a test of whether ones or zeros tend to come in groups more than they should.

```c
/* Driver for routine kendl2 */

#include <stdio.h>
#include "nr.h"
#include "nrutil.h"

#define NDAT 1000
#define IP 8
#define JP 8

int main(void)
{
    int ifunc,i=IP,j=JP,k,l,m,n,twoton;
    unsigned long iseed;
    float prob,tau,z,**tab;
    static char *txt[8]=
        {"000","001","010","011","100","101","110","111"};

    /* Look for 'ones-after-zeros' in IRBIT1 and IRBIT2 sequences */
    tab=matrix(1,IP,1,JP);
    printf("Are ones followed by zeros and vice-versa?\n");
    for (ifunc=1;ifunc<=2;ifunc++) {
        iseed=2468;
        if (ifunc == 1)
            printf("test of irbit1:\n");
        else
            printf("test of irbit2:\n");
        for (k=1;k<=i;k++)
            for (l=1;l<=j;l++) tab[k][l]=0.0;
        for (m=1;m<=NDAT;m++) {
            k=1;
            twoton=1;
            for (n=0;n<=2;n++) {
                if (ifunc == 1)
                    k += (irbit1(&iseed)*twoton);
                else
                    k += (irbit2(&iseed)*twoton);
                twoton *= 2;
```

```
                    }
                    l=1;
                    twoton=1;
                    for (n=0;n<=2;n++) {
                        if (ifunc == 1)
                            l += (irbit1(&iseed)*twoton);
                        else
                            l += (irbit2(&iseed)*twoton);
                        twoton *= 2;
                    }
                    ++tab[k][l];
                }
                kendl2(tab,i,j,&tau,&z,&prob);
                printf("      ");
                for (n=0;n<=7;n++) printf("%6s",txt[n]);
                printf("\n");
                for (n=1;n<=8;n++) {
                    printf("%3s",txt[n-1]);
                    for (m=1;m<=8;m++)
                        printf("%6d",(int) (0.5+tab[n][m]));
                    printf("\n");
                }
                printf("\n%17s %14s %16s\n",
                    "kendall tau","std. dev.","probability");
                printf("%15.6f %15.6f %15.6f\n\n",tau,z,prob);
            }
            free_matrix(tab,1,IP,1,JP);
            return 0;
        }
}
```

The output of xkendl2 looks like this:

```
Are ones followed by zeros and vice-versa?
test of irbit1:
          000    001    010    011    100    101    110    111
000        17     15      3     14     14     18     10     15
001        18     14      8     16     14     22     17     10
010        18     15      6     14     14     16     13     15
011        17     13      8     18     21     26     17     13
100        15     16     12     11     18     18     18     18
101        14     10     15     17     18     16     25     15
110        16     12     25     15     23     15     13     22
111        17     15     16     18     16     15     23     14
        kendall tau      std. dev.       probability
         0.011227        0.531589        0.595010
test of irbit2:
          000    001    010    011    100    101    110    111
000        21     12     21     16     17      9     14     15
001        14     17     15     18     10     17     20     16
010        19     17     15     16     17     17      9     14
011        12     18      9     16     13     26     13     13
100        16     18     14     16     12     19     15     11
101        12     21     20     16     17     13      8     15
110        14     18     20     17     12     17     12     19
111        19     16     15     22     19      9     19     13
        kendall tau      std. dev.       probability
        -0.003014       -0.142715        0.886516
```

The routine `ks2d1s` tests a single sample (hence "1s") of a two-dimensional distribution (hence "2d") against a theoretical model distribution, returning the significance level with which the distribution can be said to differ from the model. The model, supplied by the user, is contained in the routine `quadvl`. The main book supplies a sample `quadvl` where the model is a distribution that is uniform inside the square $|x| < 1$, $|y| < 1$, so we will use that `quadvl` in our example program.

We generate a sample that is made nonuniform by "warping" the output of the uniform random number generator by a quadratic function of adjustable amplitude (`factor`). Then, the program invokes `ks2d1s` and prints the significance of the result. This is repeated some number of times, so that you can get an idea of how variable the test is from run to run.

If you run the program, you will see that `factor` as large as 0.1 is easily detectable in a 500 point sample (i.e., it gives very small values for PROB), but not detectable in 100 points. In a 100 point sample, `factor` must be on the order of 0.5 before it becomes easily detectable.

```
/* Driver for routine ks2d1s */

#include <stdio.h>
#include "nr.h"
#include "nrutil.h"

#define NMAX 1000

int main(void)
{
    long idum;
    unsigned long j,jtrial,n1,ntrial;
    float d,factor,prob,u,v,*x1,*y1;

    x1=vector(1,NMAX);
    y1=vector(1,NMAX);
    for (;;) {
        printf("How many points?\n");
        if (scanf("%lu",&n1) == EOF) break;
        if (n1 > NMAX) {
            printf("n1 too large.\n");
            continue;
        }
        printf("What factor nonlinearity (0 to 1)?\n");
        for (;;) {
            if (scanf("%f",&factor) == EOF) {
                factor = -1.0;
                break;
            }
            if (factor < 0.0) {
                printf("factor less than 0\n");
                continue;
            }
            if (factor > 1.0) {
                printf("factor greater than 1\n");
                continue;
            }
            break;
```

```
            }
            if (factor == -1.0) break;
            printf("How many trials?\n");
            if (scanf("%lu",&ntrial) == EOF) break;
            idum = -289-ntrial-n1;
            for (jtrial=1;jtrial<=ntrial;jtrial++) {
                for (j=1;j<=n1;j++) {
                    u=ran1(&idum);
                    u=u*((1.0-factor)+u*factor);
                    x1[j]=2.0*u-1.0;
                    v=ran1(&idum);
                    v=v*((1.0-factor)+v*factor);
                    y1[j]=2.0*v-1.0;
                }
                ks2d1s(x1,y1,n1,quadvl,&d,&prob);
                printf("d, prob= %12.6f %12.6f\n",d,prob);
            }
        }
        free_vector(y1,1,NMAX);
        free_vector(x1,1,NMAX);
        printf("Normal completion\n");
        return 0;
}
```

The routine ks2d2s performs a similar statistical test, but on two samples and without the necessity of a model. The example program is quite similar to that for ks2d1s. Now, however, we generate two samples, drawn from 2-dimensional Gaussian distributions of slightly different shape (as controlled by the variable shrink).

Running the program, one finds that with, e.g., sample sizes of 100 points for the first sample and 80 points for the second sample, a value shrink = 0.5 is easily detectable as a low value for prob, while shrink = 0.8 is not.

```
/* Driver for routine ks2d2s */

#include <stdio.h>
#include "nr.h"
#include "nrutil.h"

#define NMAX 1000

int main(void)
{
        long idum;
        unsigned long j,jtrial,n1,n2,ntrial;
        float d,prob,shrink,u,v;
        float *x1,*y1,*x2,*y2;

        x1=vector(1,NMAX);
        y1=vector(1,NMAX);
        x2=vector(1,NMAX);
        y2=vector(1,NMAX);
        for (;;) {
            printf("Input N1,N2\n");
            if (scanf("%lu %lu",&n1,&n2) == EOF) break;
            if (n1 > NMAX) {
                printf("n1 too large.\n");
```

```
            continue;
        }
        if (n2 > NMAX) {
            printf("n2 too large.\n");
            continue;
        }
        printf("What shrinkage?\n");
        if (scanf("%f",&shrink) == EOF) break;
        printf("How many trials?\n");
        if (scanf("%lu",&ntrial) == EOF) break;
        if (ntrial > NMAX) {
            printf("Too many trials.\n");
            continue;
        }
        idum = -287-ntrial-n1-n2;
        for (jtrial=1;jtrial<=ntrial;jtrial++) {
            for (j=1;j<=n1;j++) {
                u=gasdev(&idum);
                v=gasdev(&idum)*shrink;
                x1[j]=u+v;
                y1[j]=u-v;
            }
            for (j=1;j<=n2;j++) {
                u=gasdev(&idum)*shrink;
                v=gasdev(&idum);
                x2[j]=u+v;
                y2[j]=u-v;
            }
            ks2d2s(x1,y1,n1,x2,y2,n2,&d,&prob);
            printf("d, prob= %12.6f %12.6f\n",d,prob);
        }
    }
    free_vector(y2,1,NMAX);
    free_vector(x2,1,NMAX);
    free_vector(y1,1,NMAX);
    free_vector(x1,1,NMAX);
    printf("Normal completion\n");
    return 0;
}
```

Sample program `xsavgol` tests `savgol` by computing Savitzky-Golay smoothing coefficients for various values of M, n_L and n_R. It then compares the answers with pre-saved values.

```
/* Driver for routine savgol */

#include <stdio.h>
#include "nr.h"
#include "nrutil.h"

#define NMAX 1000
#define NTEST 6

int main(void)
{
    int i,j,m,nl,np,nr;
    float *c,sum;
```

```
      static int mtest[NTEST+1] = {0,2,2,2,2,4,4};
      static int nltest[NTEST+1] = {0,2,3,4,5,4,5};
      static int nrtest[NTEST+1] = {0,2,1,0,5,4,5};
      static char *ans[NTEST+1] = {"",
"                 -0.086  0.343  0.486  0.343 -0.086",
"          -0.143  0.171  0.343  0.371  0.257",
"      0.086 -0.143 -0.086  0.257  0.886",
" -0.084  0.021  0.103  0.161  0.196  0.207  0.196  0.161  0.103  0.021 -0.084",
"      0.035 -0.128  0.070  0.315  0.417  0.315  0.070 -0.128  0.035",
"  0.042 -0.105 -0.023  0.140  0.280  0.333  0.280  0.140 -0.023 -0.105  0.042"};

      c=vector(1,NMAX);
      printf("M nl nr\n");
      printf("\t\t\tSample Savitzky-Golay Coefficients\n");
      for (i=1;i<=NTEST;i++) {
          m=mtest[i];
          nl=nltest[i];
          nr=nrtest[i];
          np=nl+nr+1;
          savgol(c,np,nl,nr,0,m);
          for (sum=0.0,j=1;j<=np;j++) sum += c[j];
          printf("%1d %1d %1d\n",m,nl,nr);
          for (j=nl;j<5;j++) printf("%7s"," ");
          for (j=nl+1;j>=1;j--) printf("%7.3f",c[j]);
          for (j=0;j<nr;j++) printf("%7.3f",c[np-j]);
          printf("\n");
          printf("Sum = %7.3f\n",sum);
          printf("Compare answer:\n%s\n",ans[i]);
      }
      free_vector(c,1,NMAX);
      return 0;
}
```

Appendix

File table1.dat:

Accidental Deaths by Month and Type (1979)

Month:	jan	feb	mar	apr	may	jun	jul	aug	sep	oct	nov	dec
Motor Vehicle	3298	3304	4241	4291	4594	4710	4914	4942	4861	4914	4563	4892
Falls	1150	1034	1089	1126	1142	1100	1112	1099	1114	1079	999	1181
Drowning	180	190	370	530	800	1130	1320	990	580	320	250	212
Fires	874	768	630	516	385	324	277	272	271	381	533	760
Choking	299	264	258	247	273	269	251	269	271	279	297	266
Fire-arms	168	142	122	140	153	142	147	160	162	172	266	230
Poisons	298	277	346	263	253	239	268	228	240	260	252	241
Gas-poison	267	193	144	127	70	63	55	53	60	118	150	172
Other	1264	1234	1172	1220	1547	1339	1419	1453	1359	1308	1264	1246

File table2.dat:

Average solar radiation (watts/square meter) for selected cities

Month:	jul	aug	sep	oct	nov	dec	jan	feb	mar	apr	may	jun	ave	lat
Atlanta, GA	257	246	201	166	30	102	106	140	184	236	258	271	192	34.0
Barrow, AK	208	123	56	20	0	0	0	18	87	184	248	256	100	71.0
Bismark, ND	296	251	185	132	78	60	76	121	170	217	267	284	178	47.0
Boise, ID	324	275	221	152	88	60	69	113	164	235	284	309	191	43.5

Boston, MA	240	206	165	115	70	58	67	96	142	176	228	242	150	42.5
Caribou, ME	246	218	161	102	53	51	66	111	178	194	229	232	153	47.0
Cleveland, OH	267	239	182	127	68	56	60	87	151	182	253	271	162	41.5
Dodge City, KS	311	287	239	184	138	113	123	153	202	256	275	315	216	38.0
El Paso, TX	324	309	278	224	178	151	160	209	266	317	346	353	260	32.0
Fresno, CA	323	293	243	182	117	77	90	143	212	264	308	337	216	37.0
Greensboro, NC	263	235	197	156	118	95	97	134	171	227	257	273	185	36.0
Honolulu, HI	305	293	271	245	208	176	175	200	234	262	300	297	247	21.0
Little Rock, AR	270	250	214	167	118	91	96	127	173	220	256	272	188	35.0
Miami, FL	260	246	216	188	171	154	166	201	238	263	267	257	219	26.0
New York, NY	251	238	175	127	77	62	71	102	151	183	220	255	159	41.0
Omaha, NE	275	252	192	142	96	80	99	134	172	224	248	272	182	21.0
Rapid City, SD	288	262	208	152	99	76	90	135	193	235	259	287	190	44.0
Seattle, WA	242	209	150	84	44	29	34	60	118	174	216	228	132	47.5
Tucson, AZ	304	286	281	216	172	144	151	195	264	322	358	343	253	41.0
Washington, DC	267	190	196	145	75	64	101	124	153	182	215	247	163	39.0

Chapter 15: Modeling of Data

*Chapter 15 of Numerical Recipes deals with the fitting of a model func-
tion to a set of data, in order to summarize the data in terms of a few model
parameters. Both traditional least-squares fitting and robust fitting are considered. Fits to a straight line are carried out by routine* fit. *When there
are errors in both x and y, straight-line fitting is done by* fitexy. *More general linear least-squares fits are handled by* lfit *and* covsrt. *(Remember
that the term "linear" here refers not to a linear dependence of the fitting
function on its argument, but rather to the fact that a linear combination of
the functions is fit to the data.) In cases where* lfit *fails, owing probably to
near degeneracy of some basis functions, the answer may still be found using*
svdfit *and* svdvar. *In fact, these are generally recommended in preference
to* lfit *because they never (?) fail. For nonlinear least-squares fits, the
Levenberg-Marquardt method is discussed, and is implemented in* mrqmin,
which makes use also of covsrt *and* mrqcof.

Robust estimation is discussed in several forms, and illustrated by routine
medfit *which fits a straight line to data points based on the criterion of
least absolute deviations rather than least-squared deviations.* rofunc *is an
auxiliary function for* medfit.

$$\star \quad \star \quad \star \quad \star$$

Routine fit fits a set of N data points (x[i],y[i]), with standard deviations
sig[i], to the linear model $y = A + Bx$. It uses χ^2 as the criterion for goodness-
of-fit. To demonstrate fit, we generate some noisy data in sample program xfit.
For NPT values of i we take $x = 0.1i$ and $y = -2x + 1$ plus some values drawn
from a Gaussian distribution to represent noise. Then we make two calls to fit, first
performing the fit without allowance for standard deviations sig[i], and then with
such allowance. Since sig[i] has been set to the constant value spread, it should
not affect the resulting parameter values. The values output from this routine are:

Ignoring standard deviation:

```
a  =     1.079573      uncertainty:  0.099821
b  =    -2.006663      uncertainty:  0.017161
chi-squared:          24.047955
goodness-of-fit:       1.000000
```

Including standard deviation:

```
a  =     1.079573      uncertainty:  0.100755
b  =    -2.006663      uncertainty:  0.017321
chi-squared:          96.191818
```

```
          goodness-of-fit:     0.532772
```

```c
/* Driver for routine fit */

#include <stdio.h>
#include "nr.h"
#include "nrutil.h"

#define NPT 100
#define SPREAD 0.5

int main(void)
{
    long idum=(-117);
    int i,mwt;
    float a,b,chi2,q,siga,sigb,*x,*y,*sig;

    x=vector(1,NPT);
    y=vector(1,NPT);
    sig=vector(1,NPT);
    for (i=1;i<=NPT;i++) {
        x[i]=0.1*i;
        y[i] = -2.0*x[i]+1.0+SPREAD*gasdev(&idum);
        sig[i]=SPREAD;
    }
    for (mwt=0;mwt<=1;mwt++) {
        fit(x,y,NPT,sig,mwt,&a,&b,&siga,&sigb,&chi2,&q);
        if (mwt == 0)
            printf("\nIgnoring standard deviations\n");
        else
            printf("\nIncluding standard deviations\n");
        printf("%12s %9.6f %18s %9.6f \n",
            "a  =  ",a,"uncertainty:",siga);
        printf("%12s %9.6f %18s %9.6f \n",
            "b  =  ",b,"uncertainty:",sigb);
        printf("%19s %14.6f \n","chi-squared: ",chi2);
        printf("%23s %10.6f \n","goodness-of-fit: ",q);
    }
    free_vector(sig,1,NPT);
    free_vector(y,1,NPT);
    free_vector(x,1,NPT);
    return 0;
}
```

Here is a program that puts the routine `fitexy`, which fits a straight line to data with error bars in both x and y, through its paces. The underlying straight line is taken to be

$$y = 2x - 5$$

We generate a synthetic data set that has errors in both x and y, with the errors having a magnitude that is highly variable (but known) from point to point. We then call `fitexy` and print its result. As a comparison, we set the x errors to zero and do the fit both with `fitexy` and with the simpler routine `fit`. Similarly, interchanging the roles of x and y, we can do this setting the y errors to zero. (We have to transform the answers, however, so as to be able to interpret x and y in their true roles.)

Output from this program looks like this:

```
Values of a,b,siga,sigb,chi2,q:
Fit with x and y errors gives:
 -5.058980    2.022800    0.274571    0.020892 42.586609    0.038141
Setting x errors to zero gives:
 -5.538356    2.008888    0.110675    0.009360 277.534546    0.000000
...to be compared with fit result:
 -5.538357    2.008888    0.104930    0.008537 277.534424    0.000000
Setting y errors to zero gives:
 -5.477018    2.043204    0.184778    0.012426 105.700729    0.000000
...to be compared with fit result:
 -5.494896    2.044625    0.160466    0.012380 105.687660    0.000000
```

Notice that the intercept and slope are obtained correctly, notwithstanding a somewhat unlucky value for the χ^2 probability. (This demonstrates that you should not reject a hypothesis just because its χ^2 probability is as small as a few percent: that happens a few percent of the time for correct models!)

Setting either the x or y errors to zero, however, produces significantly wrong results for the intercept, as well as misleading formal errors on the parameters and unacceptable χ^2 probabilities.

```c
/* Driver for routine fitexy */

#include <stdio.h>
#include <math.h>
#include "nr.h"
#include "nrutil.h"

#define NPT 30

int main(void)
{
    long idum=(-1);
    int j;
    float a,b,chi2,q,sa,sb,siga,sigb;
    float *x,*y,*dx,*dy,*dz;

    x=vector(1,NPT);
    y=vector(1,NPT);
    dx=vector(1,NPT);
    dy=vector(1,NPT);
    dz=vector(1,NPT);
    for (j=1;j<=NPT;j++) {
        dx[j]=0.1+ran1(&idum);
        dy[j]=0.1+ran1(&idum);
        dz[j]=0.0;
        x[j]=10.0+10.0*gasdev(&idum);
        y[j]=2.0*x[j]-5.0+dy[j]*gasdev(&idum);
        x[j] += dx[j]*gasdev(&idum);
    }
    printf("Values of a,b,siga,sigb,chi2,q:\n");
    printf("Fit with x and y errors gives:\n");
    fitexy(x,y,NPT,dx,dy,&a,&b,&siga,&sigb,&chi2,&q);
    printf("%11.6f %11.6f %11.6f %11.6f %11.6f %11.6f\n\n",
            a,b,siga,sigb,chi2,q);
```

```
        printf("Setting x errors to zero gives:\n");
        fitexy(x,y,NPT,dz,dy,&a,&b,&siga,&sigb,&chi2,&q);
        printf("%11.6f %11.6f %11.6f %11.6f %11.6f\n",
            a,b,siga,sigb,chi2,q);
        printf("...to be compared with fit result:\n");
        fit(x,y,NPT,dy,1,&a,&b,&siga,&sigb,&chi2,&q);
        printf("%11.6f %11.6f %11.6f %11.6f %11.6f %11.6f\n\n",
            a,b,siga,sigb,chi2,q);
        printf("Setting y errors to zero gives:\n");
        fitexy(x,y,NPT,dx,dz,&a,&b,&siga,&sigb,&chi2,&q);
        printf("%11.6f %11.6f %11.6f %11.6f %11.6f %11.6f\n",
            a,b,siga,sigb,chi2,q);
        printf("...to be compared with fit result:\n");
        fit(y,x,NPT,dx,1,&a,&b,&siga,&sigb,&chi2,&q);
        sa=sqrt(siga*siga+SQR(sigb*(a/b)))/b;
        sb=sigb/(b*b);
        printf("%11.6f %11.6f %11.6f %11.6f %11.6f %11.6f\n",
            -a/b,1.0/b,sa,sb,chi2,q);
        free_vector(dz,1,NPT);
        free_vector(dy,1,NPT);
        free_vector(dx,1,NPT);
        free_vector(y,1,NPT);
        free_vector(x,1,NPT);
        return 0;
}
```

`lfit` carries out the same sort of fit as `fit` but this time does a linear least-squares fit to a more general function. In sample program `xlfit` the chosen function is a linear sum of powers of x, generated by routine `funcs`. For convenience in checking the result we have generated data according to $y = 1 + 2x + 3x^2 + \cdots$. This series is truncated depending on the choice of NTERM, and some Gaussian noise is added to simulate realistic data. The `sig[i]` are taken as constant errors. `lfit` is called twice to fit the same data. The first time ia[i] is set to 1, so that `lfit` fits for all the parameters: a[1] ≈ 1.0, a[2] ≈ 2.0, a[3] ≈ 3.0,.... Then, as a test of the ia feature, which determines which parameters are to be fit, the fit is restricted to odd-numbered parameters, while even-numbered parameters are fixed. In this case the elements of the covariance matrix associated with fixed parameters should be zero. In `xlfit`, we have set NTERM=5 to fit a 5th degree polynomial.

```
/* Driver for routine lfit */

#include <stdio.h>
#include <math.h>
#include "nr.h"
#include "nrutil.h"

#define NPT 100
#define SPREAD 0.1
#define NTERM 5

void funcs(float x,float afunc[],int ma)
{
    int i;

    afunc[1]=1.0;
```

```
        afunc[2]=x;
        for (i=3;i<=ma;i++) afunc[i]=sin(i*x);
}

int main(void)
{
        long idum=(-911);
        int i,j,*ia;
        float chisq,*a,*x,*y,*sig,**covar;

        ia=ivector(1,NTERM);
        a=vector(1,NTERM);
        x=vector(1,NPT);
        y=vector(1,NPT);
        sig=vector(1,NPT);
        covar=matrix(1,NTERM,1,NTERM);

        for (i=1;i<=NPT;i++) {
                x[i]=0.1*i;
                funcs(x[i],a,NTERM);
                y[i]=0.0;
                for (j=1;j<=NTERM;j++) y[i] += j*a[j];
                y[i] += SPREAD*gasdev(&idum);
                sig[i]=SPREAD;
        }
        for (i=1;i<=NTERM;i++) ia[i]=1;
        lfit(x,y,sig,NPT,a,ia,NTERM,covar,&chisq,funcs);
        printf("\n%11s %21s\n","parameter","uncertainty");
        for (i=1;i<=NTERM;i++)
                printf("  a[%1d] = %8.6f %12.6f\n",
                        i,a[i],sqrt(covar[i][i]));
        printf("chi-squared = %12f\n",chisq);
        printf("full covariance matrix\n");
        for (i=1;i<=NTERM;i++) {
                for (j=1;j<=NTERM;j++) printf("%12f",covar[i][j]);
                printf("\n");
        }
        printf("\npress RETURN to continue...\n");
        (void) getchar();
        /* Now check results of restricting fit parameters */
        for (i=2;i<=NTERM;i+=2) ia[i]=0;
        lfit(x,y,sig,NPT,a,ia,NTERM,covar,&chisq,funcs);
        printf("\n%11s %21s\n","parameter","uncertainty");
        for (i=1;i<=NTERM;i++)
                printf("  a[%1d] = %8.6f %12.6f\n",
                        i,a[i],sqrt(covar[i][i]));
        printf("chi-squared = %12f\n",chisq);
        printf("full covariance matrix\n");
        for (i=1;i<=NTERM;i++) {
                for (j=1;j<=NTERM;j++) printf("%12f",covar[i][j]);
                printf("\n");
        }
        printf("\n");
        free_matrix(covar,1,NTERM,1,NTERM);
        free_vector(sig,1,NPT);
        free_vector(y,1,NPT);
        free_vector(x,1,NPT);
```

```
        free_vector(a,1,NTERM);
        free_ivector(ia,1,NTERM);
        return 0;
}
```

covsrt is used in conjunction with lfit (and later with the routine svdfit) to redistribute the covariance matrix covar so that it represents the true order of coefficients, rather than the order in which they were fit. In sample routine xcovsrt an artificial 10×10 covariance matrix covar[i][j] is created, which is all zeros except for the upper left 5×5 section, for which the elements are covar[i][j]=i+j-1. Then two tests are performed:

1. By setting ia[i] = 1 for $i = 1, \ldots, 10$ and MFIT=5, we simulate a fit where the first 5 out of 10 parameters are fit. covar[i][j] should be returned unchanged.

2. By setting the even elements of the ia array to 1 and the odd elements to zero, we simulate a fit where parameters 2,4,6,8 and 10 have been fit. The elements of covar[i][j] should be spread so that alternate elements are zero.

```
/* Driver for routine covsrt */

#include <stdio.h>
#include "nr.h"
#include "nrutil.h"

#define MA 10
#define MFIT 5

int main(void)
{
    int i,j,*ia;
    float **covar;

    ia=ivector(1,MA);
    covar=matrix(1,MA,1,MA);
    for (i=1;i<=MA;i++)
        for (j=1;j<=MA;j++) {
            covar[i][j]=0.0;
            if ((i <= MFIT) && (j <= MFIT))
                covar[i][j]=i+j-1;
        }
    printf("\noriginal matrix\n");
    for (i=1;i<=MA;i++) {
        for (j=1;j<=MA;j++) printf("%4.1f",covar[i][j]);
        printf("\n");
    }
    printf("press RETURN to continue...\n");
    (void) getchar();
    printf("\nTest #1 - full fitting\n");
    for (i=1;i<=MA;i++) ia[i]=1;
    covsrt(covar,MA,ia,MA);
    for (i=1;i<=MA;i++) {
        for (j=1;j<=MA;j++) printf("%4.1f",covar[i][j]);
        printf("\n");
    }
    printf("press RETURN to continue...\n");
    (void) getchar();
```

```
        printf("\nTest #2 - spread\n");
        for (i=1;i<=MA;i++)
            for (j=1;j<=MA;j++) {
                covar[i][j]=0.0;
                if ((i <= MFIT) && (j <= MFIT)) covar[i][j]=i+j-1;
            }
        for (i=1;i<=MA;i+=2) ia[i]=0;
        covsrt(covar,MA,ia,MFIT);
        for (i=1;i<=MA;i++) {
            for (j=1;j<=MA;j++) printf("%4.1f",covar[i][j]);
            printf("\n");
        }
        free_matrix(covar,1,MA,1,MA);
        free_ivector(ia,1,MA);
        return 0;
    }
```

Routine `svdfit` is recommended in preference to `lfit` for performing linear least-squares fits. The sample program `xsvdfit` puts `svdfit` to work on the data generated according to

$$F(x) = 1 + 2x + 3x^2 + 4x^3 + 5x^4 + \text{Gaussian noise.}$$

This data is fit first to a five-term polynomial sum, and then to a five-term Legendre polynomial sum. In each case `sig[i]`, the measurement fluctuation in y, is taken to be constant. For the polynomial fit, the resulting coefficients should clearly have the values `a[i]` $\approx$ i. For Legendre polynomials the expected results are:

$$a[1] \approx 3.0$$

$$a[2] \approx 4.4$$

$$a[3] \approx 4.9$$

$$a[4] \approx 1.6$$

$$a[5] \approx 1.1$$

```
/* Driver for routine svdfit */

#include <stdio.h>
#include <math.h>
#include "nr.h"
#include "nrutil.h"

#define NPT 100
#define SPREAD 0.02
#define NPOL 5

int main(void)
{
    long idum=(-911);
    int i;
    float chisq,*x,*y,*sig,*a,*w,**cvm,**u,**v;

    x=vector(1,NPT);
    y=vector(1,NPT);
```

```
        sig=vector(1,NPT);
        a=vector(1,NPOL);
        w=vector(1,NPOL);
        cvm=matrix(1,NPOL,1,NPOL);
        u=matrix(1,NPT,1,NPOL);
        v=matrix(1,NPOL,1,NPOL);
        for (i=1;i<=NPT;i++) {
            x[i]=0.02*i;
            y[i]=1.0+x[i]*(2.0+x[i]*(3.0+x[i]*(4.0+x[i]*5.0)));
            y[i] *= (1.0+SPREAD*gasdev(&idum));
            sig[i]=y[i]*SPREAD;
        }
        svdfit(x,y,sig,NPT,a,NPOL,u,v,w,&chisq,fpoly);
        svdvar(v,NPOL,w,cvm);
        printf("\npolynomial fit:\n\n");
        for (i=1;i<=NPOL;i++)
            printf("%12.6f %s %10.6f\n",a[i]," +-",sqrt(cvm[i][i]));
        printf("\nChi-squared %12.6f\n",chisq);
        svdfit(x,y,sig,NPT,a,NPOL,u,v,w,&chisq,fleg);
        svdvar(v,NPOL,w,cvm);
        printf("\nLegendre polynomial fit:\n\n");
        for (i=1;i<=NPOL;i++)
            printf("%12.6f %s %10.6f\n",a[i]," +-",sqrt(cvm[i][i]));
        printf("\nChi-squared %12.6f\n",chisq);
        free_matrix(v,1,NPOL,1,NPOL);
        free_matrix(u,1,NPT,1,NPOL);
        free_matrix(cvm,1,NPOL,1,NPOL);
        free_vector(w,1,NPOL);
        free_vector(a,1,NPOL);
        free_vector(sig,1,NPT);
        free_vector(y,1,NPT);
        free_vector(x,1,NPT);
        return 0;
    }
```

svdvar is used with svdfit to evaluate the covariance matrix cvm of a fit with MA parameters. In program xsvdvar, we provide input vector w and array v for this routine via two data statements, and calculate the covariance matrix cvm determined from them. We have also done the calculation by hand and recorded the correct results in array tru for comparison.

```
/* Driver for routine svdvar */

#include <stdio.h>
#include "nr.h"
#include "nrutil.h"

#define NP 6
#define MA 3

int main(void)
{
    int i,j;
    float **cvm,**v;
    static float vtemp[NP][NP]=
        {1.0,1.0,1.0,1.0,1.0,1.0,
        2.0,2.0,2.0,2.0,2.0,2.0,
```

```
            3.0,3.0,3.0,3.0,3.0,3.0,
            4.0,4.0,4.0,4.0,4.0,4.0,
            5.0,5.0,5.0,5.0,5.0,5.0,
            6.0,6.0,6.0,6.0,6.0,6.0};
    static float w[NP+1]=
            {0.0,0.0,1.0,2.0,3.0,4.0,5.0};
    static float tru[MA][MA]=
            {1.25,2.5,3.75,
            2.5,5.0,7.5,
            3.75,7.5,11.25};

    cvm=matrix(1,MA,1,MA);
    v=convert_matrix(&vtemp[0][0],1,NP,1,NP);
    printf("\nmatrix v\n");
    for (i=1;i<=NP;i++) {
        for (j=1;j<=NP;j++) printf("%12.6f",v[i][j]);
        printf("\n");
    }
    printf("\nvector w\n");
    for (i=1;i<=NP;i++) printf("%12.6f",w[i]);
    printf("\n");
    svdvar(v,MA,w,cvm);
    printf("\ncovariance matrix from svdvar\n");
    for (i=1;i<=MA;i++) {
        for (j=1;j<=MA;j++) printf("%12.6f",cvm[i][j]);
        printf("\n");
    }
    printf("\nexpected covariance matrix\n");
    for (i=1;i<=MA;i++) {
        for (j=1;j<=MA;j++) printf("%12.6f",tru[i-1][j-1]);
        printf("\n");
    }
    free_convert_matrix(v,1,NP,1,NP);
    free_matrix(cvm,1,MA,1,MA);
    return 0;
}
```

Routines `fpoly` and `fleg` were used above with sample program `xsvdfit` to generate the powers of x and the Legendre polynomials, respectively. In the case of `fpoly`, sample program `xfpoly` below is used to list the powers of x generated by `fpoly` so that they may be checked "by eye". For `fleg`, the generated polynomials in program `xfleg` are compared to values from routine `plgndr`.

```
/* Driver for routine fpoly */

#include <stdio.h>
#include "nr.h"
#include "nrutil.h"

#define NVAL 15
#define DX 0.1
#define NPOLY 5

int main(void)
{
    int i,j;
    float x,*afunc;
```

```
    afunc=vector(1,NPOLY);
    printf("\n%38s\n","powers of x");
    printf("%8s %10s %9s %9s %9s %9s\n",
        "x","x**0","x**1","x**2","x**3","x**4");
    for (i=1;i<=NVAL;i++) {
        x=i*DX;
        fpoly(x,afunc,NPOLY);
        printf("%10.4f",x);
        for (j=1;j<=NPOLY;j++) printf("%10.4f",afunc[j]);
        printf("\n");
    }
    free_vector(afunc,1,NPOLY);
    return 0;
}

/* Driver for routine fleg */

#include <stdio.h>
#include "nr.h"
#include "nrutil.h"

#define NVAL 5
#define DX 0.2
#define NPOLY 5

int main(void)
{
    int i,j;
    float x,*afunc;

    afunc=vector(1,NPOLY);
    printf("\n%3s\n","Legendre polynomials");
    printf("%9s %9s %9s %9s %9s\n","n=1","n=2","n=3","n=4","n=5");
    for (i=1;i<=NVAL;i++) {
        x=i*DX;
        fleg(x,afunc,NPOLY);
        printf("x =%5.2f\n",x);
        for (j=1;j<=NPOLY;j++) printf("%10.4f",afunc[j]);
        printf("  routine FLEG\n");
        for (j=1;j<=NPOLY;j++) printf("%10.4f",plgndr(j-1,0,x));
        printf("  routine PLGNDR\n\n");
    }
    free_vector(afunc,1,NPOLY);
    return 0;
}
```

mrqmin is used along with mrqcof to perform nonlinear least-squares fits with the Levenberg-Marquardt method. The artificial data used to try it in sample program xmrqmin is computed as the sum of two Gaussians plus noise:

$$y[i] = a[1] \exp\{-[(x[i] - a[2])/a[3]]^2\}$$
$$+ a[4] \exp\{-[(x[i] - a[5])/a[6]]^2\} + \text{noise}.$$

The a[i] are assigned at the beginning of the program, as are the initial guesses gues[i] for these parameters to be used in initiating the fit. Also initialized for the

fit are `ia[i]=1` for `i=1,..,mfit` to specify that all six of the parameters are to be fit.
On the first call to `mrqmin`, `alamda=-1` to initialize. Then a loop is entered in which
`mrqmin` is iterated while monitoring successive values of chi-squared `chisq`. When
`chisq` changes by less than 0.1 on two consecutive iterations, the fit is considered
complete, and `mrqmin` is called one final time with `alamda=0.0` so that array `covar`
will return the covariance matrix. Uncertainties are derived from the square roots of
the diagonal elements of `covar`. Expected results for the parameters are, of course,
the values used to generate the "data" in the first place. As a test of the `ia` feature,
the fit is redone holding fixed `a[2]=2` and `a[5]=5`.

```c
/* Driver for routine mrqmin */

#include <stdio.h>
#include <math.h>
#include "nr.h"
#include "nrutil.h"

#define NPT 100
#define MA 6
#define SPREAD 0.001

int main(void)
{
    long idum=(-911);
    int i,*ia,iter,itst,j,k,mfit=MA;
    float alamda,chisq,ochisq,*x,*y,*sig,**covar,**alpha;
    static float a[MA+1]=
        {0.0,5.0,2.0,3.0,2.0,5.0,3.0};
    static float gues[MA+1]=
        {0.0,4.5,2.2,2.8,2.5,4.9,2.8};

    ia=ivector(1,MA);
    x=vector(1,NPT);
    y=vector(1,NPT);
    sig=vector(1,NPT);
    covar=matrix(1,MA,1,MA);
    alpha=matrix(1,MA,1,MA);
    /* First try a sum of two Gaussians */
    for (i=1;i<=NPT;i++) {
        x[i]=0.1*i;
        y[i]=0.0;
        for (j=1;j<=MA;j+=3) {
            y[i] += a[j]*exp(-SQR((x[i]-a[j+1])/a[j+2]));
        }
        y[i] *= (1.0+SPREAD*gasdev(&idum));
        sig[i]=SPREAD*y[i];
    }
    for (i=1;i<=mfit;i++) ia[i]=1;
    for (i=1;i<=MA;i++) a[i]=gues[i];
    for (iter=1;iter<=2;iter++) {
        alamda = -1;
        mrqmin(x,y,sig,NPT,a,ia,MA,covar,alpha,&chisq,fgauss,&alamda);
        k=1;
        itst=0;
        for (;;) {
            printf("\n%s %2d %17s %10.4f %10s %9.2e\n","Iteration #",k,
```

```
                    "chi-squared:",chisq,"alamda:",alamda);
            printf("%8s %8s %8s %8s %8s\n",
                    "a[1]","a[2]","a[3]","a[4]","a[5]","a[6]");
            for (i=1;i<=6;i++) printf("%9.4f",a[i]);
            printf("\n");
            k++;
            ochisq=chisq;
            mrqmin(x,y,sig,NPT,a,ia,MA,covar,alpha,&chisq,fgauss,&alamda);
            if (chisq > ochisq)
                itst=0;
            else if (fabs(ochisq-chisq) < 0.1)
                itst++;
            if (itst < 4) continue;
            alamda=0.0;
            mrqmin(x,y,sig,NPT,a,ia,MA,covar,alpha,&chisq,fgauss,&alamda);
            printf("\nUncertainties:\n");
            for (i=1;i<=6;i++) printf("%9.4f",sqrt(covar[i][i]));
            printf("\n");
            printf("\nExpected results:\n");
            printf(" %7.2f %8.2f %8.2f %8.2f %8.2f %8.2f\n",
                    5.0,2.0,3.0,2.0,5.0,3.0);
            break;
        }
        if (iter == 1) {
            printf("press return to continue with constraint\n");
            (void) getchar();
            printf("holding a[2] and a[5] constant\n");
            for (j=1;j<=MA;j++) a[j] += 0.1;
            a[2]=2.0;
            ia[2]=0;
            a[5]=5.0;
            ia[5]=0;
        }
    }
    free_matrix(alpha,1,MA,1,MA);
    free_matrix(covar,1,MA,1,MA);
    free_vector(sig,1,NPT);
    free_vector(y,1,NPT);
    free_vector(x,1,NPT);
    free_ivector(ia,1,MA);
    return 0;
}
```

The nonlinear least-squares fit makes use of a vector β_k (the gradient of χ^2 in parameter-space) and α_{kl} (the Hessian of χ^2 in the same space). These quantities are produced by mrqcof, as demonstrated by sample program xmrqcof. The function is a sum of two Gaussians with noise added (the same function as in xmrqmin) and it is used twice. In the first call, ia[i]=1 and mfit=6 so all six parameters are used. In the second call, ia[i]=0 for i=1,2,3 and mfit=3 so the first three parameters are fixed and the last three, a[4]...a[6] are fit. Here is the output you should get:

```
matrix alpha
    137.2021      60.6572     148.2204     147.7027     -12.6093      90.9625
     60.6572     500.8828     366.9686     448.3022     287.4484     257.7019
    148.2204     366.9686     583.6553     659.6396     553.0109     612.6782
    147.7027     448.3022     659.6396     971.0057    1131.7701    1410.9380
```

```
     -12.6093      287.4484      553.0109     1131.7701     1911.5935     2409.3484
      90.9625      257.7019      612.6782     1410.9380     2409.3484     3341.0002
vector beta
     -24.7629     -118.2998      -76.5074     -124.1268     -104.8776     -145.4018
chi-squared:      147.9271
matrix alpha
     971.0057     1131.7701     1410.9380
    1131.7701     1911.5935     2409.3484
    1410.9380     2409.3484     3341.0002
vector beta
    -124.1268     -104.8776     -145.4018
chi-squared:      147.9271
```

```c
/* Driver for routine mrqcof */

#include <stdio.h>
#include <math.h>
#include "nr.h"
#include "nrutil.h"

#define NPT 100
#define MA 6
#define SPREAD 0.1

int main(void)
{
    long idum=(-911);
    int i,j,mfit=MA,*ia;
    float chisq,*beta,*x,*y,*sig,**covar,**alpha;
    static float a[MA+1]=
        {0.0,5.0,2.0,3.0,2.0,5.0,3.0};
    static float gues[MA+1]=
        {0.0,4.9,2.1,2.9,2.1,4.9,3.1};

    ia=ivector(1,MA);
    beta=vector(1,MA);
    x=vector(1,NPT);
    y=vector(1,NPT);
    sig=vector(1,NPT);
    covar=matrix(1,MA,1,MA);
    alpha=matrix(1,MA,1,MA);
    /* First try sum of two gaussians */
    for (i=1;i<=NPT;i++) {
        x[i]=0.1*i;
        y[i]=0.0;
        y[i] += a[1]*exp(-SQR((x[i]-a[2])/a[3]));
        y[i] += a[4]*exp(-SQR((x[i]-a[5])/a[6]));
        y[i] *= (1.0+SPREAD*gasdev(&idum));
        sig[i]=SPREAD*y[i];
    }
    for (i=1;i<=mfit;i++) ia[i]=1;
    for (i=1;i<=mfit;i++) a[i]=gues[i];
    mrqcof(x,y,sig,NPT,a,ia,MA,alpha,beta,&chisq,fgauss);
    printf("\nmatrix alpha\n");
    for (i=1;i<=MA;i++) {
        for (j=1;j<=MA;j++) printf("%12.4f",alpha[i][j]);
        printf("\n");
```

```
    }
    printf("vector beta\n");
    for (i=1;i<=MA;i++) printf("%12.4f",beta[i]);
    printf("\nchi-squared: %12.4f\n\n",chisq);
    /* Next fix one line and improve the other */
    mfit=3;
    for (i=1;i<=mfit;i++) ia[i]=0;
    for (i=1;i<=MA;i++) a[i]=gues[i];
    mrqcof(x,y,sig,NPT,a,ia,MA,alpha,beta,&chisq,fgauss);
    printf("matrix alpha\n");
    for (i=1;i<=mfit;i++) {
        for (j=1;j<=mfit;j++) printf("%12.4f",alpha[i][j]);
        printf("\n");
    }
    printf("vector beta\n");
    for (i=1;i<=mfit;i++) printf("%12.4f",beta[i]);
    printf("\nchi-squared: %12.4f\n\n",chisq);
    free_matrix(alpha,1,MA,1,MA);
    free_matrix(covar,1,MA,1,MA);
    free_vector(sig,1,NPT);
    free_vector(y,1,NPT);
    free_vector(x,1,NPT);
    free_vector(beta,1,MA);
    free_ivector(ia,1,MA);
    return 0;
}
```

fgauss is an example of the type of procedure that must be supplied to mrqfit in order to fit a user-defined function, in this case the sum of Gaussians. fgauss calculates both the function, and its derivative with respect to each adjustable parameter in a fairly compact fashion. The sample program xfgauss calculates the same quantities in a more pedantic fashion, just to be sure we got everything right.

```
/* Driver for routine fgauss */

#include <stdio.h>
#include <math.h>
#include "nr.h"
#include "nrutil.h"

#define NPT 3
#define NLIN 2
#define NA 3*NLIN

int main(void)
{
    int i,j;
    float e1,e2,f,x,y;
    static float a[NA+1]={0.0,3.0,0.2,0.5,1.0,0.7,0.3};
    float dyda[NA+1],df[NA+1];

    printf("\n%6s %8s %8s %7s %7s %7s %7s %7s\n",
        "x","y","dyda1","dyda2","dyda3","dyda4","dyda5","dyda6");
    for (i=1;i<=NPT;i++) {
        x=0.3*i;
        fgauss(x,a,&y,dyda,NA);
        e1=exp(-SQR((x-a[2])/a[3]));
```

```
        e2=exp(-SQR((x-a[5])/a[6]));
        f=a[1]*e1+a[4]*e2;
        df[1]=e1;
        df[4]=e2;
        df[2]=a[1]*e1*2.0*(x-a[2])/(a[3]*a[3]);
        df[5]=a[4]*e2*2.0*(x-a[5])/(a[6]*a[6]);
        df[3]=a[1]*e1*2.0*SQR(x-a[2])/(a[3]*a[3]*a[3]);
        df[6]=a[4]*e2*2.0*SQR(x-a[5])/(a[6]*a[6]*a[6]);
        printf("from FGAUSS\n");
        printf("%8.4f %8.4f",x,y);
        for (j=1;j<=6;j++) printf("%8.4f",dyda[j]);
        printf("\nindependent calc.\n");
        printf("%8.4f %8.4f",x,f);
        for (j=1;j<=6;j++) printf("%8.4f",df[j]);
        printf("\n\n");
    }
    return 0;
}
```

medfit is a routine illustrating a more "robust" way of fitting. It performs a fit of data to a straight line, but instead of using the least-squares criterion for figuring the merit of a fit, it uses the least-absolute-deviation. For comparison, sample routine xmedfit fits lines to a noisy linear data set, using first the least-squares routine fit, and then the least-absolute-deviation routine medfit. You may be interested to see if you can figure out what mean value of absolute deviation you expect for data with gaussian noise of amplitude SPREAD.

```
/* Driver for routine medfit */

#include <stdio.h>
#include "nr.h"
#include "nrutil.h"

#define NPT 100
#define SPREAD 0.1
#define NDATA NPT

int main(void)
{
    long idum=(-1984);
    int i,mwt=1;
    float a,abdev,b,chi2,q,siga,sigb;
    float *x,*y,*sig;

    x=vector(1,NDATA);
    y=vector(1,NDATA);
    sig=vector(1,NDATA);
    for (i=1;i<=NPT;i++) {
        x[i]=0.1*i;
        y[i] = -2.0*x[i]+1.0+SPREAD*gasdev(&idum);
        sig[i]=SPREAD;
    }
    fit(x,y,NPT,sig,mwt,&a,&b,&siga,&sigb,&chi2,&q);
    printf("\nAccording to routine FIT the result is:\n");
    printf("   a = %8.4f   uncertainty: %8.4f\n",a,siga);
    printf("   b = %8.4f   uncertainty: %8.4f\n",b,sigb);
```

```
printf("   chi-squared: %8.4f  for  %4d  points\n",chi2,NPT);
printf("   goodness-of-fit: %8.4f\n",q);
printf("\nAccording to routine MEDFIT the result is:\n");
medfit(x,y,NPT,&a,&b,&abdev);
printf("   a = %8.4f\n",a);
printf("   b = %8.4f\n",b);
printf("   absolute deviation (per data point): %8.4f\n",abdev);
printf("   (note: gaussian SPREAD is %8.4f)\n",SPREAD);
free_vector(sig,1,NDATA);
free_vector(y,1,NDATA);
free_vector(x,1,NDATA);
return 0;
}
```

rofunc is an auxiliary function for medfit. It evaluates the quantity

$$\sum_{i=1}^{N} x_i \, \text{sgn}(y_i - a - bx_i)$$

given arrays x_i and y_i. Data are communicated to and from rofunc primarily through global variables, but the value of the sum above is returned as the value of rofunc(b). abdev is the summed absolute deviation, and aa (listed below as a) is given the value that minimizes abdev. Our results for these quantities are:

b	a	ROFUNC	ABDEVT
-2.10	1.51	249.80	25.17
-2.08	1.40	247.40	20.20
-2.06	1.30	247.40	15.25
-2.04	1.21	231.20	10.36
-2.02	1.11	193.60	5.92
-2.00	1.00	-15.60	3.75
-1.98	0.89	-199.20	6.00
-1.96	0.80	-234.40	10.40
-1.94	0.71	-247.20	15.25
-1.92	0.60	-247.20	20.19
-1.90	0.51	-247.20	25.14

```
/* Driver for routine rofunc */

#include <stdio.h>
#include "nr.h"
#include "nrutil.h"

#define SPREAD 0.05
#define NDATA 100

int ndatat; /* defining declaration */
float *xt,*yt,aa,abdevt;    /* defining declaration */

int main(void)
{
    long idum=(-11);
    int i;
    float b,rf,*x,*y;

    x=vector(1,NDATA);
    y=vector(1,NDATA);
```

```
ndatat=NDATA;
xt=x;
yt=y;
for (i=1;i<=NDATA;i++) {
    x[i]=0.1*i;
    y[i] = -2.0*x[i]+1.0+SPREAD*gasdev(&idum);
}
printf("%9s %9s %12s %10s\n","b","a","ROFUNC","ABDEVT");
for (i = -5;i<=5;i++) {
    b = -2.0+0.02*i;
    rf=rofunc(b);
    printf("%10.2f %9.2f %11.2f %10.2f\n",
        b,aa,rf,abdevt);
}
free_vector(y,1,NDATA);
free_vector(x,1,NDATA);
return 0;
}
```

Chapter 16: Ordinary Differential Equations

Chapter 16 of Numerical Recipes deals with the integration of ordinary differential equations, restricting its attention specifically to initial-value problems. Three practical methods are introduced: 1) Runge-Kutta methods (rk4, rkdumb, rkqs, and odeint), 2) Richardson extrapolation and the Bulirsch-Stoer method (bsstep, mmid, stoerm, rzextr, pzextr), 3) predictor-corrector methods. In general, for applications not demanding high precision, and where convenience is paramount, Runge-Kutta with adaptive step-size control is recommended. For higher precision applications, the Bulirsch-Stoer method dominates. The predictor-corrector methods are covered because of their history of widespread use.

Stiff differential equations require special methods. Routine stiff *uses a generalization of Runge-Kutta, known as a Rosenbrock method. The semi-implicit midpoint extrapolation method is a Bulirsch-Stoer method designed for stiff equations, and implemented in* stifbs.

$$\star \quad \star \quad \star \quad \star$$

Routine rk4 advances the solution vector y[n] of a set of ordinary differential equations over a single small interval h in x using the fourth-order Runge-Kutta method. The operation is shown by sample program xrk4 for an array of four variables y[1], ..., y[4]. The first-order differential equations satisfied by these variables are specified by the accompanying routine derivs, and are simply the equations describing the first four Bessel functions $J_0(x), \ldots, J_3(x)$. The y's are initialized to the values of these functions at $x = 1.0$. Note that the values of dydx are also initialized at $x = 1.0$, because rk4 uses the values of dydx before its first call to derivs. The reason for this is discussed in the text. The sample program calls rk4 with h (the step-size) set to various values from 0.2 to 1.0, so that you can see how well rk4 can do even with quite sizeable steps.

```
/* Driver for routine rk4 */

#include <stdio.h>
#include "nr.h"
#include "nrutil.h"

#define N 4

void derivs(float x,float y[],float dydx[])
{
    dydx[1] = -y[2];
    dydx[2]=y[1]-(1.0/x)*y[2];
    dydx[3]=y[2]-(2.0/x)*y[3];
```

264

```
        dydx[4]=y[3]-(3.0/x)*y[4];
}

int main(void)
{
    int i,j;
    float h,x=1.0,*y,*dydx,*yout;

    y=vector(1,N);
    dydx=vector(1,N);
    yout=vector(1,N);
    y[1]=bessj0(x);
    y[2]=bessj1(x);
    y[3]=bessj(2,x);
    y[4]=bessj(3,x);
    derivs(x,y,dydx);
    printf("\n%16s %5s %12s %12s %12s\n",
        "Bessel function:","j0","j1","j3","j4");
    for (i=1;i<=5;i++) {
        h=0.2*i;
        rk4(y,dydx,N,x,h,yout,derivs);
        printf("\nfor a step size of: %6.2f\n",h);
        printf("%12s","rk4:");
        for (j=1;j<=4;j++) printf(" %12.6f",yout[j]);
        printf("\n%12s %12.6f %12.6f %12.6f %12.6f\n","actual:",
            bessj0(x+h),bessj1(x+h),bessj(2,x+h),bessj(3,x+h));
    }
    free_vector(yout,1,N);
    free_vector(dydx,1,N);
    free_vector(y,1,N);
    return 0;
}
```

rkdumb is an extension of rk4 that allows you to integrate over larger intervals. It is "dumb" in the sense that it has no adaptive step-size determination, and no code to estimate errors. Sample program xrkdumb works with the same functions and derivatives as the previous program, but integrates from x1=1.0 to x2=20.0, breaking the interval into NSTEP=150 equal steps. The variables vstart[1],...,vstart[4] which become the starting values of the y's, are initialized as before, but their derivatives this time are not initialized; rkdumb takes care of that. This time only the results for the fourth variable $J_3(x)$ are listed, and only every tenth value is given.

```
/* Driver for routine rkdumb */

#include <stdio.h>
#include "nr.h"
#include "nrutil.h"

#define NVAR 4
#define NSTEP 150

void derivs(float x,float y[],float dydx[])
{
    dydx[1] = -y[2];
    dydx[2]=y[1]-(1.0/x)*y[2];
    dydx[3]=y[2]-(2.0/x)*y[3];
```

```
        dydx[4]=y[3]-(3.0/x)*y[4];
}

extern float **y,*xx;    /* referencing declaration */

int main(void)
{
    int j;
    float x1=1.0,x2=20.0,*vstart;

    vstart=vector(1,NVAR);
    /* Note: The arrays xx and y must have indices up to NSTEP+1 */
    xx=vector(1,NSTEP+1);
    y=matrix(1,NVAR,1,NSTEP+1);
    vstart[1]=bessj0(x1);
    vstart[2]=bessj1(x1);
    vstart[3]=bessj(2,x1);
    vstart[4]=bessj(3,x1);
    rkdumb(vstart,NVAR,x1,x2,NSTEP,derivs);
    printf("%8s %17s %10s\n","x","integrated","bessj3");
    for (j=10;j<=NSTEP;j+=10)
        printf("%10.4f %14.6f %12.6f\n",
            xx[j],y[4][j],bessj(3,xx[j]));
    free_matrix(y,1,NVAR,1,NSTEP+1);
    free_vector(xx,1,NSTEP+1);
    free_vector(vstart,1,NVAR);
    return 0;
}
```

rkqs performs a single step of fifth-order Runge-Kutta integration, this time with monitoring of local truncation error and corresponding step-size adjustment. Its sample program xrkqs is similar to that for routine rk4, using four Bessel functions as the example, and starting the integration at $x = 1.0$. However, on each pass a value is set for eps, the desired accuracy, and the trial value htry for the interval size is set to 0.6. For the first few passes, eps is not too demanding and htry may be perfectly adequate. As eps becomes smaller, the routine will be forced to diminish h and return smaller values of hdid and hnext. Our results (in single precision) are:

eps	htry	hdid	hnext
0.367879	0.60	0.600000	3.000000
0.135335	0.60	0.600000	2.542155
0.049787	0.60	0.600000	2.081340
0.018316	0.60	0.600000	1.704057
0.006738	0.60	0.600000	1.395164
0.002479	0.60	0.600000	1.142264
0.000912	0.60	0.600000	0.935206
0.000335	0.60	0.600000	0.765682
0.000123	0.60	0.600000	0.626888
0.000045	0.60	0.506775	0.513415
0.000017	0.60	0.394677	0.421229
0.000006	0.60	0.307374	0.346002
0.000002	0.60	0.239383	0.284358
0.000001	0.60	0.186432	0.233682
0.000000	0.60	0.145193	0.191925

```
/* Driver for routine rkqs */

#include <stdio.h>
#include <math.h>
#include "nr.h"
#include "nrutil.h"

#define N 4

void derivs(float x,float y[],float dydx[])
{
    dydx[1] = -y[2];
    dydx[2]=y[1]-(1.0/x)*y[2];
    dydx[3]=y[2]-(2.0/x)*y[3];
    dydx[4]=y[3]-(3.0/x)*y[4];
}

int main(void)
{
    int i,j;
    float eps,hdid,hnext,htry,x=1.0,*y,*dydx,*dysav,*ysav,*yscal;

    y=vector(1,N);
    dydx=vector(1,N);
    dysav=vector(1,N);
    ysav=vector(1,N);
    yscal=vector(1,N);
    ysav[1]=bessj0(x);
    ysav[2]=bessj1(x);
    ysav[3]=bessj(2,x);
    ysav[4]=bessj(3,x);
    derivs(x,ysav,dysav);
    for (i=1;i<=N;i++) yscal[i]=1.0;
    htry=0.6;
    printf("%10s %11s %12s %13s\n","eps","htry","hdid","hnext");
    for (i=1;i<=15;i++) {
        eps=exp((double) -i);
        x=1.0;
        for (j=1;j<=N;j++) {
            y[j]=ysav[j];
            dydx[j]=dysav[j];
        }
        rkqs(y,dydx,N,&x,htry,eps,yscal,&hdid,&hnext,derivs);
        printf("%13f %8.2f %14.6f %12.6f \n",eps,htry,hdid,hnext);
    }
    free_vector(yscal,1,N);
    free_vector(ysav,1,N);
    free_vector(dysav,1,N);
    free_vector(dydx,1,N);
    free_vector(y,1,N);
    return 0;
}
```

The full driver routine for `rkqs`, which provides Runge-Kutta integration over large intervals with adaptive step-size control, is `odeint`. It plays the same role for `rkqs` that `rkdumb` plays for `rk4`. Integration is performed on four Bessel functions from x1=1.0 to x2=10.0, with an accuracy eps=1.0e-4. Independent of the values of step-size actually used by odeint, intermediate values will be recorded only at intervals greater than dxsav. The sample program returns values of $J_3(x)$ for checking against actual values produced by bessj. It also records the number of function evaluations used, how many steps were successful, and how many were "bad". Bad steps are redone, and indicate no extra loss in accuracy. At the same time, they do represent a loss in efficiency, so that an excessive number of bad steps should initiate an investigation.

```
/* Driver for routine odeint */

#include <stdio.h>
#include "nr.h"
#include "nrutil.h"

#define N 4

float dxsav,*xp,**yp;  /* defining declarations */
int kmax,kount;

int nrhs;   /* counts function evaluations */

void derivs(float x,float y[],float dydx[])
{
    nrhs++;
    dydx[1] = -y[2];
    dydx[2]=y[1]-(1.0/x)*y[2];
    dydx[3]=y[2]-(2.0/x)*y[3];
    dydx[4]=y[3]-(3.0/x)*y[4];
}

int main(void)
{
    int i,nbad,nok;
    float eps=1.0e-4,h1=0.1,hmin=0.0,x1=1.0,x2=10.0,*ystart;

    ystart=vector(1,N);
    xp=vector(1,200);
    yp=matrix(1,10,1,200);
    ystart[1]=bessj0(x1);
    ystart[2]=bessj1(x1);
    ystart[3]=bessj(2,x1);
    ystart[4]=bessj(3,x1);
    nrhs=0;
    kmax=100;
    dxsav=(x2-x1)/20.0;
    odeint(ystart,N,x1,x2,eps,h1,hmin,&nok,&nbad,derivs,rkqs);
    printf("\n%s %13s %3d\n","successful steps:"," ",nok);
    printf("%s %20s %3d\n","bad steps:"," ",nbad);
    printf("%s %9s %3d\n","function evaluations:"," ",nrhs);
    printf("\n%s %3d\n","stored intermediate values:    ",kount);
    printf("\n%8s %18s %15s\n","x","integral","bessj(3,x)");
    for (i=1;i<=kount;i++)
        printf("%10.4f %16.6f %14.6f\n",
            xp[i],yp[4][i],bessj(3,xp[i]));
    free_matrix(yp,1,10,1,200);
```

```
        free_vector(xp,1,200);
        free_vector(ystart,1,N);
        return 0;
}
```

The modified midpoint routine `mmid` is presented in *Numerical Recipes* primarily as a component of the more powerful Bulirsch-Stoer routine. It integrates variables over an interval `htot` through a sequence of much smaller steps. Sample routine `xmmid` takes the number of subintervals `i` to be $5, 10, 15, \ldots, 50$ so that we can witness any improvements in accuracy that may occur. The values of the four Bessel functions are compared with the results of the integrations.

```c
/* Driver for routine mmid */

#include <stdio.h>
#include "nr.h"
#include "nrutil.h"

#define NVAR 4
#define X1 1.0
#define HTOT 0.5

void derivs(float x,float y[],float dydx[])
{
    dydx[1] = -y[2];
    dydx[2]=y[1]-(1.0/x)*y[2];
    dydx[3]=y[2]-(2.0/x)*y[3];
    dydx[4]=y[3]-(3.0/x)*y[4];
}

int main(void)
{
    int i;
    float b1,b2,b3,b4,xf=X1+HTOT,*y,*yout,*dydx;

    y=vector(1,NVAR);
    yout=vector(1,NVAR);
    dydx=vector(1,NVAR);
    y[1]=bessj0(X1);
    y[2]=bessj1(X1);
    y[3]=bessj(2,X1);
    y[4]=bessj(3,X1);
    derivs(X1,y,dydx);
    b1=bessj0(xf);
    b2=bessj1(xf);
    b3=bessj(2,xf);
    b4=bessj(3,xf);
    printf("First four Bessel functions:\n");
    for (i=5;i<=50;i+=5) {
        mmid(y,dydx,NVAR,X1,HTOT,i,yout,derivs);
        printf("\n%s %5.2f %s %5.2f %s %2d %s \n",
            "x=",X1," to ",X1+HTOT," in ",i," steps");
        printf("%14s %9s\n","integration","bessj");
        printf("%12.6f %12.6f\n",yout[1],b1);
        printf("%12.6f %12.6f\n",yout[2],b2);
        printf("%12.6f %12.6f\n",yout[3],b3);
```

```
            printf("%12.6f %12.6f\n",yout[4],b4);
            printf("\nPress RETURN to continue...\n");
            (void) getchar();
        }
        free_vector(dydx,1,NVAR);
        free_vector(yout,1,NVAR);
        free_vector(y,1,NVAR);
        return 0;
    }
```

The Bulirsch-Stoer method, carried out by routine bsstep, is the integrator of choice for higher accuracy calculations on smooth functions. An interval h is broken into finer and finer steps, and the results of integration are extrapolated to zero step-size. The extrapolation is via a polynomial with pzextr. bsstep monitors local truncation error and adjusts the step-size appropriately, to keep errors below eps. From an external point of view, bsstep operates exactly as does rkqs: it has the same arguments and in the same order. Consequently it can be used in place of rkqs in routine odeint, allowing more efficient integration over large regions of x. For this reason, the sample program xbsstep is of the same form used to demonstrate rkqs. You will notice that bsstep integrates the test problem in far fewer steps than rkqs, which is typical. Because of the low accuracy requirement, however, rkqs makes fewer function evaluations.

```c
/* Driver for routine bsstep */

#include <stdio.h>
#include "nr.h"
#include "nrutil.h"

#define N 4

float dxsav,*xp,**yp;   /* defining declarations */
int kmax,kount;

int nrhs;   /* counts function evaluations */

void derivs(float x,float y[],float dydx[])
{
    nrhs++;
    dydx[1] = -y[2];
    dydx[2]=y[1]-(1.0/x)*y[2];
    dydx[3]=y[2]-(2.0/x)*y[3];
    dydx[4]=y[3]-(3.0/x)*y[4];
}

int main(void)
{
    int i,nbad,nok;
    float eps=1.0e-4,h1=0.1,hmin=0.0,x1=1.0,x2=10.0,*ystart;

    ystart=vector(1,N);
    xp=vector(1,200);
    yp=matrix(1,10,1,200);
    ystart[1]=bessj0(x1);
    ystart[2]=bessj1(x1);
```

```
        ystart[3]=bessj(2,x1);
        ystart[4]=bessj(3,x1);
        nrhs=0;
        kmax=100;
        dxsav=(x2-x1)/20.0;
        odeint(ystart,N,x1,x2,eps,h1,hmin,&nok,&nbad,derivs,bsstep);
        printf("\n%s %13s %3d\n","successful steps:"," ",nok);
        printf("%s %20s %3d\n","bad steps:"," ",nbad);
        printf("%s %9s %3d\n","function evaluations:"," ",nrhs);
        printf("\n%s %3d\n","stored intermediate values:    ",kount);
        printf("\n%8s %18s %15s\n","x","integral","bessj(3,x)");
        for (i=1;i<=kount;i++)
            printf("%10.4f %16.6f %14.6f\n",
                xp[i],yp[4][i],bessj(3,xp[i]));
        free_matrix(yp,1,10,1,200);
        free_vector(xp,1,200);
        free_vector(ystart,1,N);
        return 0;
}
```

pzextr performs a polynomial extrapolation for **bsstep**. It takes a sequence of interval lengths and corresponding integrated values, and extrapolates to the value the integral would have if the interval length were zero. Sample routine **xpzextr** works with a known function

$$F_n = \frac{1 - x + x^3}{(x+1)^n} \qquad n = 1, .., 4$$

We extrapolate the vector **yest** $= (F_1, F_2, F_3, F_4)$ given a sequence of ten values. The ten values are labelled **iest**=1, ..., 10 and are evaluated at **xest**=1.0/iest. A call to **pzextr** produces extrapolated values **yz**, and estimated errors **dy**, and compares to the true values $(1.0, 1.0, 1.0, 1.0)$ at **xest**=0.0.

```
/* Driver for routine pzextr */

#include <stdio.h>
#include "nr.h"
#include "nrutil.h"

#define NV 4
#define IMAXX 10

float **d,*x;    /* defining declaration */

int main(void)
{
    int i,iest,j;
    float dum,xest,*dy,*yest,*yz;

    dy=vector(1,NV);
    yest=vector(1,NV);
    yz=vector(1,NV);
    x=vector(1,IMAXX);
    d=matrix(1,IMAXX,1,IMAXX);
    /* Feed values from a rational function */
    /* fn(x)=(1-x+x**3)/(x+1)**n */
    for (i=1;i<=IMAXX;i++) {
```

```
        iest=i;
        xest=1.0/i;
        dum=1.0-xest+xest*xest*xest;
        for (j=1;j<=NV;j++) {
            dum /= (xest+1.0);
            yest[j]=dum;
        }
        pzextr(iest,xest,yest,yz,dy,NV);
        printf("\ni = %2d",i);
        printf("\nExtrap. function:");
        for (j=1;j<=NV;j++) printf("%12.6f",yz[j]);
        printf("\nEstimated error: ");
        for (j=1;j<=NV;j++) printf("%12.6f",dy[j]);
        printf("\n");
    }
    printf("\nactual values: %14.6f %11.6f %11.6f %11.6f\n",
        1.0,1.0,1.0,1.0);
    free_matrix(d,1,IMAXX,1,IMAXX);
    free_vector(x,1,IMAXX);
    free_vector(yz,1,NV);
    free_vector(yest,1,NV);
    free_vector(dy,1,NV);
    return 0;
}
```

rzextr performs a diagonal rational function extrapolation for bsstep. It is usually a less efficient standby for rzextr, to be used primarily when some problem crops up with the extrapolation. The sample program xrzextr is identical to that for pzextr.

```
/* Driver for routine rzextr */

#include <stdio.h>
#include "nr.h"
#include "nrutil.h"

#define NV 4
#define IMAXX 10

float **d,*x;    /* defining declaration */

int main(void)
{
    int i,iest,j;
    float dum,xest,*dy,*yest,*yz;

    dy=vector(1,NV);
    yest=vector(1,NV);
    yz=vector(1,NV);
    x=vector(1,IMAXX);
    d=matrix(1,IMAXX,1,IMAXX);
    /* Feed values from a rational function */
    /* fn(x)=(1-x+x**3)/(x+1)**n */
    for (i=1;i<=IMAXX;i++) {
        iest=i;
        xest=1.0/i;
        dum=1.0-xest+xest*xest*xest;
```

```
            for (j=1;j<=NV;j++) {
                dum /= (xest+1.0);
                yest[j]=dum;
            }
            rzextr(iest,xest,yest,yz,dy,NV);
            printf("\n%s %2d %s %8.4f\n",
                "iest = ",i,"    xest =",xest);
            printf("Extrap. function: ");
            for (j=1;j<=NV;j++) printf("%12.6f",yz[j]);
            printf("\nEstimated error:  ");
            for (j=1;j<=NV;j++) printf("%12.6f",dy[j]);
            printf("\n");
        }
        printf("\nActual values: %15.6f %11.6f %11.6f %11.6f \n",
            1.0,1.0,1.0,1.0);
        free_matrix(d,1,IMAXX,1,IMAXX);
        free_vector(x,1,IMAXX);
        free_vector(yz,1,NV);
        free_vector(yest,1,NV);
        free_vector(dy,1,NV);
        return 0;
}
```

Stoermer's method is a special-purpose integration method for second-order ODEs in which the first derivative term is absent. It replaces `mmid` in the Bulirsch-Stoer method. Sample program `xstoerm` tests it out on the system

$$y_1'' = x - y_1$$
$$y_2'' = x^2 - y_2$$

The system is integrated from $x = 0$ to $x = \pi/2$ with initial conditions

$$y_1 = 0, \quad y_2 = -1, \quad y_1' = 2, \quad y_2' = 0$$

The program compares the result of integrating with varying number of substeps to the exact solution

$$y_1 = x + \sin x$$
$$y_2 = x^2 + \cos x - 2$$

Of course, in real life you would use `stoerm` as a replacement for `mmid` in a modified version of `bsstep`, as described in *Numerical Recipes*.

```
/* Driver for routine stoerm */

#include <stdio.h>
#include <math.h>
#include "nr.h"
#include "nrutil.h"

#define NVAR 4
#define X1 0.0
#define HTOT 1.570796

float d2y1(float x)
{
    return sin(x)+x;
```

```
}

float d2y2(float x)
{
    return cos(x)+x*x-2;
}

void derivs(float x,float y[],float dydx[])
{
    dydx[1]=x-y[1];
    dydx[2]=x*x-y[2];
}

int main(void)
{
    int i;
    float a1,a2,xf,*y,*yout,*d2y;

    y=vector(1,NVAR);
    yout=vector(1,NVAR);
    d2y=vector(1,NVAR);
    y[1]=0.0;
    y[2] = -1.0;
    y[3]=2.0;
    y[4]=0.0;
    derivs(X1,y,d2y);
    xf=X1+HTOT;
    a1=d2y1(xf);
    a2=d2y2(xf);
    printf("Stoermer's Rule:\n");
    for (i=5;i<=45;i+=10) {
        stoerm(y,d2y,NVAR,X1,HTOT,i,yout,derivs);
        printf("\n%s %5.2f %s %5.2f %s %2d %s \n",
            "x=",X1," to ",X1+HTOT," in ",i," steps");
        printf("%14s %9s\n","integration","answers");
        printf("%12.6f %12.6f\n",yout[1],a1);
        printf("%12.6f %12.6f\n",yout[2],a2);
    }
    free_vector(d2y,1,NVAR);
    free_vector(yout,1,NVAR);
    free_vector(y,1,NVAR);
    return 0;
}
```

Numerical Recipes provides two routines for stiff systems. The first, `stiff`, is a fourth-order Rosenbrock method. We try it out on the sample problem described in §16.6. Note that for best results one should change the line

```
        yscal[i]=fabs(y[i])+fabs(dydx[i]*h)+TINY;
```

in odeint to

```
        yscal[i]=FMAX(1.0,fabs(y[i]));
```

```
/* Driver for routine stiff */

#include <stdio.h>
```

```
#include "nr.h"
#include "nrutil.h"

int kmax,kount;            /* defining declarations */
float *xp,**yp,dxsav;

int main(void)
{
    float eps,hstart,x1=0.0,x2=50.0,y[4];
    int nbad,nok;

    for (;;) {
        printf("Enter eps,hstart\n");
        if (scanf("%f %f",&eps,&hstart) == EOF) break;
        kmax=0;
        y[1]=y[2]=1.0;
        y[3]=0.0;
        odeint(y,3,x1,x2,eps,hstart,0.0,&nok,&nbad,derivs,stiff);
        printf("\n%s %13s %3d\n","successful steps:"," ",nok);
        printf("%s %20s %3d\n","bad steps:"," ",nbad);
        printf("Y(END) = %12.6f %12.6f %12.6f\n",y[1],y[2],y[3]);
    }
    printf("Normal completion\n");
    return 0;
}
```

The second method for stiff equations is a version of Bulirsch-Stoer integration that uses the semi-implicit midpoint rule `simpr` instead of `mmid`. Demonstration program `xsimpr` tries it out on the same test problem as above, exploring the effect of different stepsizes.

```
/* Driver for routine simpr */

#include <stdio.h>
#include "nr.h"
#include "nrutil.h"

#define NVAR 3
#define X1 0.0
#define HTOT 50.0

int main(void)
{
    int i;
    float a1=0.5976,a2=1.4023,a3=0.0,*y,*yout,*dfdx,**dfdy,*dydx;

    y=vector(1,NVAR);
    yout=vector(1,NVAR);
    dfdx=vector(1,NVAR);
    dfdy=matrix(1,NVAR,1,NVAR);
    dydx=vector(1,NVAR);
    y[1]=y[2]=1.0;
    y[3]=0.0;
    derivs(X1,y,dydx);
    jacobn(X1,y,dfdx,dfdy,NVAR);
    printf("Test Problem:\n");
    for (i=5;i<=50;i+=5) {
```

```
        simpr(y,dydx,dfdx,dfdy,NVAR,X1,HTOT,i,yout,derivs);
        printf("\n%s %5.2f %s %5.2f %s %2d %s \n",
            "x=",X1," to ",X1+HTOT," in ",i," steps");
        printf("%14s %9s\n","integration","bessj");
        printf("%12.6f %12.6f\n",yout[1],a1);
        printf("%12.6f %12.6f\n",yout[2],a2);
        printf("%12.6f %12.6f\n",yout[3],a3);
    }
    free_vector(dydx,1,NVAR);
    free_matrix(dfdy,1,NVAR,1,NVAR);
    free_vector(dfdx,1,NVAR);
    free_vector(yout,1,NVAR);
    free_vector(y,1,NVAR);
    return 0;
}
```

Finally, sample program `xstifbs` solves the same problem with the full machinery of the Bulirsch-Stoer method via routine `stifbs`. Once again you should modify the scaling in `odeint` as in `xstiff` for best results.

```
/* Driver for routine stifbs */

#include <stdio.h>
#include "nr.h"
#include "nrutil.h"

int kmax,kount;          /* defining declarations */
float *xp,**yp,dxsav;

int main(void)
{
    float eps,hstart,x1=0.0,x2=50.0,y[4];
    int nbad,nok;

    for (;;) {
        printf("Enter eps,hstart\n");
        if (scanf("%f %f",&eps,&hstart) == EOF) break;
        kmax=0;
        y[1]=y[2]=1.0;
        y[3]=0.0;
        odeint(y,3,x1,x2,eps,hstart,0.0,&nok,&nbad,derivs,stifbs);
        printf("\n%s %13s %3d\n","successful steps:"," ",nok);
        printf("%s %20s %3d\n","bad steps:"," ",nbad);
        printf("Y(END) = %12.6f %12.6f %12.6f\n",y[1],y[2],y[3]);
    }
    printf("Normal completion\n");
    return 0;
}
```

Chapter 17: Two-Point Boundary Value Problems

Two-point boundary value problems, and their iterative solution, are the substance of Chapter 17 of Numerical Recipes. The first step in solving them is to cast the problem as a set of N coupled first-order ordinary differential equations, satisfying n_1 conditions at one boundary point, and $n_2 = N - n_1$ conditions at the other boundary point. We apply two general methods to the solutions. First are the shooting methods, typified by routines shoot and shootf, which enforce the n_1 conditions at one boundary and set n_2 conditions freely. Then they integrate across the interval to find discrepancies with the n_2 conditions to be enforced at the other end. The Newton-Raphson method is used to reduce these discrepancies by adjusting the variable parameters.

The other approach is the relaxation method in which the differential equations are replaced by finite difference equations on a grid that covers the range of interest. Routine solvde demonstrates this method, and is demonstrated "in action" by program sfroid, which uses it to compute eigenvalues of spheroidal harmonics. The programs sphoot and sphfpt in Numerical Recipes similarly demonstrate the use of shoot and shootf on the same problem. Since these three programs are essentially self-contained, we do not need separate demonstration programs here. The file xsphfpt.c on the diskette merely contains a copy of the routine derivs needed by sphfpt. We reproduce it here, along with typical output from the three test programs.

$$\star \quad \star \quad \star \quad \star$$

```
/* Auxiliary routine for sphfpt */

extern int m,n;
extern float c2,dx,gmma;

void derivs(float x,float y[],float dydx[])
{
    dydx[1]=y[2];
    dydx[2]=(2.0*x*(m+1.0)*y[2]-(y[3]-c2*x*x)*y[1])/(1.0-x*x);
    dydx[3]=0.0;
}
```

In the following sample output, note that λ (lamda) and μ (mu) are merely two different conventions for expressing the same result, related by

$$\mu(m,n) = \lambda(m,n) - m(m+1)$$

Typical output for sfroid.c:

```
enter m n
3 5
Enter c**2 or 999 to end.
16.0
   Iter.       Error        FAC
    1       0.271212     1.000000
   Iter.       Error        FAC
    2       0.080674     1.000000
   Iter.       Error        FAC
    3       0.006255     1.000000
   Iter.       Error        FAC
    4       0.000039     1.000000
   Iter.       Error        FAC
    5       0.000000     1.000000
  m =  3    n =  5   c**2 =  16.000  lamda =  35.461006
Enter c**2 or 999 to end.
20.0
   Iter.       Error        FAC
    1       0.075860     1.000000
   Iter.       Error        FAC
    2       0.004998     1.000000
   Iter.       Error        FAC
    3       0.000025     1.000000
   Iter.       Error        FAC
    4       0.000000     1.000000
  m =  3    n =  5   c**2 =  20.000  lamda =  36.771168
Enter c**2 or 999 to end.
-16.0
   Iter.       Error        FAC
    1       0.517707     1.000000
   Iter.       Error        FAC
    2       0.835383     1.000000
   Iter.       Error        FAC
    3       0.375960     1.000000
   Iter.       Error        FAC
    4       0.172314     1.000000
   Iter.       Error        FAC
    5       0.020227     1.000000
   Iter.       Error        FAC
    6       0.000824     1.000000
   Iter.       Error        FAC
    7       0.000001     1.000000
  m =  3    n =  5   c**2 = -16.000  lamda =  24.310978
Enter c**2 or 999 to end.
-20.0
   Iter.       Error        FAC
    1       0.052214     1.000000
   Iter.       Error        FAC
    2       0.003544     1.000000
   Iter.       Error        FAC
    3       0.000022     1.000000
   Iter.       Error        FAC
    4       0.000000     1.000000
  m =  3    n =  5   c**2 = -20.000  lamda =  22.893074
Enter c**2 or 999 to end.
999
```

Typical output from `sphoot.c`:

```
input m,n,c-squared (999 to end)
3 5 1.5
     mu(m,n)
     18.524508
input m,n,c-squared (999 to end)
3 5 0.0
     mu(m,n)
     18.000002
input m,n,c-squared (999 to end)
3 5 999
```

Typical output from `sphfpt.c`:

```
input m,n,c-squared (999 to end)
3 5 1.5
     mu(m,n)
     18.524508
input m,n,c-squared (999 to end)
3 5 -1.5
     mu(m,n)
     17.473288
input m,n,c-squared (999 to end)
3 5 0.0
     mu(m,n)
     18.000004
input m,n,c-squared (999 to end)
3, 5, 999
```

Chapter 18: Integral Equations and Inverse Theory

Linear Fredholm equations of the second kind can be solved with the routine fred2. *The accompanying routine* fredin *interpolates the solution returned by* fred2 *on the quadrature points to arbitrary abscissas.* voltra *handles linear Volterra equations of the second kind. Routine* wwghts *constructs quadrature weights for an arbitrarily singular kernel. In Numerical Recipes a sample program demonstrating its use is given in* fredex, *so we do not give another sample program here.*

Fredholm equations of the first kind are closely related to so-called inverse problems. While no general-purpose routines are given in Numerical Recipes for such problems, there is an extensive discussion in Chapter 18.

$$\star \quad \star \quad \star \quad \star$$

Our test problem for routine fred2 is Example 4.2.5b from *Computational Methods for Integral Equations* by Delves and Mohamed:

$$f(t) = \int_0^{\pi/2} (ts)^{3/4} f(s)\, ds + t^{1/2} - \left(\frac{\pi}{2}\right)^{9/4} \frac{t^{3/4}}{9/4}$$

with exact solution

$$f(t) = t^{1/2}$$

Using only an 8-point Gaussian quadrature gives reasonable accuracy in the solution.

```
/* Driver for routine fred2 */

#include <stdio.h>
#include <math.h>
#include "nr.h"
#include "nrutil.h"

#define N 8
#define PI 3.1415927

float g(float t)
{
	return sqrt(t)-pow(PI/2.0,2.25)*pow(t,0.75)/2.25;
}

float ak(float t,float s)
{
	return pow(t*s,0.75);
}
```

```
int main(void)
{
    int i;
    float a=0.0,b=PI/2.0,*f;
    float *t,*w;

    t=vector(1,N);
    f=vector(1,N);
    w=vector(1,N);
    fred2(N,a,b,t,f,w,g,ak);
    /* Compare with exact solution */
    printf("Abscissa, Calc soln, True soln\n");
    for (i=1;i<=N;i++) printf("%10.6f %10.6f %10.6f\n",t[i],f[i],sqrt(t[i]));
    free_vector(w,1,N);
    free_vector(f,1,N);
    free_vector(t,1,N);
    return 0;
}
```

Routine `fredin` interpolates the solution obtained with `fred2` onto an arbitrary point t using Nystrom interpolation. Sample program `xfredin` demonstrates its use on the same problem solved in `xfred2`.

```
/* Driver for routine fredin */

#include <stdio.h>
#include <math.h>
#include "nr.h"
#include "nrutil.h"

#define N 8
#define PI 3.1415927

float g(float t)
{
    return sqrt(t)-pow(PI/2.0,2.25)*pow(t,0.75)/2.25;
}

float ak(float t,float s)
{
    return pow(t*s,0.75);
}

int main(void)
{
    float a=0.0,ans,b=PI/2.0,x,*f;
    float *t,*w;

    t=vector(1,N);
    f=vector(1,N);
    w=vector(1,N);
    fred2(N,a,b,t,f,w,g,ak);
    for (;;) {
        printf("Enter T between 0 and PI/2\n");
        if (scanf("%f",&x) == EOF) break;
        ans=fredin(x,N,a,b,t,f,w,g,ak);
        printf("T, Calculated answer, True answer\n");
```

```
        printf("%10.6f %10.6f %10.6f\n",x,ans,sqrt(x));
    }
    free_vector(w,1,N);
    free_vector(f,1,N);
    free_vector(t,1,N);
    return 0;
}
```

Our test system of Volterra equations comes from p. 151 of Delves and Mohamed's book:

$$f_1(t) = -\int_0^t e^{t-s} f_1(s)\,ds - \int_0^t \cos(t-s) f_2(s)\,ds + \cosh t + t\sin t$$

$$f_2(t) = -\int_0^t e^{t+s} f_1(s)\,ds - \int_0^t t\cos s\, f_2(s)\,ds + 2\sin t + t(\sin^2 t + e^t)$$

with exact solution

$$f_1(t) = e^{-t}$$
$$f_2(t) = 2\sin t$$

```
/* Driver for routine voltra */
#include <stdio.h>
#include <math.h>
#include "nr.h"
#include "nrutil.h"

float g(int k,float t)
{
    return (k == 1 ? cosh(t)+t*sin(t) : 2.0*sin(t)+t*(SQR(sin(t))+exp(t)));
}

float ak(int k,int l,float t,float s)
{
    return ((k == 1) ?
        (l == 1 ? -exp(t-s) : -cos(t-s)) :
        (l == 1 ? -exp(t+s) : -t*cos(s)));
}

#define N 30
#define H 0.05
#define M 2

int main(void)
{
    int nn;
    float t0=0.0,*t,**f;

    t=vector(1,N);
    f=matrix(1,M,1,N);
    voltra(N,M,t0,H,t,f,g,ak);
    /* exact soln is f[1]=exp(-t), f[2]=2sin(t) */
    printf(" abscissa, voltra answer1, real answer1,");
    printf(" voltra answer2, real answer2\n");
    for (nn=1;nn<=N;nn++)
        printf("%12.6f %12.6f %12.6f %12.6f %12.6f\n",
            t[nn],f[1][nn],exp(-t[nn]),f[2][nn],2.0*sin(t[nn]));
```

```
    free_vector(t,1,N);
    free_matrix(f,1,M,1,N);
    return 0;
}
```

Chapter 19: Partial Differential Equations

Several methods for solving partial differential equations by numerical means are treated in Chapter 19 of Numerical Recipes. All are finite differencing methods, including forward time centered space differencing, the Lax method, staggered leapfrog differencing, the two-step Lax-Wendroff scheme, the Crank-Nicholson method, Fourier analysis and cyclic reduction (FACR), Jacobi's method, the Gauss-Seidel method, successive over-relaxation (SOR) with and without Chebyshev acceleration, operator splitting methods as exemplified by the alternating direction implicit (ADI) method, and multigrid methods. There are so many methods, in fact, that we have not provided each topic with a routine of its own. In many cases the nature of such routines follows naturally from the description. In some cases, you will have to consult other references. The routines that do appear in the chapter, sor, mglin and mgfas, show two of the more useful and efficient methods for elliptic equations in application.

$$\star \quad \star \quad \star \quad \star$$

Procedure sor incorporates successive over-relaxation with Chebyshev acceleration to solve an elliptic partial differential equation. As input it accepts six arrays of coefficients, an estimate of the spectral radius of Jacobi iteration, and a trial solution which is often just set to zero over the solution grid. In program xsor the method is applied to the model problem

$$\frac{\partial^2 u}{\partial x^2} + \frac{\partial^2 u}{\partial y^2} = \rho$$

which is treated as the relaxation problem

$$\frac{\partial u}{\partial t} = \frac{\partial^2 u}{\partial x^2} + \frac{\partial^2 u}{\partial y^2} - \rho$$

Using FTCS differencing, this becomes

$$u_{j+1,l}^n + u_{j-1,l}^n + u_{j,l+1}^n + u_{j,l-1}^n - 4u_{j,l}^{n+1} = \rho_{jl}\Delta^2$$

(The notation is explained in Chapter 19 of *Numerical Recipes*.) This is a simple form of the general difference equation to which sor may be applied, with

$$A_{jl} = B_{jl} = C_{jl} = D_{jl} = 1.0 \text{ and } E_{jl} = -4.0$$

for all j and l. The starting guess for u is $u_{jl} = 0.0$ for all j, l. We choose the density ρ to be zero everywhere except directly in the center of the grid where $\rho = 2$. For a 32 × 32 grid, $\Delta = 1/32$ and so $F_{j,l} = 0.0$ except in the center of the grid where

$F(\text{midl}, \text{midl}) = 2/32^2 \approx 0.0020$. The value of ρ_{Jacobi}, which is called `rjac`, is taken from equation (19.5.24) of *Numerical Recipes*,

$$\rho_{\text{Jacobi}} = \frac{\cos\dfrac{\pi}{J} + \left(\dfrac{\Delta x}{\Delta y}\right)^2 \cos\dfrac{\pi}{L}}{1 + \left(\dfrac{\Delta x}{\Delta y}\right)^2}$$

In this case, j=l=JMAX and $\Delta x = \Delta y$ so $\text{rjac} = \cos(\pi/\text{JMAX})$. A call to `sor` leads to the solution shown below. To fit the output on the page, the solution at only every fourth grid point is printed out. As a test that this is indeed a solution to the finite difference equation, the program plugs the result back into that equation, calculating

$$F_{j,l} = u_{j+1,l}^n + u_{j-1,l}^n + u_{j,l+1}^n + u_{j,l-1}^n - 4u_{j,l}^{n+1}$$

The test is whether $F_{j,l}$ is almost everywhere zero, but equal to 0.0020 at the very centerpoint of the grid.

```
SOR solution:
  0.0000  0.0000  0.0000  0.0000  0.0000  0.0000  0.0000  0.0000  0.0000
  0.0000  0.0000 -0.0001 -0.0001 -0.0001 -0.0001 -0.0001  0.0000  0.0000
  0.0000 -0.0001 -0.0001 -0.0002 -0.0002 -0.0002 -0.0001 -0.0001  0.0000
  0.0000 -0.0001 -0.0002 -0.0003 -0.0005 -0.0003 -0.0002 -0.0001  0.0000
  0.0000 -0.0001 -0.0002 -0.0005 -0.0014 -0.0005 -0.0002 -0.0001  0.0000
  0.0000 -0.0001 -0.0002 -0.0003 -0.0005 -0.0003 -0.0002 -0.0001  0.0000
  0.0000 -0.0001 -0.0001 -0.0002 -0.0002 -0.0002 -0.0001 -0.0001  0.0000
  0.0000  0.0000 -0.0001 -0.0001 -0.0001 -0.0001 -0.0001  0.0000  0.0000
  0.0000  0.0000  0.0000  0.0000  0.0000  0.0000  0.0000  0.0000  0.0000
```

```c
/* Driver for routine sor */

#include <stdio.h>
#include <math.h>
#include "nr.h"
#include "nrutil.h"

#define NSTEP 4
#define JMAX 33
#define PI 3.1415926

int main(void)
{
    int i,j,midl;
    double **a,**b,**c,**d,**e,**f,**u,rjac;

    a=dmatrix(1,JMAX,1,JMAX);
    b=dmatrix(1,JMAX,1,JMAX);
    c=dmatrix(1,JMAX,1,JMAX);
    d=dmatrix(1,JMAX,1,JMAX);
    e=dmatrix(1,JMAX,1,JMAX);
    f=dmatrix(1,JMAX,1,JMAX);
    u=dmatrix(1,JMAX,1,JMAX);
    for (i=1;i<=JMAX;i++)
        for (j=1;j<=JMAX;j++) {
            a[i][j]=b[i][j]=c[i][j]=d[i][j]=1.0;
```

```
                e[i][j]=(-4.0);
                f[i][j]=u[i][j]=0.0;
            }
        midl=JMAX/2+1;
        f[midl][midl]=2.0/((JMAX-1)*(JMAX-1));
        rjac=cos(PI/JMAX);
        sor(a,b,c,d,e,f,u,JMAX,rjac);
        printf("SOR solution:\n");
        for (i=1;i<=JMAX;i+=NSTEP) {
            for (j=1;j<=JMAX;j+=NSTEP) printf("%8.4f",u[i][j]);
            printf("\n");
        }
        printf("\n Test that solution satisfies difference equations:\n");
        for (i=NSTEP+1;i<JMAX;i+=NSTEP) {
            for (j=NSTEP+1;j<JMAX;j+=NSTEP)
                f[i][j]=u[i+1][j]+u[i-1][j]+u[i][j+1]+u[i][j-1]
                    -4.0*u[i][j];
            printf("%7s"," ");
            for (j=NSTEP+1;j<JMAX;j+=NSTEP) printf("%8.4f",f[i][j]);
            printf("\n");
        }
        free_dmatrix(u,1,JMAX,1,JMAX);
        free_dmatrix(f,1,JMAX,1,JMAX);
        free_dmatrix(e,1,JMAX,1,JMAX);
        free_dmatrix(d,1,JMAX,1,JMAX);
        free_dmatrix(c,1,JMAX,1,JMAX);
        free_dmatrix(b,1,JMAX,1,JMAX);
        free_dmatrix(a,1,JMAX,1,JMAX);
        return 0;
    }
```

Multigrid methods are generally much more efficient than SOR. Routine `mglin` solves linear elliptic equations using the full multigrid method. Program `xmglin` solves the same model problem as `xsor` and prints out the solution, which may be compared with the copy listed above. However, because of the way `mglin` is set up, when the solution is substituted into the difference equations, it should give back ρ and not $\rho\Delta^2$. Here is the output on a 32×32 grid:

```
Test that solution satisfies difference equations:
        0.0001 -0.0001 -0.0004  0.0008 -0.0004 -0.0001  0.0001
       -0.0001 -0.0002 -0.0008  0.0028 -0.0008 -0.0002 -0.0001
       -0.0004 -0.0008 -0.0027  0.0048 -0.0027 -0.0008 -0.0004
        0.0008  0.0028  0.0048  2.0171  0.0048  0.0028  0.0008
       -0.0004 -0.0008 -0.0027  0.0048 -0.0027 -0.0008 -0.0004
       -0.0001 -0.0002 -0.0008  0.0028 -0.0008 -0.0002 -0.0001
        0.0001 -0.0001 -0.0004  0.0008 -0.0004 -0.0001  0.0001
```

```
/* Driver for routine mglin */

#include <stdio.h>
#include <math.h>
#include "nr.h"
#include "nrutil.h"

#define NSTEP 4
#define JMAX 33
```

```
int main(void)
{
    int i,j,midl=JMAX/2+1;
    double **f,**u;

    f=dmatrix(1,JMAX,1,JMAX);
    u=dmatrix(1,JMAX,1,JMAX);
    for (i=1;i<=JMAX;i++)
        for (j=1;j<=JMAX;j++)
            u[i][j]=0.0;
    u[midl][midl]=2.0;
    mglin(u,JMAX,2);
    printf("MGLIN solution:\n");
    for (i=1;i<=JMAX;i+=NSTEP) {
        for (j=1;j<=JMAX;j+=NSTEP) printf("%8.4f",u[i][j]);
        printf("\n");
    }
    printf("\n Test that solution satisfies difference equations:\n");
    for (i=NSTEP+1;i<JMAX;i+=NSTEP) {
        for (j=NSTEP+1;j<JMAX;j+=NSTEP)
            f[i][j]=u[i+1][j]+u[i-1][j]+u[i][j+1]+u[i][j-1]
                -4.0*u[i][j];
        printf("%7s"," ");
        for (j=NSTEP+1;j<JMAX;j+=NSTEP) printf("%8.4f",f[i][j]*(JMAX-1)*(JMAX-1));
        printf("\n");
    }
    free_dmatrix(u,1,JMAX,1,JMAX);
    free_dmatrix(f,1,JMAX,1,JMAX);
    return 0;
}
```

Numerical Recipes provides `mgfas` as a prototype for solving nonlinear elliptic problems with the Full Approximation Storage (FAS) scheme. Program `xmgfas` solves the simple problem

$$\nabla^2 u + u^2 = \rho$$

as described in *Numerical Recipes* with the same choice of ρ made in `xsor` and `xmglin`. The solution is checked once again by substituting into the difference equation. Here is the output for the 32×32 grid:

```
MGFAS solution:
  0.0000  0.0000  0.0000  0.0000  0.0000  0.0000  0.0000  0.0000
  0.0000  0.0000 -0.0001 -0.0001 -0.0001 -0.0001 -0.0001  0.0000  0.0000
  0.0000 -0.0001 -0.0001 -0.0002 -0.0002 -0.0002 -0.0001 -0.0001  0.0000
  0.0000 -0.0001 -0.0002 -0.0004 -0.0004 -0.0004 -0.0002 -0.0001  0.0000
  0.0000 -0.0001 -0.0002 -0.0004 -0.0013 -0.0004 -0.0002 -0.0001  0.0000
  0.0000 -0.0001 -0.0002 -0.0004 -0.0004 -0.0004 -0.0002 -0.0001  0.0000
  0.0000 -0.0001 -0.0001 -0.0002 -0.0002 -0.0002 -0.0001 -0.0001  0.0000
  0.0000  0.0000 -0.0001 -0.0001 -0.0001 -0.0001 -0.0001  0.0000  0.0000
  0.0000  0.0000  0.0000  0.0000  0.0000  0.0000  0.0000  0.0000  0.0000
 Test that solution satisfies difference equations:
          0.0000  0.0008 -0.0009  0.0000 -0.0009  0.0008  0.0000
          0.0008  0.0033  0.0007 -0.0035  0.0007  0.0033  0.0008
         -0.0009  0.0007  0.0051 -0.0328  0.0051  0.0007 -0.0009
          0.0000 -0.0035 -0.0328  1.8660 -0.0328 -0.0035  0.0000
         -0.0009  0.0007  0.0051 -0.0328  0.0051  0.0007 -0.0009
```

```
      0.0008   0.0033   0.0007  -0.0035   0.0007   0.0033   0.0008
      0.0000   0.0008  -0.0009   0.0000  -0.0009   0.0008   0.0000
```

You might be a little puzzled by this output at first. For example, the center value of $\nabla^2 u + u^2$ for the solution is 1.8660, while the true value is 2. Both `sor` and `mglin` get closer, with 2.0171, yet the nonlinear correction u^2 is only $(-.0013)^2$. Here's what's going on: `mgfas` iterates until the iteration error is smaller than the truncation error. These errors are r.m.s. errors, computed with routine `anorm2`. To check the performance of `mgfas`, we did a run with the parameter `ALPHA` set to 10^{-6} and with `maxcyc = 6`. This high accuracy solution is essentially the exact solution of the finite difference equations, and is in fact very close to the solution obtained by `sor` and `mglin`. We next computed the truncation error by applying the routine `lop` to this accurate solution and then finding the difference between the answer and the original right-hand side, ρ. The norm of this error was 6×10^{-2}. Next, we applied `lop` to the solution obtained in the demonstration program. (This gives the quantity printed above as "Test that solution satisfies difference equations:".) The difference between this quantity and `lop` of the "exact" finite difference solution is the iteration error. Its norm is 6×10^{-3}. Thus `mgfas` is solving the problem to better than the truncation error, as advertised. Since in this problem ρ is so sharply peaked, the r.m.s. error is perhaps not the best measure of the error — the maximum deviation might be better. For a smoother ρ, these error measures are not that different and `mgfas` would give more intuitive looking results. For this problem, however, you need a smaller value of `ALPHA` if you want the maximum deviation below the truncation error; `ALPHA = 0.1` will do the trick.

```c
/* Driver for routine mgfas */

#include <stdio.h>
#include <math.h>
#include "nr.h"
#include "nrutil.h"

#define NSTEP 4
#define JMAX 33

int main(void)
{
    int i,j,midl=JMAX/2+1;
    double **f,**u;

    f=dmatrix(1,JMAX,1,JMAX);
    u=dmatrix(1,JMAX,1,JMAX);
    for (i=1;i<=JMAX;i++)
        for (j=1;j<=JMAX;j++)
            u[i][j]=0.0;
    u[midl][midl]=2.0;
    mgfas(u,JMAX,2);
    printf("MGFAS solution:\n");
    for (i=1;i<=JMAX;i+=NSTEP) {
        for (j=1;j<=JMAX;j+=NSTEP) printf("%8.4f",u[i][j]);
        printf("\n");
    }
    printf("\n Test that solution satisfies difference equations:\n");
    for (i=NSTEP+1;i<JMAX;i+=NSTEP) {
```

```
        for (j=NSTEP+1;j<JMAX;j+=NSTEP)
            f[i][j]=u[i+1][j]+u[i-1][j]+u[i][j+1]+u[i][j-1]-
                4.0*u[i][j]+u[i][j]*u[i][j]/((JMAX-1)*(JMAX-1));
        printf("%7s"," ");
        for (j=NSTEP+1;j<JMAX;j+=NSTEP) printf("%8.4f",f[i][j]*(JMAX-1)*(JMAX-1));
        printf("\n");
    }
    free_dmatrix(u,1,JMAX,1,JMAX);
    free_dmatrix(f,1,JMAX,1,JMAX);
    return 0;
}
```

Chapter 20: Less-Numerical Algorithms

> *Chapter 20 of Numerical Recipes collects together an idiosyncratic group
> of "less-numerical" algorithms whose usefulness lies either in tasks peripheral
> to numerical work, or in elucidating some less obvious applications of numeri-
> cal techniques (for example, the use of Fast Fourier Transforms for performing
> extended precision arithmetic).*
>
> *The routine* machar *is a useful aid in understanding the limits of floating
> point arithmetic on your particular computer.* igray *is a simple function that
> converts integers into their Gray code, or back again. The combination of* icrc
> *and its captive routine* icrc1 *is used to construct cyclic redundancy check-
> sums, in some standard conventions. The main book discusses two rather
> different methods for compressing information, Huffman coding, realized in
> the routines* hufmak, hufapp, hufenc, *and* hufdec, *and arithmetic coding,
> realized as* arcmak, arcode, *and* arcsum. *Finally, a whole group of routines
> for multiple precision arithmetic are given, with names of the form* mp...,
> *culminating in the program* mppi *which is used to compute* π *to as many as
> several thousand decimal digits.*

$$\star \quad \star \quad \star \quad \star$$

The sample program for machar does nothing more than run the routine and print
the results. machar, by some fairly subtle arithmetic processes, feels out the floating
point characteristics of your computer. The main book explains the meaning of the
parameters that are reported. The most important ones are eps, xmin, and xmax,
which give respectively the floating point precision, and the smallest and largest (in
magnitude) numbers that are representable at full precision. (Most ANSI C compilers
already provide values for these quantities, under different names, in the standard
header file float.h.)

```
/* Driver for routine machar */

#include <stdio.h>
#include "nr.h"

int main(void)
{
    int ibeta,iexp,irnd,it,machep,maxexp,minexp,negep,ngrd;
    float eps,epsneg,xmax,xmin;

    machar(&ibeta,&it,&irnd,&ngrd,&machep,&negep,&iexp,&minexp,&maxexp,
        &eps,&epsneg,&xmin,&xmax);
    printf("ibeta = %d\n",ibeta);
    printf("it = %d\n",it);
```

```
        printf("irnd = %d\n",irnd);
        printf("ngrd = %d\n",ngrd);
        printf("machep = %d\n",machep);
        printf("negep = %d\n",negep);
        printf("iexp = %d\n",iexp);
        printf("minexp = %d\n",minexp);
        printf("maxexp = %d\n",maxexp);
        printf("eps = %12.6g\n",eps);
        printf("epsneg = %12.6g\n",epsneg);
        printf("xmin = %12.6g\n",xmin);
        printf("xmax = %12.6g\n",xmax);
        return 0;
}
```

The function `igray` returns the Gray code of its first argument, or else the inverse Gray code (if the second argument is -1). The following example program verifies, for some range of integers, that the function, followed by its inverse, does in fact produce the original input value. It also prints out a few values within the chosen range, and verifies, for those values, that the Gray code of two consecutive integers differ in only one bit: the last column printed should always be a power of 2; half the time it will simply be the value 1, since the results differ only in the least significant bit.

```
/* Driver for routine igray */

#include <stdio.h>
#include "nr.h"

int main(void)
{
    unsigned long jp,n,ng,nmax,nmin,nni,nxor;

    for (;;) {
        printf("input nmin,nmax: \n");
        if (scanf("%lu %lu",&nmin,&nmax) == EOF) break;
        jp=(nmax-nmin)/11;
        if (jp < 1) jp=1;
        printf("n, Gray[n], Gray(Gray[n]), Gray[n] ^ Gray[n+1]\n");
        for (n=nmin;n<=nmax;n++) {
            ng=igray(n,1);
            nni=igray(ng,-1);
            if (nni != n) printf("WRONG ! AT %d %d %d\n",n,ng,nni);
            if (!((n-nmin) % jp)) {
                    nxor=ng ^ igray(n+1,1);
                    printf("%lu %lu %lu %lu\n",n,ng,nni,nxor);
            }
        }
    }
    printf("Normal completion\n");
    return 0;
}
```

Cyclic redundancy checksums are constructed, according to any one of several standard conventions, by the routine `icrc`. The following test program, with the input lines

```
1 T
17 CatMouse987654321
```

will reproduce the test values that are given in the Table in the main book. Notice how some conventions (e.g., X.25), require that the packet (message and checksum together) be made from the bitwise complement of the CRC bytes, while others use the bytes themselves. Note also that the desired checksum of the full packet is zero in some conventions, but some particular value (here hexadecimal F0B8) in others.

```
/* Driver for routine icrc */

#include <stdio.h>
#include "nr.h"

#define LOBYTE(x) ((unsigned char)((x) & 0xFF))
#define HIBYTE(x) ((unsigned char)((x) >> 8))

int main(void)
{
        unsigned char lin[256];
        unsigned short i1,i2;
        unsigned long n;

        for (;;) {
            printf("Enter length and string: \n");
            if (scanf("%lu %s",&n,&lin[1]) == EOF) break;
            lin[n+1]=0;
            printf("%s\n",&lin[1]);
            i1=icrc(0,lin,n,(short)0,1);
            lin[n+1]=HIBYTE(i1);
            lin[n+2]=LOBYTE(i1);
            i2=icrc(i1,lin,n+2,(short)0,1);
            printf("    XMODEM: String CRC, Packet CRC=    0x%x    0x%x\n",i1,i2);
            i1=icrc(i2,lin,n,(short)0xff,-1);
            lin[n+1] = ~LOBYTE(i1);
            lin[n+2] = ~HIBYTE(i1);
            i2=icrc(i1,lin,n+2,(short)0xff,-1);
            printf("     X.25: String CRC, Packet CRC=    0x%x    0x%x\n",i1,i2);
            i1=icrc(i2,lin,n,(short)0,-1);
            lin[n+1]=LOBYTE(i1);
            lin[n+2]=HIBYTE(i1);
            i2=icrc(i1,lin,n+2,(short)0,-1);
            printf(" CRC-CCITT: String CRC, Packet CRC=    0x%x    0x%x\n",i1,i2);
        }
        printf("Normal completion\n");
        return 0;
}
```

The example program for the decimal check digit routine decchk first verifies the claim in the main book that the method finds about 95% of jump transpositions ($acb \rightarrow bca$). (The exact answer is that it finds all but 52 out of 900.) The program then prompts the user for test strings (terminated by "x") and prints out the string with a following check digit, and verification that the augmented string passes the check-digit test. You might test this program with, e.g., telephone numbers in various formats, such as "1-800-433-7300" or "1 (800) 241-6522," to verify that decchk in fact ignores all input characters except decimal digits in computing its check digit.

```c
/* Driver for routine decchk */

#include <stdio.h>
#include "nr.h"

#define MAXLINE 128

int main(void)
{
    int j,k,l,n,nbad=0,ntot=0;
    int iok,jok;
    char lin[MAXLINE+2],ch,chh;

    /* test all jump transpositions of the form 86jlk41 */
    lin[0]='8';
    lin[1]='6';
    lin[5]='4';
    lin[6]='1';
    for (j=48;j<=57;j++) {
        for (k=48;k<=57;k++) {
            for (l=48;l<=57;l++) {
                lin[3]=l;
                if (j != k) {
                    ntot++;
                    lin[2]=j;
                    lin[4]=k;
                    iok=decchk(lin,7,&ch);
                    lin[7]=ch;
                    iok=decchk(lin,8,&chh);
                    lin[2]=k;
                    lin[4]=j;
                    jok=decchk(lin,8,&chh);
                    if (!iok || jok) nbad++;
                }
            }
        }
    }
    printf("%s %14s %3d\n","Total tries:"," ",ntot);
    printf("%s %16s %3d\n","Bad tries:"," ",nbad);
    printf("%s %11s %4.2f\n","Fraction good:"," ",((float)(ntot-nbad))/ntot);
    for (;;) {
        printf("enter string terminated by x:\n");
        if (gets(lin) == NULL) break;
        for (j=0;j<MAXLINE;j++)
            if (lin[j] == 'x') break;
        n=j;
        if (!n) break;
        iok=decchk(lin,n,&ch);
        lin[n]=ch;
        jok=decchk(lin,n+1,&chh);
        lin[n+1]=0;
        printf("%s checks as %c\n",lin,jok ? 'T' : 'F');
    }
    printf("Normal completion\n");
    return 0;
```

```
}
```

The example program that demonstrates Huffman compression of alphabetic text has several separate tasks to perform. First, it must construct a statistical profile of the kind of text to be compressed, in the form of a table of letter frequencies. It does this using the file text.dat which, as supplied on the diskette, contains a particularly racy passage from Thomas Hardy's *Far from the Madding Crowd* (New York: Harper, 1895).

Next, the Huffman code itself is constructed from the frequency count. The demonstration program prints out the entire code, allowing you to see how more common letters are encoded in fewer bits than less common ones.

The program then prompts repeatedly for an input line of text, compresses that line, decompresses it, and prints out the result (hopefully the same as you input) and the lengths of the line in uncompressed and compressed forms. Notice how a compressed line is terminated by encoding a (spurious) rare character, while *not* incrementing the message byte count. This ensures that on decoding, the spurious character will not be fully decoded. In addition to trying ordinary text as input to the program, you might try a line of all lower case e's, or all upper case X's.

```c
/* Driver for routines hufmak, hufenc, hufdec */

#include <stdio.h>
#include <stdlib.h>
#include <string.h>
#include "nr.h"
#include "nrutil.h"

#define MC 512
#define MQ (2*MC-1)

#define MAXLINE 256

int main(void)
{
    int k;
    unsigned long i,j,ilong,n,nb,nh,nlong,nt,lcode=MAXLINE,nch,nfreq[257];
    unsigned char *code,mess[MAXLINE],ness[MAXLINE];
    huffcode hcode;
    FILE *fp;

    code=cvector(0,MAXLINE);
    hcode.icod=lvector(1,MQ);
    hcode.ncod=lvector(1,MQ);
    hcode.left=lvector(1,MQ);
    hcode.right=lvector(1,MQ);
    for (j=1;j<=MQ;j++) hcode.icod[j]=hcode.ncod[j]=0;
    /* construct a letter frequency table from the file text.dat */
    if ((fp = fopen("text.dat","r")) == NULL)
        nrerror("Input file text.dat not found.\n");
    for (j=1;j<=256;j++) nfreq[j]=0;
    while ((k=getc(fp)) != EOF) {
        if ((k -= 31) >= 1) nfreq[k]++;
    }
    fclose(fp);
```

```
nch=96;
/* here is the initialization that constructs the code */
hufmak(nfreq,nch,&ilong,&nlong,&hcode);
printf("ind char  nfreq  ncod   icod\n");
for (j=1;j<=nch;j++)
    if (nfreq[j]) printf("%3lu %c %6lu %6lu %6lu\n",
        j,(char)(j+31),nfreq[j],hcode.ncod[j],hcode.icod[j]);
for (;;) {
    /* now ready to prompt for lines to encode */
    printf("Enter a line:\n");
    if (gets((char *)&mess[1]) == NULL) break;
    n=strlen((char *)&mess[1]);
    /* shift from 256 character alphabet to 96 printing characters */
    for (j=1;j<=n;j++) mess[j] -= 32;
    /* here we Huffman encode mess[1..n] */
    nb=0;
    for (j=1;j<=n;j++) hufenc((unsigned long)mess[j],&code,&lcode,&nb,&hcode);
    nh=(nb>>3)+1;
    /* message termination (encode a single long character) */
    hufenc(ilong,&code,&lcode,&nb,&hcode);
    /* here we decode the message, hopefully to get the original back */
    nb=0;
    for (j=1;j<=MAXLINE;j++) {
        hufdec(&i,code,nh,&nb,&hcode);
        if (i == nch) break;
        else ness[j]=(unsigned char)i;
    }
    if (j > lcode) nrerror("Huffman coding: Never get here");
    nt=j-1;
    printf("Length of line input,coded= %lu %lu\n",n,nh);
    printf("Decoded output:\n");
    for (j=1;j<=nt;j++) printf("%c",(char)(ness[j]+32));
    printf("\n");
    if (nt != n) printf("Error! :  n decoded != n input\n");
    if (nt-n == 1) printf("May be harmless spurious character.\n");
}
free_cvector(code,0,MAXLINE);
free_lvector(hcode.right,1,MQ);
free_lvector(hcode.left,1,MQ);
free_lvector(hcode.ncod,1,MQ);
free_lvector(hcode.icod,1,MQ);
printf("Normal completion\n");
return 0;
}
```

The demonstration program for arithmetic compression is very similar to that for Huffman compression. As before, we construct a letter frequency table from a corpus that is statistically similar to the text that we are expecting to compress, then initialize the code, and finally encode/decode messages.

There are several characters' worth of output associated with message initialization and message termination in arithmetic coding, so one would not normally begin a new compression for each input line (as is done in the following demonstration). Rather, a whole file would normally be compressed in a single pass.

```
/* Driver for routines arcmak and arcode */

#include <stdio.h>
#include <stdlib.h>
#include <string.h>
#include "nr.h"
#include "nrutil.h"

#define MC 512

#define NWK 20

#define MAXLINE 256

int main(void)
{
    int k;
    unsigned long i,j,lc,lcode=MAXLINE,n,nch,nrad,nt,nfreq[257],tmp,zero=0;
    unsigned char *code,mess[MAXLINE],ness[MAXLINE];
    arithcode acode;
    FILE *fp;

    code=cvector(0,MAXLINE);
    acode.ilob=lvector(1,NWK);
    acode.iupb=lvector(1,NWK);
    acode.ncumfq=lvector(1,MC+2);
    if ((fp = fopen("text.dat","r")) == NULL)
        nrerror("Input file text.dat not found.\n");
    for (j=1;j<=256;j++) nfreq[j]=0;
    while ((k=getc(fp)) != EOF) {
        if ((k -= 31) >= 1) nfreq[k]++;
    }
    fclose(fp);
    nch=96;
    nrad=256;
    /* here is the initialization that constructs the code */
    arcmak(nfreq,(int)nch,(int)nrad,&acode);
    /* now ready to prompt for lines to encode */
    for (;;) {
        printf("Enter a line:\n");
        if (gets((char *)&mess[1]) == NULL) break;
        n=strlen((char *)&mess[1]);
        /* shift from 256 character alphabet to 96 printing characters */
        for (j=1;j<=n;j++) mess[j] -= 32;
        /* message initialization */
        lc=1;
        arcode(&zero,&code,&lcode,&lc,0,&acode);
        /* here we arithmetically encode mess(1:n) */
        for (j=1;j<=n;j++) {
            tmp=mess[j];
            arcode(&tmp,&code,&lcode,&lc,1,&acode);
        }
        /* message termination */
        arcode(&nch,&code,&lcode,&lc,1,&acode);
        printf("Length of line input, coded= %lu %lu\n",n,lc-1);
        /* here we decode the message, hopefully to get the original back */
        lc=1;
```

```
        arcode(&zero,&code,&lcode,&lc,0,&acode);
        for (j=1;j<=lcode;j++) {
            arcode(&i,&code,&lcode,&lc,-1,&acode);
            if (i == nch) break;
            else ness[j]=(unsigned char)i;
        }
        if (j > lcode) nrerror("Arith. coding: Never get here");
        nt=j-1;
        printf("Decoded output:\n");
        for (j=1;j<=nt;j++) printf("%c",(char)(ness[j]+32));
        printf("\n");
        if (nt != n) printf("Error ! j decoded != n input.\n");
    }
    free_cvector(code,0,MAXLINE);
    free_lvector(acode.ncumfq,1,MC+2);
    free_lvector(acode.iupb,1,NWK);
    free_lvector(acode.ilob,1,NWK);
    printf("Normal completion\n");
    return 0;
}
```

The *Numerical Recipes* book ends on a note of numerical denouement with the calculation of π to some 2398 decimal places. The following demonstration program, and its one additional function, repeats that calculation, and also uses the various multiple precision routines to calculate a couple of simpler quantities at high precision, $\sqrt{2}$ and $2 - \sqrt{2}$. You might wonder why bother with the latter quantity. The answer is that the multiple precision integer divide routine mpdiv happens not to be used at all in the calculation of π (although the routine for reciprocals, mpinv, is used). To give it some exercise, we therefore calculate $2 - \sqrt{2}$ by the following nutty scheme that does make use of the integer divide:

$$\frac{2 \times 2^M}{[\sqrt{2} \times 2^M]} = 1 \text{ remainder } \left(2 - \sqrt{2}\right) \times 2^M$$

at least to the same approximation that $\sqrt{2}$ is computed. Here square brackets indicate integer part, and M is the number of binary digits in the precision that you supply when prompted.

Before you go off and compute a million digits of π, be sure to read the main book's note about the routine mp2dfr for converting from base 256 to decimal fractions. Unless you are satisfied with knowing π in base 256 (or binary, octal, or hexadecimal), you will need to rewrite that routine to make it faster.

```
/* Driver for mp routines */

#include <stdio.h>
#include <math.h>
#include "nr.h"
#include "nrutil.h"

#define IAOFF 48
#define NMAX 1024

void mpsqr2(int n)
{
```

```
    int j,m;
    unsigned char *x,*y,*t,*q,*r,*s;

    x=cvector(1,NMAX);
    y=cvector(1,NMAX);
    t=cvector(1,NMAX);
    q=cvector(1,NMAX);
    r=cvector(1,NMAX);
    s=cvector(1,3*NMAX);
    t[1]=2;
    for (j=2;j<=n;j++) t[j]=0;
    mpsqrt(x,x,t,n,n);
    mpmov(y,x,n);
    printf("sqrt(2)=\n");
    s[1]=y[1]+IAOFF;
    s[2]='.';
    /* caution: next step is N**2! omit it for large N */
    mp2dfr(&y[1],&s[2],n-1,&m);
    s[m+3]=0;
    printf(" %64s\n",&s[1]);
    printf("Result rounded to 1 less base-256 place:\n");
    /* use s as scratch space */
    mpsad(s,x,n,128);
    mpmov(y,&s[1],n-1);
    s[1]=y[1]+IAOFF;
    s[2]='.';
    /* caution: next step is N**2! omit it for large N */
    mp2dfr(&y[1],&s[2],n-2,&m);
    s[m+3]=0;
    printf(" %64s\n",&s[1]);
    printf("2-sqrt(2)=\n");
    /* Calculate this the hard way to exercise the mpdiv function */
    mpdiv(q,r,t,x,n,n);
    s[1]=r[1]+IAOFF;
    s[2]='.';
    /* caution: next step is N**2! omit it for large N */
    mp2dfr(&r[1],&s[2],n-1,&m);
    s[m+3]=0;
    printf(" %64s\n",&s[1]);
    free_cvector(s,1,3*NMAX);
    free_cvector(r,1,NMAX);
    free_cvector(q,1,NMAX);
    free_cvector(t,1,NMAX);
    free_cvector(y,1,NMAX);
    free_cvector(x,1,NMAX);
}

int main(void)
{
    int n;

    for (;;) {
        printf("Input n\n");
        if (scanf("%d",&n) == EOF) break;
        mpsqr2(n);
        mppi(n);
    }
```

```
        printf("Normal completion\n");
        return 0;
}
```

Appendix A: Header Files

The Numerical Recipes functions, and the example routines in this book, make use of the following header files. The files listed here are abbreviated forms of those found on the Numerical Recipes diskettes, in that they do not show the alternative constructions for use with compilers that do not support the full prototype specifications of the ANSI C standard. We list the full ANSI prototypes because they are more helpful in checking the data types expected by each function. They can easily be abbreviated to the traditional K&R form.

$\star$ $\quad$ $\star$ $\quad$ $\star$ $\quad$ $\star$

File `nrutil.h` contains a number of macros and utility routines, which are discussed in Chapter 1 of the main book. The macros are used for common mathematical operations such as squaring a quantity, finding the maximum of a pair of numbers, and so on. The utility functions are used for error reporting, dynamic allocation and deallocation of memory for vectors and matrices, and the creation of references to submatrices. Source code for the utility functions is in the file `nrutil.c`, listed in Appendix B. Note that `nrutil.h` and `nrutil.c` are not copyrighted, and you may copy them freely for any purpose.

```
#ifndef _NR_UTILS_H_
#define _NR_UTILS_H_

static float sqrarg;
#define SQR(a) ((sqrarg=(a)) == 0.0 ? 0.0 : sqrarg*sqrarg)

static double dsqrarg;
#define DSQR(a) ((dsqrarg=(a)) == 0.0 ? 0.0 : dsqrarg*dsqrarg)

static double dmaxarg1,dmaxarg2;
#define DMAX(a,b) (dmaxarg1=(a),dmaxarg2=(b),(dmaxarg1) > (dmaxarg2) ?\
        (dmaxarg1) : (dmaxarg2))

static double dminarg1,dminarg2;
#define DMIN(a,b) (dminarg1=(a),dminarg2=(b),(dminarg1) < (dminarg2) ?\
        (dminarg1) : (dminarg2))

static float maxarg1,maxarg2;
#define FMAX(a,b) (maxarg1=(a),maxarg2=(b),(maxarg1) > (maxarg2) ?\
        (maxarg1) : (maxarg2))

static float minarg1,minarg2;
#define FMIN(a,b) (minarg1=(a),minarg2=(b),(minarg1) < (minarg2) ?\
        (minarg1) : (minarg2))
```

```
static long lmaxarg1,lmaxarg2;
#define LMAX(a,b) (lmaxarg1=(a),lmaxarg2=(b),(lmaxarg1) > (lmaxarg2) ?\
        (lmaxarg1) : (lmaxarg2))

static long lminarg1,lminarg2;
#define LMIN(a,b) (lminarg1=(a),lminarg2=(b),(lminarg1) < (lminarg2) ?\
        (lminarg1) : (lminarg2))

static int imaxarg1,imaxarg2;
#define IMAX(a,b) (imaxarg1=(a),imaxarg2=(b),(imaxarg1) > (imaxarg2) ?\
        (imaxarg1) : (imaxarg2))

static int iminarg1,iminarg2;
#define IMIN(a,b) (iminarg1=(a),iminarg2=(b),(iminarg1) < (iminarg2) ?\
        (iminarg1) : (iminarg2))

#define SIGN(a,b) ((b) >= 0.0 ? fabs(a) : -fabs(a))

#if defined(__STDC__) || defined(ANSI) || defined(NRANSI) /* ANSI */

void nrerror(char error_text[]);
float *vector(long nl, long nh);
int *ivector(long nl, long nh);
unsigned char *cvector(long nl, long nh);
unsigned long *lvector(long nl, long nh);
double *dvector(long nl, long nh);
float **matrix(long nrl, long nrh, long ncl, long nch);
double **dmatrix(long nrl, long nrh, long ncl, long nch);
int **imatrix(long nrl, long nrh, long ncl, long nch);
float **submatrix(float **a, long oldrl, long oldrh, long oldcl, long oldch,
        long newrl, long newcl);
float **convert_matrix(float *a, long nrl, long nrh, long ncl, long nch);
float ***f3tensor(long nrl, long nrh, long ncl, long nch, long ndl, long ndh);
void free_vector(float *v, long nl, long nh);
void free_ivector(int *v, long nl, long nh);
void free_cvector(unsigned char *v, long nl, long nh);
void free_lvector(unsigned long *v, long nl, long nh);
void free_dvector(double *v, long nl, long nh);
void free_matrix(float **m, long nrl, long nrh, long ncl, long nch);
void free_dmatrix(double **m, long nrl, long nrh, long ncl, long nch);
void free_imatrix(int **m, long nrl, long nrh, long ncl, long nch);
void free_submatrix(float **b, long nrl, long nrh, long ncl, long nch);
void free_convert_matrix(float **b, long nrl, long nrh, long ncl, long nch);
void free_f3tensor(float ***t, long nrl, long nrh, long ncl, long nch,
        long ndl, long ndh);

#else /* ANSI */
/* traditional - K&R */

void nrerror();


void free_f3tensor();

#endif /* ANSI */
```

```
#endif /* _NR_UTILS_H_ */
```

The `complex.h` header file contains prototypes for a set of routines that perform arithmetic functions with complex variables. Complex data types, and predefined arithmetic operations on complex numbers, are not part of the ANSI C standard. However, if your C library provides such facilities, you may use them in place of the ones we provide. Source code for these functions may be found in Appendix C.

```
#ifndef _FCOMPLEX_DECLARE_T_
typedef struct FCOMPLEX {float r,i;} fcomplex;
#define _FCOMPLEX_DECLARE_T_
#endif /* _FCOMPLEX_DECLARE_T_ */

#if defined(__STDC__) || defined(ANSI) || defined(NRANSI) /* ANSI */

fcomplex Cadd(fcomplex a, fcomplex b);
fcomplex Csub(fcomplex a, fcomplex b);
fcomplex Cmul(fcomplex a, fcomplex b);
fcomplex Complex(float re, float im);
fcomplex Conjg(fcomplex z);
fcomplex Cdiv(fcomplex a, fcomplex b);
float Cabs(fcomplex z);
fcomplex Csqrt(fcomplex z);
fcomplex RCmul(float x, fcomplex a);

#else /* ANSI */

fcomplex Cadd();

    .

    .

fcomplex RCmul();

#endif /* ANSI */
```

The following is a list of prototypes for all functions in the Numerical Recipes software collection. This is a section of the file `nr.h`, a header file included in virtually all of the example routines in this book. As an alternative to including `nr.h` in your programs, you may cull from this file the references to the specific Recipes that you will be using, and incorporate them as individual declarations.

```
#ifndef _NR_H_
#define _NR_H_

#ifndef _FCOMPLEX_DECLARE_T_
typedef struct FCOMPLEX {float r,i;} fcomplex;
#define _FCOMPLEX_DECLARE_T_
#endif /* _FCOMPLEX_DECLARE_T_ */

#ifndef _ARITHCODE_DECLARE_T_
typedef struct {
        unsigned long *ilob,*iupb,*ncumfq,jdif,nc,minint,nch,ncum,nrad;
} arithcode;
#define _ARITHCODE_DECLARE_T_
#endif /* _ARITHCODE_DECLARE_T_ */

#ifndef _HUFFCODE_DECLARE_T_
```

```
typedef struct {
        unsigned long *icod,*ncod,*left,*right,nch,nodemax;
} huffcode;
#define _HUFFCODE_DECLARE_T_
#endif /* _HUFFCODE_DECLARE_T_ */


#include <stdio.h>


#if defined(__STDC__) || defined(ANSI) || defined(NRANSI) /* ANSI */

void addint(double **uf, double **uc, double **res, int nf);
void airy(float x, float *ai, float *bi, float *aip, float *bip);
void amebsa(float **p, float y[], int ndim, float pb[],          float *yb,
        float ftol, float (*funk)(float []), int *iter, float temptr);
void amoeba(float **p, float y[], int ndim, float ftol,
        float (*funk)(float []), int *iter);
float amotry(float **p, float y[], float psum[], int ndim,
        float (*funk)(float []), int ihi, float fac);
float amotsa(float **p, float y[], float psum[], int ndim, float pb[],
        float *yb, float (*funk)(float []), int ihi, float *yhi, float fac);
void anneal(float x[], float y[], int iorder[], int ncity);
double anorm2(double **a, int n);
void arcmak(unsigned long nfreq[], unsigned long nchh, unsigned long nradd,
        arithcode *acode);
void arcode(unsigned long *ich, unsigned char **codep, unsigned long *lcode,
        unsigned long *lcd, int isign, arithcode *acode);
void arcsum(unsigned long iin[], unsigned long iout[], unsigned long ja,
        int nwk, unsigned long nrad, unsigned long nc);
void asolve(unsigned long n, double b[], double x[], int itrnsp);
void atimes(unsigned long n, double x[], double r[], int itrnsp);
void avevar(float data[], unsigned long n, float *ave, float *var);
void balanc(float **a, int n);
void banbks(float **a, unsigned long n, int m1, int m2, float **al,
        unsigned long indx[], float b[]);
void bandec(float **a, unsigned long n, int m1, int m2, float **al,
        unsigned long indx[], float *d);
void banmul(float **a, unsigned long n, int m1, int m2, float x[], float b[]);
void bcucof(float y[], float y1[], float y2[], float y12[], float d1,
        float d2, float **c);
void bcuint(float y[], float y1[], float y2[], float y12[],
        float x1l, float x1u, float x2l, float x2u, float x1,
        float x2, float *ansy, float *ansy1, float *ansy2);
void beschb(double x, double *gam1, double *gam2, double *gampl,
        double *gammi);
float bessi(int n, float x);
float bessi0(float x);
float bessi1(float x);
void bessik(float x, float xnu, float *ri, float *rk, float *rip,
        float *rkp);
float bessj(int n, float x);
float bessj0(float x);
float bessj1(float x);
void bessjy(float x, float xnu, float *rj, float *ry, float *rjp,
        float *ryp);
float bessk(int n, float x);
float bessk0(float x);
float bessk1(float x);
```

```
float bessy(int n, float x);
float bessy0(float x);
float bessy1(float x);
float beta(float z, float w);
float betacf(float a, float b, float x);
float betai(float a, float b, float x);
float bico(int n, int k);
void bksub(int ne, int nb, int jf, int k1, int k2, float ***c);
float bnldev(float pp, int n, long *idum);
float brent(float ax, float bx, float cx,
        float (*f)(float), float tol, float *xmin);
void broydn(float x[], int n, int *check,
        void (*vecfunc)(int, float [], float []));
void bsstep(float y[], float dydx[], int nv, float *xx, float htry,
        float eps, float yscal[], float *hdid, float *hnext,
        void (*derivs)(float, float [], float []));
void caldat(long julian, int *mm, int *id, int *iyyy);
void chder(float a, float b, float c[], float cder[], int n);
float chebev(float a, float b, float c[], int m, float x);
void chebft(float a, float b, float c[], int n, float (*func)(float));
void chebpc(float c[], float d[], int n);
void chint(float a, float b, float c[], float cint[], int n);
float chixy(float bang);
void choldc(float **a, int n, float p[]);
void cholsl(float **a, int n, float p[], float b[], float x[]);
void chsone(float bins[], float ebins[], int nbins, int knstrn,
        float *df, float *chsq, float *prob);
void chstwo(float bins1[], float bins2[], int nbins, int knstrn,
        float *df, float *chsq, float *prob);
void cisi(float x, float *ci, float *si);
void cntab1(int **nn, int ni, int nj, float *chisq,
        float *df, float *prob, float *cramrv, float *ccc);
void cntab2(int **nn, int ni, int nj, float *h, float *hx, float *hy,
        float *hygx, float *hxgy, float *uygx, float *uxgy, float *uxy);
void convlv(float data[], unsigned long n, float respns[], unsigned long m,
        int isign, float ans[]);
void copy(double **aout, double **ain, int n);
void correl(float data1[], float data2[], unsigned long n, float ans[]);
void cosft(float y[], int n, int isign);
void cosft1(float y[], int n);
void cosft2(float y[], int n, int isign);
void covsrt(float **covar, int ma, int ia[], int mfit);
void crank(unsigned long n, float w[], float *s);
void cyclic(float a[], float b[], float c[], float alpha, float beta,
        float r[], float x[], unsigned long n);
void daub4(float a[], unsigned long n, int isign);
float dawson(float x);
float dbrent(float ax, float bx, float cx,
        float (*f)(float), float (*df)(float), float tol, float *xmin);
void ddpoly(float c[], int nc, float x, float pd[], int nd);
int decchk(char string[], int n, char *ch);
void derivs(float x, float y[], float dydx[]);
float df1dim(float x);
void dfour1(double data[], unsigned long nn, int isign);
void dfpmin(float p[], int n, float gtol, int *iter, float *fret,
        float (*func)(float []), void (*dfunc)(float [], float []));
float dfridr(float (*func)(float), float x, float h, float *err);
```

```
void dftcor(float w, float delta, float a, float b, float endpts[],
        float *corre, float *corim, float *corfac);
void dftint(float (*func)(float), float a, float b, float w,
        float *cosint, float *sinint);
void difeq(int k, int k1, int k2, int jsf, int is1, int isf,
        int indexv[], int ne, float **s, float **y);
void dlinmin(float p[], float xi[], int n, float *fret,
        float (*func)(float []), void (*dfunc)(float [], float[]));
double dpythag(double a, double b);
void drealft(double data[], unsigned long n, int isign);
void dsprsax(double sa[], unsigned long ija[], double x[], double b[],
        unsigned long n);
void dsprstx(double sa[], unsigned long ija[], double x[], double b[],
        unsigned long n);
void dsvbksb(double **u, double w[], double **v, int m, int n, double b[],
        double x[]);
void dsvdcmp(double **a, int m, int n, double w[], double **v);
void eclass(int nf[], int n, int lista[], int listb[], int m);
void eclazz(int nf[], int n, int (*equiv)(int, int));
float ei(float x);
void eigsrt(float d[], float **v, int n);
float elle(float phi, float ak);
float ellf(float phi, float ak);
float ellpi(float phi, float en, float ak);
void elmhes(float **a, int n);
float erfcc(float x);
float erff(float x);
float erffc(float x);
void eulsum(float *sum, float term, int jterm, float wksp[]);
float evlmem(float fdt, float d[], int m, float xms);
float expdev(long *idum);
float expint(int n, float x);
float f1(float x);
float f1dim(float x);
float f2(float y);
float f3(float z);
float factln(int n);
float factrl(int n);
void fasper(float x[], float y[], unsigned long n, float ofac, float hifac,
        float wk1[], float wk2[], unsigned long nwk, unsigned long *nout,
        unsigned long *jmax, float *prob);
void fdjac(int n, float x[], float fvec[], float **df,
        void (*vecfunc)(int, float [], float []));
void fgauss(float x, float a[], float *y, float dyda[], int na);
void fill0(double **u, int n);
void fit(float x[], float y[], int ndata, float sig[], int mwt,
        float *a, float *b, float *siga, float *sigb, float *chi2, float *q);
void fitexy(float x[], float y[], int ndat, float sigx[], float sigy[],
        float *a, float *b, float *siga, float *sigb, float *chi2, float *q);
void fixrts(float d[], int m);
void fleg(float x, float pl[], int nl);
void flmoon(int n, int nph, long *jd, float *frac);
float fmin(float x[]);
void four1(float data[], unsigned long nn, int isign);
void fourew(FILE *file[5], int *na, int *nb, int *nc, int *nd);
void fourfs(FILE *file[5], unsigned long nn[], int ndim, int isign);
void fourn(float data[], unsigned long nn[], int ndim, int isign);
```

```
void fpoly(float x, float p[], int np);
void fred2(int n, float a, float b, float t[], float f[], float w[],
        float (*g)(float), float (*ak)(float, float));
float fredin(float x, int n, float a, float b, float t[], float f[], float w[],
        float (*g)(float), float (*ak)(float, float));
void frenel(float x, float *s, float *c);
void frprmn(float p[], int n, float ftol, int *iter, float *fret,
        float (*func)(float []), void (*dfunc)(float [], float []));
void ftest(float data1[], unsigned long n1, float data2[], unsigned long n2,
        float *f, float *prob);
float gamdev(int ia, long *idum);
float gammln(float xx);
float gammp(float a, float x);
float gammq(float a, float x);
float gasdev(long *idum);
void gaucof(int n, float a[], float b[], float amu0, float x[], float w[]);
void gauher(float x[], float w[], int n);
void gaujac(float x[], float w[], int n, float alf, float bet);
void gaulag(float x[], float w[], int n, float alf);
void gauleg(float x1, float x2, float x[], float w[], int n);
void gaussj(float **a, int n, float **b, int m);
void gcf(float *gammcf, float a, float x, float *gln);
float golden(float ax, float bx, float cx, float (*f)(float), float tol,
        float *xmin);
void gser(float *gamser, float a, float x, float *gln);
void hpsel(unsigned long m, unsigned long n, float arr[], float heap[]);
void hpsort(unsigned long n, float ra[]);
void hqr(float **a, int n, float wr[], float wi[]);
void hufapp(unsigned long index[], unsigned long nprob[], unsigned long n,
        unsigned long i);
void hufdec(unsigned long *ich, unsigned char *code, unsigned long lcode,
        unsigned long *nb, huffcode *hcode);
void hufenc(unsigned long ich, unsigned char **codep, unsigned long *lcode,
        unsigned long *nb, huffcode *hcode);
void hufmak(unsigned long nfreq[], unsigned long nchin, unsigned long *ilong,
        unsigned long *nlong, huffcode *hcode);
void hunt(float xx[], unsigned long n, float x, unsigned long *jlo);
void hypdrv(float s, float yy[], float dyyds[]);
fcomplex hypgeo(fcomplex a, fcomplex b, fcomplex c, fcomplex z);
void hypser(fcomplex a, fcomplex b, fcomplex c, fcomplex z,
        fcomplex *series, fcomplex *deriv);
unsigned short icrc(unsigned short crc, unsigned char *bufptr,
        unsigned long len, short jinit, int jrev);
unsigned short icrc1(unsigned short crc, unsigned char onech);
unsigned long igray(unsigned long n, int is);
void iindexx(unsigned long n, long arr[], unsigned long indx[]);
void indexx(unsigned long n, float arr[], unsigned long indx[]);
void interp(double **uf, double **uc, int nf);
int irbit1(unsigned long *iseed);
int irbit2(unsigned long *iseed);
void jacobi(float **a, int n, float d[], float **v, int *nrot);
void jacobn(float x, float y[], float dfdx[], float **dfdy, int n);
long julday(int mm, int id, int iyyy);
void kendl1(float data1[], float data2[], unsigned long n, float *tau, float *z,
        float *prob);
void kendl2(float **tab, int i, int j, float *tau, float *z, float *prob);
void kermom(double w[], double y, int m);
```

```
void ks2d1s(float x1[], float y1[], unsigned long n1,
        void (*quadvl)(float, float, float *, float *, float *, float *),
        float *d1, float *prob);
void ks2d2s(float x1[], float y1[], unsigned long n1, float x2[], float y2[],
        unsigned long n2, float *d, float *prob);
void ksone(float data[], unsigned long n, float (*func)(float), float *d,
        float *prob);
void kstwo(float data1[], unsigned long n1, float data2[], unsigned long n2,
        float *d, float *prob);
void laguer(fcomplex a[], int m, fcomplex *x, int *its);
void lfit(float x[], float y[], float sig[], int ndat, float a[], int ia[],
        int ma, float **covar, float *chisq, void (*funcs)(float, float [], int));
void linbcg(unsigned long n, double b[], double x[], int itol, double tol,
        int itmax, int *iter, double *err);
void linmin(float p[], float xi[], int n, float *fret,
        float (*func)(float []));
void lnsrch(int n, float xold[], float fold, float g[], float p[], float x[],
        float *f, float stpmax, int *check, float (*func)(float []));
void load(float x1, float v[], float y[]);
void load1(float x1, float v1[], float y[]);
void load2(float x2, float v2[], float y[]);
void locate(float xx[], unsigned long n, float x, unsigned long *j);
void lop(double **out, double **u, int n);
void lubksb(float **a, int n, int *indx, float b[]);
void ludcmp(float **a, int n, int *indx, float *d);
void machar(int *ibeta, int *it, int *irnd, int *ngrd,
        int *machep, int *negep, int *iexp, int *minexp, int *maxexp,
        float *eps, float *epsneg, float *xmin, float *xmax);
void matadd(double **a, double **b, double **c, int n);
void matsub(double **a, double **b, double **c, int n);
void medfit(float x[], float y[], int ndata, float *a, float *b, float *abdev);
void memcof(float data[], int n, int m, float *xms, float d[]);
int metrop(float de, float t);
void mgfas(double **u, int n, int maxcyc);
void mglin(double **u, int n, int ncycle);
float midexp(float (*funk)(float), float aa, float bb, int n);
float midinf(float (*funk)(float), float aa, float bb, int n);
float midpnt(float (*func)(float), float a, float b, int n);
float midsql(float (*funk)(float), float aa, float bb, int n);
float midsqu(float (*funk)(float), float aa, float bb, int n);
void miser(float (*func)(float []), float regn[], int ndim, unsigned long npts,
        float dith, float *ave, float *var);
void mmid(float y[], float dydx[], int nvar, float xs, float htot,
        int nstep, float yout[], void (*derivs)(float, float[], float[]));
void mnbrak(float *ax, float *bx, float *cx, float *fa, float *fb,
        float *fc, float (*func)(float));
void mnewt(int ntrial, float x[], int n, float tolx, float tolf);
void moment(float data[], int n, float *ave, float *adev, float *sdev,
        float *var, float *skew, float *curt);
void mp2dfr(unsigned char a[], unsigned char s[], int n, int *m);
void mpadd(unsigned char w[], unsigned char u[], unsigned char v[], int n);
void mpdiv(unsigned char q[], unsigned char r[], unsigned char u[],
        unsigned char v[], int n, int m);
void mpinv(unsigned char u[], unsigned char v[], int n, int m);
void mplsh(unsigned char u[], int n);
void mpmov(unsigned char u[], unsigned char v[], int n);
void mpmul(unsigned char w[], unsigned char u[], unsigned char v[], int n,
```

```
        int m);
void mpneg(unsigned char u[], int n);
void mppi(int n);
void mprove(float **a, float **alud, int n, int indx[], float b[],
        float x[]);
void mpsad(unsigned char w[], unsigned char u[], int n, int iv);
void mpsdv(unsigned char w[], unsigned char u[], int n, int iv, int *ir);
void mpsmu(unsigned char w[], unsigned char u[], int n, int iv);
void mpsqrt(unsigned char w[], unsigned char u[], unsigned char v[], int n,
        int m);
void mpsub(int *is, unsigned char w[], unsigned char u[], unsigned char v[],
        int n);
void mrqcof(float x[], float y[], float sig[], int ndata, float a[],
        int ia[], int ma, float **alpha, float beta[], float *chisq,
        void (*funcs)(float, float [], float *, float [], int));
void mrqmin(float x[], float y[], float sig[], int ndata, float a[],
        int ia[], int ma, float **covar, float **alpha, float *chisq,
        void (*funcs)(float, float [], float *, float [], int), float *alamda);
void newt(float x[], int n, int *check,
        void (*vecfunc)(int, float [], float []));
void odeint(float ystart[], int nvar, float x1, float x2,
        float eps, float h1, float hmin, int *nok, int *nbad,
        void (*derivs)(float, float [], float []),
        void (*rkqs)(float [], float [], int, float *, float, float,
        float [], float *, float *, void (*)(float, float [], float [])));
void orthog(int n, float anu[], float alpha[], float beta[], float a[],
        float b[]);
void pade(double cof[], int n, float *resid);
void pccheb(float d[], float c[], int n);
void pcshft(float a, float b, float d[], int n);
void pearsn(float x[], float y[], unsigned long n, float *r, float *prob,
        float *z);
void period(float x[], float y[], int n, float ofac, float hifac,
        float px[], float py[], int np, int *nout, int *jmax, float *prob);
void piksr2(int n, float arr[], float brr[]);
void piksrt(int n, float arr[]);
void pinvs(int ie1, int ie2, int je1, int jsf, int jc1, int k,
        float ***c, float **s);
float plgndr(int l, int m, float x);
float poidev(float xm, long *idum);
void polcoe(float x[], float y[], int n, float cof[]);
void polcof(float xa[], float ya[], int n, float cof[]);
void poldiv(float u[], int n, float v[], int nv, float q[], float r[]);
void polin2(float x1a[], float x2a[], float **ya, int m, int n,
        float x1, float x2, float *y, float *dy);
void polint(float xa[], float ya[], int n, float x, float *y, float *dy);
void powell(float p[], float **xi, int n, float ftol, int *iter, float *fret,
        float (*func)(float []));
void predic(float data[], int ndata, float d[], int m, float future[], int nfut);
float probks(float alam);
void psdes(unsigned long *lword, unsigned long *irword);
void pwt(float a[], unsigned long n, int isign);
void pwtset(int n);
float pythag(float a, float b);
void pzextr(int iest, float xest, float yest[], float yz[], float dy[],
        int nv);
float qgaus(float (*func)(float), float a, float b);
```

```
void qrdcmp(float **a, int n, float *c, float *d, int *sing);
float qromb(float (*func)(float), float a, float b);
float qromo(float (*func)(float), float a, float b,
        float (*choose)(float (*)(float), float, float, int));
void qroot(float p[], int n, float *b, float *c, float eps);
void qrsolv(float **a, int n, float c[], float d[], float b[]);
void qrupdt(float **r, float **qt, int n, float u[], float v[]);
float qsimp(float (*func)(float), float a, float b);
float qtrap(float (*func)(float), float a, float b);
float quad3d(float (*func)(float, float, float), float x1, float x2);
void quadct(float x, float y, float xx[], float yy[], unsigned long nn,
        float *fa, float *fb, float *fc, float *fd);
void quadmx(float **a, int n);
void quadvl(float x, float y, float *fa, float *fb, float *fc, float *fd);
float ran0(long *idum);
float ran1(long *idum);
float ran2(long *idum);
float ran3(long *idum);
float ran4(long *idum);
void rank(unsigned long n, unsigned long indx[], unsigned long irank[]);
void ranpt(float pt[], float regn[], int n);
void ratint(float xa[], float ya[], int n, float x, float *y, float *dy);
void ratlsq(double (*fn)(double), double a, double b, int mm, int kk,
        double cof[], double *dev);
double ratval(double x, double cof[], int mm, int kk);
float rc(float x, float y);
float rd(float x, float y, float z);
void realft(float data[], unsigned long n, int isign);
void rebin(float rc, int nd, float r[], float xin[], float xi[]);
void red(int iz1, int iz2, int jz1, int jz2, int jm1, int jm2, int jmf,
        int ic1, int jc1, int jcf, int kc, float ***c, float **s);
void relax(double **u, double **rhs, int n);
void relax2(double **u, double **rhs, int n);
void resid(double **res, double **u, double **rhs, int n);
float revcst(float x[], float y[], int iorder[], int ncity, int n[]);
void reverse(int iorder[], int ncity, int n[]);
float rf(float x, float y, float z);
float rj(float x, float y, float z, float p);
void rk4(float y[], float dydx[], int n, float x, float h, float yout[],
        void (*derivs)(float, float [], float []));
void rkck(float y[], float dydx[], int n, float x, float h,
        float yout[], float yerr[], void (*derivs)(float, float [], float []));
void rkdumb(float vstart[], int nvar, float x1, float x2, int nstep,
        void (*derivs)(float, float [], float []));
void rkqs(float y[], float dydx[], int n, float *x,
        float htry, float eps, float yscal[], float *hdid, float *hnext,
        void (*derivs)(float, float [], float []));
void rlft3(float ***data, float **speq, unsigned long nn1,
        unsigned long nn2, unsigned long nn3, int isign);
float rofunc(float b);
void rotate(float **r, float **qt, int n, int i, float a, float b);
void rsolv(float **a, int n, float d[], float b[]);
void rstrct(double **uc, double **uf, int nc);
float rtbis(float (*func)(float), float x1, float x2, float xacc);
float rtflsp(float (*func)(float), float x1, float x2, float xacc);
float rtnewt(void (*funcd)(float, float *, float *), float x1, float x2,
        float xacc);
```

```
float rtsafe(void (*funcd)(float, float *, float *), float x1, float x2,
    float xacc);
float rtsec(float (*func)(float), float x1, float x2, float xacc);
void rzextr(int iest, float xest, float yest[], float yz[], float dy[], int nv);
void savgol(float c[], int np, int nl, int nr, int ld, int m);
void score(float xf, float y[], float f[]);
void scrsho(float (*fx)(float));
float select(unsigned long k, unsigned long n, float arr[]);
float selip(unsigned long k, unsigned long n, float arr[]);
void shell(unsigned long n, float a[]);
void shoot(int n, float v[], float f[]);
void shootf(int n, float v[], float f[]);
void simp1(float **a, int mm, int ll[], int nll, int iabf, int *kp,
    float *bmax);
void simp2(float **a, int n, int l2[], int nl2, int *ip, int kp, float *q1);
void simp3(float **a, int i1, int k1, int ip, int kp);
void simplx(float **a, int m, int n, int m1, int m2, int m3, int *icase,
    int izrov[], int iposv[]);
void simpr(float y[], float dydx[], float dfdx[], float **dfdy,
    int n, float xs, float htot, int nstep, float yout[],
    void (*derivs)(float, float [], float []));
void sinft(float y[], int n);
void slvsm2(double **u, double **rhs);
void slvsml(double **u, double **rhs);
void sncndn(float uu, float emmc, float *sn, float *cn, float *dn);
double snrm(unsigned long n, double sx[], int itol);
void sobseq(int *n, float x[]);
void solvde(int itmax, float conv, float slowc, float scalv[],
    int indexv[], int ne, int nb, int m, float **y, float ***c, float **s);
void sor(double **a, double **b, double **c, double **d, double **e,
    double **f, double **u, int jmax, double rjac);
void sort(unsigned long n, float arr[]);
void sort2(unsigned long n, float arr[], float brr[]);
void sort3(unsigned long n, float ra[], float rb[], float rc[]);
void spctrm(FILE *fp, float p[], int m, int k, int ovrlap);
void spear(float data1[], float data2[], unsigned long n, float *d, float *zd,
    float *probd, float *rs, float *probrs);
void sphbes(int n, float x, float *sj, float *sy, float *sjp, float *syp);
void splie2(float x1a[], float x2a[], float **ya, int m, int n, float **y2a);
void splin2(float x1a[], float x2a[], float **ya, float **y2a, int m, int n,
    float x1, float x2, float *y);
void spline(float x[], float y[], int n, float yp1, float ypn, float y2[]);
void splint(float xa[], float ya[], float y2a[], int n, float x, float *y);
void spread(float y, float yy[], unsigned long n, float x, int m);
void sprsax(float sa[], unsigned long ija[], float x[], float b[],
    unsigned long n);
void sprsin(float **a, int n, float thresh, unsigned long nmax, float sa[],
    unsigned long ija[]);
void sprspm(float sa[], unsigned long ija[], float sb[], unsigned long ijb[],
    float sc[], unsigned long ijc[]);
void sprstm(float sa[], unsigned long ija[], float sb[], unsigned long ijb[],
    float thresh, unsigned long nmax, float sc[], unsigned long ijc[]);
void sprstp(float sa[], unsigned long ija[], float sb[], unsigned long ijb[]);
void sprstx(float sa[], unsigned long ija[], float x[], float b[],
    unsigned long n);
void stifbs(float y[], float dydx[], int nv, float *xx,
    float htry, float eps, float yscal[], float *hdid, float *hnext,
```

```
        void (*derivs)(float, float [], float []));
void stiff(float y[], float dydx[], int n, float *x,
        float htry, float eps, float yscal[], float *hdid, float *hnext,
        void (*derivs)(float, float [], float []));
void stoerm(float y[], float d2y[], int nv, float xs,
        float htot, int nstep, float yout[],
        void (*derivs)(float, float [], float []));
void svbksb(float **u, float w[], float **v, int m, int n, float b[],
        float x[]);
void svdcmp(float **a, int m, int n, float w[], float **v);
void svdfit(float x[], float y[], float sig[], int ndata, float a[],
        int ma, float **u, float **v, float w[], float *chisq,
        void (*funcs)(float, float [], int));
void svdvar(float **v, int ma, float w[], float **cvm);
void toeplz(float r[], float x[], float y[], int n);
void tptest(float data1[], float data2[], unsigned long n, float *t, float *prob);
void tqli(float d[], float e[], int n, float **z);
float trapzd(float (*func)(float), float a, float b, int n);
void tred2(float **a, int n, float d[], float e[]);
void tridag(float a[], float b[], float c[], float r[], float u[],
        unsigned long n);
float trncst(float x[], float y[], int iorder[], int ncity, int n[]);
void trnspt(int iorder[], int ncity, int n[]);
void ttest(float data1[], unsigned long n1, float data2[], unsigned long n2,
        float *t, float *prob);
void tutest(float data1[], unsigned long n1, float data2[], unsigned long n2,
        float *t, float *prob);
void twofft(float data1[], float data2[], float fft1[], float fft2[],
        unsigned long n);
void vander(double x[], double w[], double q[], int n);
void vegas(float regn[], int ndim, float (*fxn)(float [], float), int init,
        unsigned long ncall, int itmx, int nprn, float *tgral, float *sd,
        float *chi2a);
void voltra(int n, int m, float t0, float h, float *t, float **f,
        float (*g)(int, float), float (*ak)(int, int, float, float));
void wt1(float a[], unsigned long n, int isign,
        void (*wtstep)(float [], unsigned long, int));
void wtn(float a[], unsigned long nn[], int ndim, int isign,
        void (*wtstep)(float [], unsigned long, int));
void wwghts(float wghts[], int n, float h,
        void (*kermom)(double [], double ,int));
int zbrac(float (*func)(float), float *x1, float *x2);
void zbrak(float (*fx)(float), float x1, float x2, int n, float xb1[],
        float xb2[], int *nb);
float zbrent(float (*func)(float), float x1, float x2, float tol);
void zrhqr(float a[], int m, float rtr[], float rti[]);
float zriddr(float (*func)(float), float x1, float x2, float xacc);
void zroots(fcomplex a[], int m, fcomplex roots[], int polish);

#else /* ANSI */
/* traditional - K&R */


void addint();



void zroots();
```

```
#endif /* ANSI */

#endif /* _NR_H_ */
```

Appendix B: Numerical Recipes Utility Functions

> The utility functions listed below are used by many Recipes and Examples in the Numerical Recipes collection. Along with the short error-reporting function, there are functions that we have found indispensable in the handling of vectors and matrices of different data types. A full discussion of the use of matrices and vectors within the Numerical Recipes collection may be found in Chapter 1 of Numerical Recipes, along with a description of other programming standards adopted for the use of C in scientific programming.

★ ★ ★ ★

```c
#include <stdio.h>
#include <stddef.h>
#include <stdlib.h>
#define NR_END 1
#define FREE_ARG char*

void nrerror(char error_text[])
/* Numerical Recipes standard error handler */
{
        fprintf(stderr,"Numerical Recipes run-time error...\n");
        fprintf(stderr,"%s\n",error_text);
        fprintf(stderr,"...now exiting to system...\n");
        exit(1);
}

float *vector(long nl, long nh)
/* allocate a float vector with subscript range v[nl..nh] */
{
        float *v;

        v=(float *)malloc((size_t) ((nh-nl+1+NR_END)*sizeof(float)));
        if (!v) nrerror("allocation failure in vector()");
        return v-nl+NR_END;
}

int *ivector(long nl, long nh)
/* allocate an int vector with subscript range v[nl..nh] */
{
        int *v;

        v=(int *)malloc((size_t) ((nh-nl+1+NR_END)*sizeof(int)));
        if (!v) nrerror("allocation failure in ivector()");
        return v-nl+NR_END;
}
```

```
unsigned char *cvector(long nl, long nh)
/* allocate an unsigned char vector with subscript range v[nl..nh] */
{
        unsigned char *v;

        v=(unsigned char *)malloc((size_t) ((nh-nl+1+NR_END)*sizeof(unsigned char)));
        if (!v) nrerror("allocation failure in cvector()");
        return v-nl+NR_END;
}


unsigned long *lvector(long nl, long nh)
/* allocate an unsigned long vector with subscript range v[nl..nh] */
{
        unsigned long *v;

        v=(unsigned long *)malloc((size_t) ((nh-nl+1+NR_END)*sizeof(long)));
        if (!v) nrerror("allocation failure in lvector()");
        return v-nl+NR_END;
}


double *dvector(long nl, long nh)
/* allocate a double vector with subscript range v[nl..nh] */
{
        double *v;

        v=(double *)malloc((size_t) ((nh-nl+1+NR_END)*sizeof(double)));
        if (!v) nrerror("allocation failure in dvector()");
        return v-nl+NR_END;
}


float **matrix(long nrl, long nrh, long ncl, long nch)
/* allocate a float matrix with subscript range m[nrl..nrh][ncl..nch] */
{
        long i, nrow=nrh-nrl+1,ncol=nch-ncl+1;
        float **m;

        /* allocate pointers to rows */
        m=(float **) malloc((size_t)((nrow+NR_END)*sizeof(float*)));
        if (!m) nrerror("allocation failure 1 in matrix()");
        m += NR_END;
        m -= nrl;

        /* allocate rows and set pointers to them */
        m[nrl]=(float *) malloc((size_t)((nrow*ncol+NR_END)*sizeof(float)));
        if (!m[nrl]) nrerror("allocation failure 2 in matrix()");
        m[nrl] += NR_END;
        m[nrl] -= ncl;

        for(i=nrl+1;i<=nrh;i++) m[i]=m[i-1]+ncol;

        /* return pointer to array of pointers to rows */
        return m;
}


double **dmatrix(long nrl, long nrh, long ncl, long nch)
/* allocate a double matrix with subscript range m[nrl..nrh][ncl..nch] */
{
```

```
        long i, nrow=nrh-nrl+1,ncol=nch-ncl+1;
        double **m;

        /* allocate pointers to rows */
        m=(double **) malloc((size_t)((nrow+NR_END)*sizeof(double*)));
        if (!m) nrerror("allocation failure 1 in matrix()");
        m += NR_END;
        m -= nrl;

        /* allocate rows and set pointers to them */
        m[nrl]=(double *) malloc((size_t)((nrow*ncol+NR_END)*sizeof(double)));
        if (!m[nrl]) nrerror("allocation failure 2 in matrix()");
        m[nrl] += NR_END;
        m[nrl] -= ncl;

        for(i=nrl+1;i<=nrh;i++) m[i]=m[i-1]+ncol;

        /* return pointer to array of pointers to rows */
        return m;
}

int **imatrix(long nrl, long nrh, long ncl, long nch)
/* allocate a int matrix with subscript range m[nrl..nrh][ncl..nch] */
{
        long i, nrow=nrh-nrl+1,ncol=nch-ncl+1;
        int **m;

        /* allocate pointers to rows */
        m=(int **) malloc((size_t)((nrow+NR_END)*sizeof(int*)));
        if (!m) nrerror("allocation failure 1 in matrix()");
        m += NR_END;
        m -= nrl;


        /* allocate rows and set pointers to them */
        m[nrl]=(int *) malloc((size_t)((nrow*ncol+NR_END)*sizeof(int)));
        if (!m[nrl]) nrerror("allocation failure 2 in matrix()");
        m[nrl] += NR_END;
        m[nrl] -= ncl;

        for(i=nrl+1;i<=nrh;i++) m[i]=m[i-1]+ncol;

        /* return pointer to array of pointers to rows */
        return m;
}

float **submatrix(float **a, long oldrl, long oldrh, long oldcl, long oldch,
        long newrl, long newcl)
/* point a submatrix [newrl..][newcl..] to a[oldrl..oldrh][oldcl..oldch] */
{
        long i,j,nrow=oldrh-oldrl+1,ncol=oldcl-newcl;
        float **m;

        /* allocate array of pointers to rows */
        m=(float **) malloc((size_t) ((nrow+NR_END)*sizeof(float*)));
        if (!m) nrerror("allocation failure in submatrix()");
        m += NR_END;
```

```
        m -= newrl;

        /* set pointers to rows */
        for(i=oldrl,j=newrl;i<=oldrh;i++,j++) m[j]=a[i]+ncol;

        /* return pointer to array of pointers to rows */
        return m;
}

float **convert_matrix(float *a, long nrl, long nrh, long ncl, long nch)
/* allocate a float matrix m[nrl..nrh][ncl..nch] that points to the matrix
declared in the standard C manner as a[nrow][ncol], where nrow=nrh-nrl+1
and ncol=nch-ncl+1. The routine should be called with the address
&a[0][0] as the first argument. */
{
        long i,j,nrow=nrh-nrl+1,ncol=nch-ncl+1;
        float **m;

        /* allocate pointers to rows */
        m=(float **) malloc((size_t) ((nrow+NR_END)*sizeof(float*)));
        if (!m) nrerror("allocation failure in convert_matrix()");
        m += NR_END;
        m -= nrl;

        /* set pointers to rows */
        m[nrl]=a-ncl;
        for(i=1,j=nrl+1;i<nrow;i++,j++) m[j]=m[j-1]+ncol;
        /* return pointer to array of pointers to rows */
        return m;
}

float ***f3tensor(long nrl, long nrh, long ncl, long nch, long ndl, long ndh)
/* allocate a float 3tensor with range t[nrl..nrh][ncl..nch][ndl..ndh] */
{
        long i,j,nrow=nrh-nrl+1,ncol=nch-ncl+1,ndep=ndh-ndl+1;
        float ***t;

        /* allocate pointers to pointers to rows */
        t=(float ***) malloc((size_t)((nrow+NR_END)*sizeof(float**)));
        if (!t) nrerror("allocation failure 1 in f3tensor()");
        t += NR_END;
        t -= nrl;

        /* allocate pointers to rows and set pointers to them */
        t[nrl]=(float **) malloc((size_t)((nrow*ncol+NR_END)*sizeof(float*)));
        if (!t[nrl]) nrerror("allocation failure 2 in f3tensor()");
        t[nrl] += NR_END;
        t[nrl] -= ncl;

        /* allocate rows and set pointers to them */
        t[nrl][ncl]=(float *) malloc((size_t)((nrow*ncol*ndep+NR_END)*sizeof(float)));
        if (!t[nrl][ncl]) nrerror("allocation failure 3 in f3tensor()");
        t[nrl][ncl] += NR_END;
        t[nrl][ncl] -= ndl;

        for(j=ncl+1;j<=nch;j++) t[nrl][j]=t[nrl][j-1]+ndep;
        for(i=nrl+1;i<=nrh;i++) {
```

```
                     t[i]=t[i-1]+ncol;
                     t[i][ncl]=t[i-1][ncl]+ncol*ndep;
                     for(j=ncl+1;j<=nch;j++) t[i][j]=t[i][j-1]+ndep;
          }

          /* return pointer to array of pointers to rows */
          return t;
}

void free_vector(float *v, long nl, long nh)
/* free a float vector allocated with vector() */
{
          free((FREE_ARG) (v+nl-NR_END));
}

void free_ivector(int *v, long nl, long nh)
/* free an int vector allocated with ivector() */
{
          free((FREE_ARG) (v+nl-NR_END));
}

void free_cvector(unsigned char *v, long nl, long nh)
/* free an unsigned char vector allocated with cvector() */
{
          free((FREE_ARG) (v+nl-NR_END));
}

void free_lvector(unsigned long *v, long nl, long nh)
/* free an unsigned long vector allocated with lvector() */
{
          free((FREE_ARG) (v+nl-NR_END));
}

void free_dvector(double *v, long nl, long nh)
/* free a double vector allocated with dvector() */
{
          free((FREE_ARG) (v+nl-NR_END));
}

void free_matrix(float **m, long nrl, long nrh, long ncl, long nch)
/* free a float matrix allocated by matrix() */
{
          free((FREE_ARG) (m[nrl]+ncl-NR_END));
          free((FREE_ARG) (m+nrl-NR_END));
}

void free_dmatrix(double **m, long nrl, long nrh, long ncl, long nch)
/* free a double matrix allocated by dmatrix() */
{
          free((FREE_ARG) (m[nrl]+ncl-NR_END));
          free((FREE_ARG) (m+nrl-NR_END));
}

void free_imatrix(int **m, long nrl, long nrh, long ncl, long nch)
/* free an int matrix allocated by imatrix() */
{
          free((FREE_ARG) (m[nrl]+ncl-NR_END));
```

```
        free((FREE_ARG) (m+nrl-NR_END));
}


void free_submatrix(float **b, long nrl, long nrh, long ncl, long nch)
/* free a submatrix allocated by submatrix() */
{
        free((FREE_ARG) (b+nrl-NR_END));
}


void free_convert_matrix(float **b, long nrl, long nrh, long ncl, long nch)
/* free a matrix allocated by convert_matrix() */
{
        free((FREE_ARG) (b+nrl-NR_END));
}


void free_f3tensor(float ***t, long nrl, long nrh, long ncl, long nch,
        long ndl, long ndh)
/* free a float f3tensor allocated by f3tensor() */
{
        free((FREE_ARG) (t[nrl][ncl]+ndl-NR_END));
        free((FREE_ARG) (t[nrl]+ncl-NR_END));
        free((FREE_ARG) (t+nrl-NR_END));
}
```

Appendix C: Functions for Complex Arithmetic

Complex data types, and arithmetic operations on complex numbers, are not a standard part of C. We have therefore included the following functions for use with the small number of Numerical Recipes routines that use complex variables. When using C libraries that support the use of complex numbers, references to these routines may be replaced by equivalent library functions.

$\star$ $\star$ $\star$ $\star$

```
#include <math.h>

typedef struct FCOMPLEX {float r,i;} fcomplex;

fcomplex Cadd(fcomplex a, fcomplex b)
{
        fcomplex c;
        c.r=a.r+b.r;
        c.i=a.i+b.i;
        return c;
}

fcomplex Csub(fcomplex a, fcomplex b)
{
        fcomplex c;
        c.r=a.r-b.r;
        c.i=a.i-b.i;
        return c;
}


fcomplex Cmul(fcomplex a, fcomplex b)
{
        fcomplex c;
        c.r=a.r*b.r-a.i*b.i;
        c.i=a.i*b.r+a.r*b.i;
        return c;
}

fcomplex Complex(float re, float im)
{
        fcomplex c;
        c.r=re;
        c.i=im;
        return c;
}

fcomplex Conjg(fcomplex z)
```

```
{
        fcomplex c;
        c.r=z.r;
        c.i = -z.i;
        return c;
}

fcomplex Cdiv(fcomplex a, fcomplex b)
{
        fcomplex c;
        float r,den;
        if (fabs(b.r) >= fabs(b.i)) {
                r=b.i/b.r;
                den=b.r+r*b.i;
                c.r=(a.r+r*a.i)/den;
                c.i=(a.i-r*a.r)/den;
        } else {
                r=b.r/b.i;
                den=b.i+r*b.r;
                c.r=(a.r*r+a.i)/den;
                c.i=(a.i*r-a.r)/den;
        }
        return c;
}


float Cabs(fcomplex z)
{
        float x,y,ans,temp;
        x=fabs(z.r);
        y=fabs(z.i);
        if (x == 0.0)
                ans=y;
        else if (y == 0.0)
                ans=x;
        else if (x > y) {
                temp=y/x;
                ans=x*sqrt(1.0+temp*temp);
        } else {
                temp=x/y;
                ans=y*sqrt(1.0+temp*temp);
        }
        return ans;
}

fcomplex Csqrt(fcomplex z)
{
        fcomplex c;
        float x,y,w,r;
        if ((z.r == 0.0) && (z.i == 0.0)) {
                c.r=0.0;
                c.i=0.0;
                return c;
        } else {
                x=fabs(z.r);
                y=fabs(z.i);
                if (x >= y) {
                        r=y/x;
```

```
                    w=sqrt(x)*sqrt(0.5*(1.0+sqrt(1.0+r*r)));
            } else {
                    r=x/y;
                    w=sqrt(y)*sqrt(0.5*(r+sqrt(1.0+r*r)));
            }
            if (z.r >= 0.0) {
                    c.r=w;
                    c.i=z.i/(2.0*w);
            } else {
                    c.i=(z.i >= 0) ? w : -w;
                    c.r=z.i/(2.0*c.i);
            }
            return c;
        }
}

fcomplex RCmul(float x, fcomplex a)
{
        fcomplex c;
        c.r=x*a.r;
        c.i=x*a.i;
        return c;
}
```

Index of Demonstrated Routines

Following is a brief explanation of each *Numerical Recipes* product, plus two order forms (one for North American residents, one for all other), which may be used to order these items directly from the publisher if you cannot obtain them from your local bookstore.

Numerical Recipes in FORTRAN, Second Edition and *Numerical Recipes in C, Second Edition* represent the main text and reference component of the *Numerical Recipes* package. Each book contains over 300 programs, in the language of the reader's choice, and constitutes a complete subroutine library for scientific computation. Both versions contain equivalent tutorial, mathematical, and practical discussions.

There are two example books containing FORTRAN or C source programs respectively that exercise and demonstrate all of the *Numerical Recipes, Second Edition* programs. Each example program contains comments and is prefaced by a short description. Input and output data are supplied in many cases. The example books are designed to help readers incorporate procedures and subroutines and conduct simple validation tests.

The programs contained in both the second edition main books and the example books are available in several machine-readable formats that will save users hours of tedious keyboarding. Diskettes for IBM PC compatible machines are available in either 5¼ inch high density or 3½ inch format and operate on DOS 2.0 or later. Diskettes for the Apple Macintosh are 3½ inch single-sided disks.

Some selected first edition products are also still available:

Numerical Recipes in Pascal contains the original 200 *Numerical Recipes* routines translated into Pascal along with the tutorial text. The Pascal example book contains the example programs for these routines. Diskettes for each book are available in 5¼ inch double-sided, double-density IBM and 3½ inch Apple Macintosh format.

Numerical Recipes Routines and Examples in BASIC contains all the routines from the original Numerical Recipes plus the exercise programs from the example book, all translated into BASIC, along with the text from the example book. The BASIC routines and programs are also available on 5¼ inch diskette for IBM PC compatible machines.

Instructions

To obtain the books or the latest version of the disks, please order from your bookstore or complete the information on the order form in this book and mail it to Cambridge University Press in Port Chester, New York or Cambridge, England. Please note that there is a separate order form for each location. All orders must be prepaid. Ordinary postage for shipping is paid by the publisher.

NB: Technical questions, corrections, and requests for information on mainframe and workstation licenses should be directed to Numerical Recipes Software, P.O. Box 243, Cambridge, MA 02238, U.S.A. Please do not write the publisher.

There are no cash refunds for diskettes. Only diskettes with manufacturing defects may be returned to the publisher for replacement.

ORDER FORM (United States and Canada)

Order from your bookstore or mail to
Cambridge University Press, Order Department, 110 Midland Avenue, Port Chester,
New York 10573
Call for current prices: 1-800-431-1580

.... 43064-X   Numerical Recipes in FORTRAN: The Art of Scientific Computing,
                     Second Edition
.... 43717-2   FORTRAN Diskette (IBM, 5¼ inch/1.2M), second edition
.... 43719-9   FORTRAN Diskette (IBM, 3½ inch/720K), second edition
.... 43716-4   FORTRAN Diskette (Macintosh), second edition
.... 43721-0   FORTRAN Example Book, second edition

.... 43108-5   Numerical Recipes in C: The Art of Scientific Computing, Second Edition
.... 43714-8   C Diskette (IBM, 5¼ inch/1.2M), second edition
.... 43724-5   C Diskette (IBM, 3½ inch/720K), second edition
.... 43715-6   C Diskette (Macintosh), second edition
.... 43720-2   C Example Book, second edition

The following first edition products are still available:

.... 37516-9   Numerical Recipes in Pascal: The Art of Scientific Computing
.... 37532-0   Pascal Diskette (IBM)
.... 38766-3   Pascal Diskette (Macintosh)
.... 37675-0   Pascal Example Book
.... 37533-9   Pascal Example Diskette (IBM)
.... 38767-1   Pascal Example Diskette (Macintosh)

.... 40689-7   Numerical Recipes Routines and Examples in BASIC
.... 40688-9   BASIC Diskette (IBM)

Please indicate method of payment: check ...., Mastercard ...., or Visa ....

Name ....

Address ....

....

....

Card No. .... Expiration date ....

Signature ....

.... Please indicate the total number of items ordered,

.... total price,

.... tax, if applicable (NY and CA residents)

.... total enclosed

ORDER FORM (Outside North America)

Order from your bookstore or mail to
Customer Services Department, Cambridge University Press, Edinburgh Building,
Shaftesbury Road, Cambridge CB2 2RU, U.K.

.... 43064-X   Numerical Recipes in FORTRAN: The Art of Scientific Computing,
Second Edition   £35.00
.... 43717-2   FORTRAN Diskette (IBM, 5¼ inch/1.2M), second edition   £24.95
.... 43719-9   FORTRAN Diskette (IBM, 3½ inch/720K), second edition   £24.95
.... 43716-4   FORTRAN Diskette (Macintosh), second edition   £24.95
.... 43721-0   FORTRAN Example Book, second edition   £19.95

.... 43108-5   Numerical Recipes in C: The Art of Scientific Computing, Second Edition
£35.00
.... 43714-8   C Diskette (IBM, 5¼ inch/1.2M), second edition   £24.95
.... 43724-5   C Diskette (IBM, 3½ inch/720K), second edition   £24.95
.... 43715-6   C Diskette (Macintosh), second edition   £24.95
.... 43720-2   C Example Book, second edition   £19.95

The following first edition products are still available:

.... 37516-9   Numerical Recipes in Pascal: The Art of Scientific Computing   £30.00
.... 37532-0   Pascal Diskette (IBM)   £21.50
.... 38766-3   Pascal Diskette (Macintosh)   £26.50
.... 37675-0   Pascal Example Book   £19.50
.... 37533-9   Pascal Example Diskette (IBM)   £21.50
.... 38767-1   Pascal Example Diskette (Macintosh)   £26.50
.... 40689-7   Numerical Recipes Routines and Examples in BASIC   £19.50
.... 40688-9   BASIC Diskette (IBM)   £21.50

Name .... (Block capitals please)

Address ....

....

....

Please accept my payment by cheque or money order in pounds sterling:
I enclose (circle one) a Cheque (made payable to Cambridge University Press)/UK Postal
Order/International Money Order/Bank Draft/Post Office Giro.

Please accept my payment by credit card:
Charge my (circle one) Barclaycard/VISA/Eurocard/Access/Mastercard/Bank Americard/
any other credit card bearing the Interbank symbol (please specify).

Card No. .... Expiry date: ....

Signature .... Date: ....

Address as registered by card company: ....

....

....

Prices of diskettes do not include V.A.T., which should be added to all U.K. purchases.